Climate Change Management

Series editor

Walter Leal Filho, Hamburg, Germany

More information about this series at http://www.springer.com/series/8740

Walter Leal Filho · Johanna Nalau
Editors

Limits to Climate Change Adaptation

Springer

Editors
Walter Leal Filho
Hamburg University of Applied Sciences
Hamburg
Germany

Johanna Nalau
Griffith Institute for Tourism
Griffith University
Gold Coast
Australia

ISSN 1610-2010 ISSN 1610-2002 (electronic)
Climate Change Management
ISBN 978-3-319-64598-8 ISBN 978-3-319-64599-5 (eBook)
https://doi.org/10.1007/978-3-319-64599-5

Library of Congress Control Number: 2017947853

Printed on acid-free paper

This Springer imprint is published by Springer Nature
The registered company is Springer International Publishing AG
The registered company address is: Gewerbestrasse 11, 6330 Cham, Switzerland

Contents

1 **Introduction: Limits to Adaptation** 1
Johanna Nalau and Walter Leal Filho

Part I Limits to Climate Change Adaptation in Asia

2 **Strategies and Barriers to Adaptation of Hazard-Prone Rural Households in Bangladesh** 11
G. M. Monirul Alam, Khorshed Alam, Shahbaz Mushtaq, Most Nilufa Khatun and Walter Leal Filho

3 **Governance Limits to Adaptation in Cambodia's Health Sector** 25
Daniel Gilfillan

4 **Land-Based Strategic Model by Integrating Diverse Policies for Climate Change Adaptation in Nepal** 41
Adish Khezri, Arbind Man Tuladhar and Jaap Zevenbergen

5 **Climate Change and Migration in Bangladesh: Empirically Derived Lessons and Opportunities for Policy Makers and Practitioners** 59
Johannes Luetz

Part II Limits to Climate Change Adaptation in Africa

6 **Limits to Climate Change Adaptation in Zimbabwe: Insights, Experiences and Lessons** 109
Nelson Chanza

7 **Pastoralists Shifting Strategies and Perceptions of Risk: Post-crisis Recovery in Damergou, Niger** 129
Karen Marie Greenough

8 **Political Limits to Climate Change Adaptation Practices: Insights from the Johannesburg Case** 143
Karen Hetz

9 **Constraints and Limits to Climate Change Adaptation Efforts in Nigeria** 159
Idowu O. Ologeh, Joshua B. Akarakiri and Francis A. Adesina

10 **Adapting Agriculture to Climate Change: Suitability of Banana Crop Production to Future Climate Change Over Uganda** 175
Geoffrey Sabiiti, Joseph Mwalichi Ininda, Laban Ayieko Ogallo, Jully Ouma, Guleid Artan, Charles Basalirwa, Franklin Opijah, Alex Nimusiima, Saul Daniel Ddumba, Jasper Batureine Mwesigwa, George Otieno and Jamiat Nanteza

11 **Local Adaptation to Climate Extremes in Domboshawa: Opportunities and Limitations** 191
Vincent Itai Tanyanyiwa and Rejoice Madobi

Part III Limits to Climate Change Adaptation in Australia, North-America and Europe

12 **The Limits of Imagination** 211
Liese Coulter

13 **Climate Change Adaptation in Coastal East Arctic Ecosystems: Complexity and Challenges of Monitoring and Evaluation** 227
Moktar Lamari, Line Poulin-Larivière and Johann L. Jacob

14 **Limits to Adaptation on Climate Change in Bosnia and Herzegovina: Insights and Experiences** 245
Goran Trbic, Davorin Bajic, Vladimir Djurdjevic, Vladan Ducic, Raduska Cupac, Đorđe Markez, Goran Vukmir, Radoslav Dekić and Tatjana Popov

Part IV Limits to Climate Change Adaptation in the Pacific Region

15 **Climate Change Adaptation Limits in Small Island Developing States** 263
Stacy-ann Robinson

16 **Limits to Coastal Adaptation in Samoa: Insights and Experiences** 283
Richard Crichton and Miguel Esteban

17 **Limits to Capital Works Adaptation in the Coastal Zones and Islands: Lessons for the Pacific** 301
Brendan Mackey and Daniel Ware

18 A 'Cost Barrier' Perspective to Adaptation on a Coral Triangle Initiative (CTI) and Mangrove Rehabilitation Projects (MRP) in Solomon Islands 325
Michael Otoara Ha'apio, Walter Leal Filho and Morgan Wairiu

19 Customary Land and Climate Change Induced Relocation: A Case Study of Vunidogoloa Village, Vanua Levu, Fiji 345
Dhrishna Charan, Manpreet Kaur and Priyatma Singh

20 Limits to Adapting to Climate Change Through Relocations in Papua-New Guinea and Fiji 359
Dalila Gharbaoui and Julia Blocher

21 Atoll Habitability Thresholds 381
Mark H. N. Stege

22 Conclusions: Overcoming the Limits to Adaptation 401
Walter Leal Filho and Johanna Nalau

Contributors

Francis A. Adesina Department of Geography, Obafemi Awolowo University, Ile-Ife, Nigeria

Joshua B. Akarakiri African Institute of Science Policy and Innovation (AISPI), Obafemi Awolowo University, Ile-Ife, Nigeria

G. M. Monirul Alam School of Commerce, Faculty of Business, Education, Law and Arts, University of Southern Queensland, Queensland, Australia; Bangabandhu Sheikh Mujibur Rahman Agricultural University, Joydebpur, Bangladesh

Khorshed Alam School of Commerce, Faculty of Business, Education, Law and Arts, University of Southern Queensland, Queensland, Australia

Guleid Artan Inter-Governmental Authority on Development's (IGAD) Climate Prediction and Applications Centre (ICPAC), Nairobi, Kenya

Davorin Bajic Faculty of Science, University of Banja Luka, Banja Luka, Bosnia and Herzegovina

Charles Basalirwa Department of Geography, Geo-Informatics and Climatic Sciences, Makerere University, Kampala, Uganda

Julia Blocher Hugo Observatory for Environmental Migration University of Liège, Liège, Belgium; United Nations University, New York, NY, USA

Nelson Chanza Department of Geography, Bindura University of Science Education, Bindura, Zimbabwe

Dhrishna Charan Department of Science, The University of Fiji, Lautoka, Fiji

Liese Coulter Griffith University, Nathan, QLD, Australia

Richard Crichton University of Tokyo, Kashiwa, Chiba Prefecture, Japan

Raduska Cupac Faculty of Science, University of Banja Luka, Banja Luka, Bosnia and Herzegovina

Saul Daniel Ddumba Department of Geography, Geo-Informatics and Climatic Sciences, Makerere University, Kampala, Uganda

Radoslav Dekić Faculty of Science, University of Banja Luka, Banja Luka, Bosnia and Herzegovina

Vladimir Djurdjevic Faculty of Science, University of Banja Luka, Banja Luka, Bosnia and Herzegovina

Vladan Ducic Faculty of Science, University of Banja Luka, Banja Luka, Bosnia and Herzegovina

Miguel Esteban University of Tokyo, Kashiwa, Chiba Prefecture, Japan

Dalila Gharbaoui Hugo Observatory for Environmental Migration University of Liège, Liège, Belgium; Macmillan Brown Center for Pacific Studies (MBC), University of Canterbury, Christchurch, New Zealand

Daniel Gilfillan Fenner School of Environment and Society, The Australian National University, Acton, Australia

Karen Marie Greenough WASCAL, Ouagadougou, Burkina Faso

Michael Otoara Ha'apio Pacific Centre for Environment and Sustainable Development, The University of the South Pacific, Suva, Fiji

Karen Hetz Geography Institute, Humboldt Universität zu Berlin, Berlin, Germany; School of Architecture and Planning, University of the Witwatersrand, Johannesburg, South Africa; Environmental and Regional Sciences Department, University Trier, Trier, Germany

Joseph Mwalichi Ininda Department of Meteorology, University of Nairobi, Nairobi, Kenya

Johann L. Jacob École National d'administration Publique, University of Quebec, Quebec, Canada

Manpreet Kaur Department of Language and Literature, The University of Fiji, Lautoka, Fiji

Most Nilufa Khatun Bangabandhu Sheikh Mujibur Rahman Agricultural University, Joydebpur, Bangladesh

Adish Khezri Faculty of Geo-Information Science and Earth Observation (ITC), University of Twente, Enschede, The Netherlands

Moktar Lamari École National d'administration Publique, University of Quebec, Quebec, Canada

Walter Leal Filho School of Science and the Environment, Manchester Metropolitan University, Manchester, UK

Johannes Luetz CHC Higher Education, Brisbane, Carindale, QLD, Australia

Brendan Mackey Griffith Climate Change Response Program, Griffith University, Gold Coast, QLD, Australia

Rejoice Madobi Department of Geography and Environmental Studies, Faculty of Science and Technology, Zimbabwe Open University, Harare, Zimbabwe

Đorđe Markez Faculty of Science, University of Banja Luka, Banja Luka, Bosnia and Herzegovina

Shahbaz Mushtaq International Centre for Applied Climate Sciences, University of Southern Queensland, Queensland, Australia

Jasper Batureine Mwesigwa Inter-Governmental Authority on Development's (IGAD) Climate Prediction and Applications Centre (ICPAC), Nairobi 00100, Kenya; Institute for Climate Change and Adaptation, University of Nairobi, Nairobi 00100, Kenya

Johanna Nalau Griffith Institute for Tourism and Griffith Climate Change Response Program, Griffith Business School, Griffith University, Queensland, Australia

Jamiat Nanteza Department of Geography, Geo-Informatics and Climatic Sciences, Makerere University, Kampala, Uganda

Alex Nimusiima Department of Geography, Geo-Informatics and Climatic Sciences, Makerere University, Kampala, Uganda

Laban Ayieko Ogallo Department of Meteorology, University of Nairobi, Nairobi 00100, Kenya; Inter-Governmental Authority on Development's (IGAD) Climate Prediction and Applications Centre (ICPAC), Nairobi 00100, Kenya

Idowu O. Ologeh African Institute of Science Policy and Innovation (AISPI), Obafemi Awolowo University, Ile-Ife, Nigeria

Franklin Opijah Department of Meteorology, University of Nairobi, Nairobi, Kenya

George Otieno Inter-Governmental Authority on Development's (IGAD) Climate Prediction and Applications Centre (ICPAC), Nairobi, Kenya

Jully Ouma Inter-Governmental Authority on Development's (IGAD) Climate Prediction and Applications Centre (ICPAC), Nairobi, Kenya

Tatjana Popov Faculty of Science, University of Banja Luka, Banja Luka, Bosnia and Herzegovina

Line Poulin-Larivière École National d'administration Publique, University of Quebec, Quebec, Canada

Stacy-ann Robinson Institute at Brown for Environment and Society, Brown University, Providence, RI, USA

Geoffrey Sabiiti Department of Geography, Geo-Informatics and Climatic Sciences, Makerere University, Kampala, Uganda; Inter Governmental Authority on Development's (IGAD) Climate Prediction and Applications Centre (ICPAC), Nairobi 00100, Kenya

Priyatma Singh Department of Science, The University of Fiji, Lautoka, Fiji

Mark H. N. Stege Executive Council, Maloelap Atoll Local Government, Taroa Maloelap, MH, Marshall Islands

Vincent Itai Tanyanyiwa Department of Geography and Environmental Studies, Faculty of Science and Technology, Zimbabwe Open University, Harare, Zimbabwe

Goran Trbic Faculty of Science, University of Banja Luka, Banja Luka, Bosnia and Herzegovina

Arbind Man Tuladhar Faculty of Geo-Information Science and Earth Observation (ITC), University of Twente, Enschede, The Netherlands

Goran Vukmir Faculty of Science, University of Banja Luka, Banja Luka, Bosnia and Herzegovina

Morgan Wairiu Pacific Centre for Environment and Sustainable Development, The University of the South Pacific, Suva, Fiji

Daniel Ware Griffith Climate Change Response Program, Griffith University, Gold Coast 4222, QLD, Australia; Centre for Coastal Management, Griffith University, Gold Coast 4222, QLD, Australia

Jaap Zevenbergen Faculty of Geo-Information Science and Earth Observation (ITC), University of Twente, Enschede, The Netherlands

Chapter 1
Introduction: Limits to Adaptation

Johanna Nalau and Walter Leal Filho

> *- Humanity is at the crossroads. We have ample evidence that Earth is headed for disaster, and for the first time in history we have the ability to prevent that disaster from wiping us out. Whether the disaster is caused by humans or by nature, it is inevitable. But our doom is not*
>
> (Newitz 2013, p. 1).

The Intergovernmental Panel on Climate Change (IPCC) in its Fifth Assessment Report (2014) already noted the many projected changes if the world does not quickly act on reducing greenhouse gases. The Paris Agreement in 2015 signalled great ambitions by most of the countries to take this task more seriously. The COP 22 in Marrakesh in 2016 saw somewhat slower progress on many of the issues from the Paris Agreement but the adaptation policy space is now very much globally recognised.

Yet, just a few recent events demonstrate how close we might be in starting to reach particular thresholds and limits. In January 2017, the year 2016 was declared the warmest year on record globally since the earth surface temperature records began in 1880 (NASA 2017). 2016 also saw extreme coral bleaching in the Great Barrier Reef in Australia with 93% of the reefs impacted (ARC Coral Reef Studies 2016). Although some of these messages might be getting lost in the constant flood of information, it seems that the discussion on how to adapt to global climate change is beginning to become a reality to more than just a few Pacific islands, which have already been facing increased storm surges and sea level rise.

J. Nalau (✉)
Griffith Institute for Tourism and Griffith Climate Change Response Program, Griffith Business School, Griffith University, Queensland, Australia
e-mail: j.nalau@griffith.edu.au

W. Leal Filho
School of Science and the Environment, Manchester Metropolitan University, Chester Street, Manchester M1 5GD, UK
e-mail: w.leal@mmu.ac.uk

W. Leal Filho and J. Nalau (eds.), *Limits to Climate Change Adaptation*,
Climate Change Management, https://doi.org/10.1007/978-3-319-64599-5_1

The concept of limits to adaptation is by no means new. We have long known of past societies reaching their limits, with vivid stories and historical reconstructions of the downfalls of even the most magnificent societies that were supposed to flourish. People's interest in societal change has spurred popular literature such as Jared Diamond's famous book 'Collapse' (Diamond 2005) and more recently Annalee Newitz's book 'Scatter, Adapt, Remember' (Newitz 2013) that explores the very reasons why particular limits have been reached and what can be done about them. Terms such as thresholds, planetary boundaries, extinctions, and limits all imply something permanent; something that once crossed cannot be easily reversed if ever. Adaptation therefore is seen as a process of guiding the society away from such thresholds or at least in minimizing potential losses and damages arising from living too close to such tipping points.

While it took the scientific and policy community time to start considering adaptation (Pielke 1998), most of the focus since then has been broadly on adaptation options, definitions, frameworks, and case studies to demonstrate adaptation processes and its implementation. Yet, given the international level negotiations through United Nations Framework Convention on Climate Change (UNFCCC) on mitigation, adaptation, and associated Loss and Damage (Warner and Afifi 2014; Warner and van der Geest 2013; Warner 2015), the concept of limits to adaptation is gathering interest in policy and research communities in trying to understand why limits to adaptation emerge, how these can be dealt with, and the kinds of ethical considerations that need to be taken into account (Adger et al. 2009; Barnett et al. 2013, 2015; Klein et al. 2014).

Limits come indeed in many shapes and sizes, and are greatly influenced by underlying cultural, social, economic, historical, and environmental contexts and processes (Adger et al. 2009; Dow et al. 2013; Klein et al. 2014). However, what exactly is an adaptation limit is still proving in many cases elusive as most of the discussions focus on potential future limits. The evidence is still in the making and many adaptation limits remain unknown and poorly understood partly due to the complexity of interacting processes and factors, which together constitute a potential limit and partly due to the changing nature of soft limits as more options become available and hence open up new pathways that enable shifts in what constitutes a limit to adaptation.

This book aims to contribute to this discussion through detailed case studies from Africa, the Pacific, Asia, and Europe, and through a broader scale synthesis, which collates these insights from across sectors and countries. The book pays special attention to evidences, which illustrate how and where such limits have arisen or are in the process of becoming limits, and indicate future trends and contexts that might prove helpful in understanding adaptation limits. In particular, the book provides an overview of some of the challenges and opportunities surrounding adaptation limits at different temporal, jurisdictional, and spatial scales, also outlining examples of case studies, projects and best practice which show how they may be addressed.

Defining Limits and Constraints to Adaptation

In the recent Intergovernmental Panel on Climate Change (IPCC)'s Fifth Assessment Report (hereafter AR5), limits to adaptation were considered mostly in terms of defining conceptual linkages between, for example, limits and transformational adaptation, and discussing particular ethical considerations and cross-linkages between concepts (Klein et al. 2014). The AR5 identifies currently three kinds of limits: biophysical, sociocultural and economic (Klein et al. 2014, p. 923). Literature to date has been more interested in examining adaptation limits within a particular sector, ecosystem or at specific species level (Klein et al. 2014). For example, the AR5 found that studies looking at adaptation limits in the AR5's regional chapters were mostly reporting on biophysical limits, with majority of information on Europe and Open oceans, and small number of studies in Africa, Asia, Australasia, Polar regions and Small islands (Klein et al. 2014, p. 923). Similarly, some socio-cultural research on adaptation limits had been conducted in Africa, Europe and Australasia, with focus on freshwater, terrestrial, urban and rural areas, human health and human security. Economic limits research has so far focused on food systems, human health and rural areas, and had only been conducted in Europe (Klein et al. 2014).

Dow et al. (2013) define an adaptation limit as "the point at which an actor's objectives cannot be secured from intolerable risks through adaptive actions" (Dow et al. 2013). A limit is reached in other words when adaptation efforts are unable to provide an acceptable level of security from risks to the existing objectives and values and prevent the loss of the key attributes, components, or services of ecosystems" (Klein et al. 2014, p. 919). Two kinds of limits can be recognised: hard limits where there are no more options available and soft limits where options might not be currently available but could be in the future (Barnett et al. 2015; Dow et al. 2013; Klein et al. 2014). Many authors note a difference between an adaptation constraint and adaptation limit: constraints can be overcome but a limit represents a point where radical change is required as there are no other options available to e.g. maintain a particular livelihood or live in a particular place (Barnett et al. 2013, 2015; Dow et al. 2013; Moser and Ekstrom 2010). Islam et al. (2014) regard limits as physical characteristics such as more intense cyclones and treat constraints as socially constructed obstacles such as lack of access to credit. Barnett et al. (2013) argue that limits to adaptation "involve irreversible losses of things individuals care about, either due to climate change impacts or as outcomes of climate change policies". Warner (2015) on the other hand argues that limits at the household scale can be for example conceptualised as tradeoffs marked by irreversibility and suffering with uncertainty.

It has long been recognised that many, although not all, adaptation limits are essentially socially constructed (Adger et al. 2009; Klein et al. 2014; O'Brien 2009). As Barnett et al. (2013) note, "the limits to adaptation are, to some extent, in the eye of the beholder" as perceptions as to what is a limit is based on values, which naturally differ among groups and individuals (Adger et al. 2009; Klein et al.

2014). This of course differs within ecological systems where for example some ecosystems reach limits regardless of how humans value a system or species habitat. What is perceived to be at risk, therefore, guides also our notions of limits to adaptation, and 'intolerable risk' (Dow et al. 2013; Klein et al. 2014).

In the policy and adaptation literature, framing an issue in a particular way impacts on what people perceive to be possible and what should be done. For example, if climate change is defined as a complex unbounded problem, such conceptualisation greatly impacts on the kinds of management interventions and options that are necessary to address it (Nalau and Handmer 2015). Perceptions do not equal action however. Even if the problem can be perceived as 'intolerable risk' but the means to address it might not be sufficient, a social compromise has to be made between other values/interests and eventually the actual response might be less comprehensive than what the actual risk category would require in reality (Nalau and Handmer 2015).

As Klein et al. (2014) reiterate, limits are multi-scalar spanning across different jurisdictions and times, and are impacted to a great extent by the interplay between both public and private actors and institutions. Hence, questions of governance and values sit also at the heart of any discussion on adaptation limits as the choices that are made or will be made very much reflect the current context, which in turn is influenced by the past and impacts on the future. Dow et al. (2013) note that planned and well-thought decision-making can change what is possible and in such sense shift an adaptation limit. The authors consider risk-based decision-making and adaptation limits as two parallel spheres of research, which are closely connected. However, not everyone agrees with research and policy focusing on the concept of adaptation limits. Bradsley (2015) for instance argues that a strong focus on constraints and limits to adaptation can hinder actors from taking adaptive responses, and that too much focus on adaptation limits can sidetrack from using a broader perspective of sustainability. This means taking into consideration other processes than just climate change, which are already impacting ecosystems and/or increasing community vulnerability.

Chapters in This Book

Given the significant gap in knowledge regarding adaptation limits as noted by the IPCC, this book aims to shed light on the kinds of limits that have and will most likely emerge with increased anthropogenic climate change. The chapters in this book demonstrate the variety and interconnectedness of factors that together constitute limits and constraints to adaptation. Many of the chapters also explore the question what adaptation limits look like in practice and how these differ potentially from adaptation constraints. All of the authors were asked to build on the knowledge base from the IPCC Fifth Assessment Report, in particular the Working Group II Chapter 16 on Adaptation Limits, Constraints and Opportunities (Klein et al. 2014). The disciplinary breadth is vast and includes contributions from psychology, social science, agriculture, political science, migration studies, coastal management, history, urban planning, and marine science. In order to be able to also contribute to region-specific adaptation limits literature, the chapters have therefore been organised around geographic regions.

The first set of chapters focus on various experiences regarding adaptation limits in Asia. Alam et al. provide a case study from Bangladesh where they explore the vulnerability of households in erosion prone riverbank areas and the perceived constraints to adopting adaptation measures. The authors find both similarities and differences in the adaptation constraints across farming groups highlighting that factors such as wealth often equal in better access to credit and services. Gilfillan in turn presents an analysis of climate change adaptation limits in the Cambodian Health sector. The chapter focuses in particular on two adaptation limits: coordination and capacity development, which are seen to significantly impede progress on adaptation in Cambodia. Khezri et al. investigate the potential of land-based strategic model in delivering cross-sectoral adaptation in Nepal. The authors find a variety of strategies to overcome adaptation constraints and limits mainly by coordinating and targeting land-based adaptation services. Luetz explores the environmental and non-environmental push and pull factors, which lead to migration in Bangladesh from rural to urban areas. He finds several tipping points that represent intolerable risks for well-being and examines climate change adaptation in the context of rural-urban linkages and slums in Dhaka.

The second set of chapters introduces a variety of adaptation limits in the African context. Chanza examines the adaptation landscape in Zimbabwe across multiple sectors such as agriculture, water, energy, human settlements and biodiversity. Chanza's analysis shows how many adaptation limits are tightly tied to existing contextual variables, including past economic performance and general level of development. Greenough uses oral histories in understanding environmental changes and coping strategies of pastoralists communities in Niger to drought and famine. She also explores how gender can be a limiting factor in resource entitlements and adaptation choices, and compares post–1984 and post–2010 recovery strategies to show how increased tolerance for certain risks helps some households take advantage of these rare opportunities. Hertz looks at the case of Johannesburg in South Africa and its urban planning practices through the lens of adaptation limits. Her research provides insights into why particular political limits emerge that impede climate adaptation including entrenched development expectations. Hertz also challenges the synergies between development and climate adaptation based on her observations of the Johannesburg urban planning processes and issues. Ologeh et al. in turn investigate how adaptation limits emerge in the context of the agricultural sector in Nigeria. The authors examine the kinds of adaptation limits, which seem particularly strong in explaining the lack of progress in adapting livelihoods in the agricultural sector including social and cultural limits in technology use. Sabiiti et al. examine the limits of banana farming in Uganda using the Intergovernmental Panel on Climate Change's (IPCC) Representative Concentration Pathways (RCPs, 2.6, 4.5, 6.0 and 8.5) and Special Report on Emission Scenarios (SRES, A1B and A2) across the period 2011–2090. The authors specifically investigate the conditions under which banana farming can maintain favourable crop growth and where banana farming is likely to face limits.

The next set of chapters span geographically different regions including Australia, Canada, East Arctic, and Eastern Europe. In her chapter on limits to imagination Liese Coultier examines how professionals who work with climate change information in Australia and Canada personally perceive the future (next two decades), and how the concept of personal vulnerability is understood. Coultier notes that behaviour to address near-term climate impacts may be limited by several factors, including lack of sense of urgency, and emphasis on expected societal interventions. Coultier's research poses interesting questions in how our imagination and perceptions can pose limits on our ability to choose and implement adaptation actions both at personal and professional levels. Lamari et al. examine the current use of Monitoring and Evaluation (M&E) indicators for climate change adaptation in Nunavut in the East Arctic. The authors are particularly interested in knowledge-related issues that potentially pose limits to adaptation including the underlying policy priorities and societal context, which might impede the gathering and utilisation of particular knowledge for effective adaptation. Trbic et al. provide a national level analysis of adaptation limits in Bosnia and Herzegovina. Trbic and colleagues note several constraints and limits to adaptation, including the legislative frameworks in Bosnia and Herzegovina, which to date do not provide any requirement in planning to even consider impacts of climate change or adaptation for that matter.

The last set of chapters focus specifically on the Small Island Developing States (SIDS), which to date have been at the forefront in recording and experiencing climate change related impacts. Robinson provides a broader regional perspective on SIDS from the Caribbean, Pacific and Mediterranean regions. She examines 19 National Communications to United Nations Framework Convention on Climate Change (UNFCCC) and provides an analysis how adaptation limits have been defined and identified by the select countries. She finds strong variability in the extent of limits but notes that institutional limits and physical and ecological limits dominate the reporting. Crichton and Esteban examine the case of Samoa and to date implemented adaptation projects in the country. Their chapter provides interesting examples of how different adaptation options might interact and in the mix give rise to adaptation limits in the process. Mackey and Ware undertake a review of the limits of capital works as a form of climate adaptation in the Pacific Islands. The authors note in particular the financial limits that come with capital works, and explore a range of factors influencing in particular the Pacific context. The chapter by Ha'apio et al. focuses on the Coral Triangle Initiative in the Solomon Islands. The authors explore the main adaptation constraints in implementing Marine Protected Areas (MPAs) and present data on the costs and benefits perceived to be accruing from conservation programs at village level.

Three of these chapters examine specifically relocation in the context of adaptation limits. Charan et al. examine the concept of climate related relocation in Vunidogoloa village in Fiji, which relocated to higher ground due to increased inundation. The authors explore the intricacies surrounding the socio-cultural aspects of the relocation process and what this means for limits to adaptation. Gbharbaoui and Blocher investigate the concept of climate-induced relocation in Papua New Guinea and Fiji, some of the largest Pacific islands in terms of populations. The notion of limits is explored through the experiences of many communities who have already faced limits to their current adaptive capacity both in social and environmental terms. The last chapter by Stege explores the concept of 'atoll habitability thresholds' as adaptation limits in Marshall Islands and focuses in particular on the question of 'when is time to move'. His chapter examines and introduces the *Reimaanlok* framework for relocation that is being trialled in the Marshall islands as a homegrown concept for planned relocation. Stege stresses the importance of integrating indigenous knowledge with science in order to adequately understand contextual adaptation limits.

Conclusions

Even though our current understanding of adaptation limits might still be conceptually fuzzy and inadequate, more and more examples are being put forth in how we can start understanding and imagining what such concepts entail in practice. This raises also philosophical and ethical issues as to whose responsibility it is to counteract or subsume to the soft and hard limits of adaptation. There is also the more complex issue of the timing of reacting to a limit: at what point do people accept an adaptation limit, and in which cases bringing adaptation options to the table pushes the limit further (the limit is retained and adaptation options simply buy time), or whether adaptation options can 'erase' or dissolve a particular limit. The contributions in this book point to a variety of experiences and answers. Given that the Paris Agreement has instituted both the Warsaw Mechanism on Loss and Damage (Article 8), and climate change adaptation (Article 7) as separate institutional items, this book offers timely reflections on how some of those institutional modalities could potentially be employed in practice in those regions already facing thresholds and limits.

References

Adger, W. N., Dessai, S., Goulden, M., Hulme, M., Lorenzoni, I., Nelson, D. R., et al. (2009). Are there social limits to adaptation to climate change? *Climatic Change, 93,* 335–354.

ARC Coral Reef Studies. (2016). Only 7% of the Great Barrier Reef has avoided coral bleaching. https://www.coralcoe.org.au/media-releases/only-7-of-the-great-barrier-reef-has-avoided-coral-bleaching. Accessed 20 April 2016.

Bardsley, D. K. (2015). Limits to adaptation or a second modernity? Responses to climate change risk in the context of failing socio-ecosystems. *Environment, Development and Sustainability, 17*(1), 41–55.

Barnett, J., Evans, L. S., Gross, C., Kiem, A. S., Kingsford, R. T., Palutikof, J. P., et al. (2015). From barriers to limits to climate change adaptation: Path dependency and the speed of change. *Ecology and Society, 20*(3), 11.

Barnett, J., Mortreux, C., & Adger, W. N. (2013). Barriers and limits to adaptation: Cautionary notes. In *Natural disasters and adaptation to climate change* (pp. 223–235).

Diamond, J. M. (2005). *Collapse: How societies choose to fail or succeed.* New York, NY: Viking.

Dow, K., Berkhout, F., & Preston, B. L. (2013). Limits to adaptation to climate change: A risk approach. *Current Opinion in Environmental Sustainability, 5,* 384–391.

Islam, M. M., Sallu, S., Hubacek, K., & Paavola, J. (2014). Limits and barriers to adaptation to climate variability and change in Bangladeshi coastal fishing communities. *Marine Policy, 43,* 208–216.

Klein, R. J. T., Midgley, G. F., Preston, B. L., Alam, M., Berkhout, F. G. H., Dow, K., et al. (2014). Adaptation opportunities, constraints and limits. In C. B. Field, V. R. Barros, D. J. Dokken, et al. (Eds.), *Climate change 2014: Impacts, adaptation, and vulnerability. Part A: Global and sectoral aspects: Contribution of the working group II to the fifth assessment report of the intergovernmental panel on climate change* (pp. 899–943). Cambridge: Cambridge University Press.

Moser, S., & Ekstrom, J. (2010). A framework to diagnose barriers to climate change adaptation. *Proceedings of the National Academy of Sciences of the United States of America, 107*(51), 22026–22031.

Nalau, J., & Handmer, J. (2015). When is transformation a viable policy alternative? *Environmental Science and Policy, 54,* 349–356.

NASA. (2017). NASA, NOAA data show 2016 warmest year on record globally. https://www.nasa.gov/press-release/nasa-noaa-data-show-2016-warmest-year-on-record-globally. Accessed 18 January 2017.

Newitz, A. (2013). *Scatter, adapt, remember: How humans will survive a mass extinction.* Collingwood: Black Inc.

O'Brien, K. (2009). Do values subjectively define the limits to climate change adaptation? In N. W. Adger, I. Lorenzoni, & K. O'Brien (Eds.), Adapting to Climate Change: Thresholds, Values, Governance (pp. 164–180). Cambridge: Cambridge University Press.

Pielke, R. A., Jr. (1998). Rethinking the role of adaptation in climate policy. *Global Environmental Change, 8*(2), 159–170.

Warner, B. P. (2015). Understanding actor-centered adaptation limits in smallholder agriculture in the Central American dry tropics. *Agriculture and Human Values, 33,* 1–13.

Warner, K., & Afifi, T. (2014). Where the rain falls: Evidence from 8 countries on how vulnerable households use migration to manage the risk of rainfall variability and food insecurity. *Climate and Development, 6*(1), 1–17.

Warner, K., & Van der Geest, K. (2013). Loss and damage from climate change: Local-level evidence from nine vulnerable countries. *International Journal of Global Warming, 5*(4), 367–386.

Part I
Limits to Climate Change Adaptation in Asia

Chapter 2
Strategies and Barriers to Adaptation of Hazard-Prone Rural Households in Bangladesh

G. M. Monirul Alam, Khorshed Alam, Shahbaz Mushtaq, Most Nilufa Khatun and Walter Leal Filho

Introduction

The issues of climate change and hazards are an ongoing part of human civilization. However, poor people in developing countries whose subsistence livelihood depend upon the utilisation of natural resources are the first and most affected by the climate change which increases their vulnerability (Alam 2016; IPCC 2014; Bardsley and Wiseman 2012; McDowell et al. 2013; Salick et al. 2009; Thomas and Twyman 2005). Climate change can manifest in four main ways: (i) slow change in mean climate conditions, (ii) increased inter annual and seasonal variability, (iii) increased frequency of extreme events, and (iv) rapid climate changes causing catastrophic shifts in ecosystems (Tompkins and Adger 2004).

Bangladesh is most vulnerable to climate-driven hazards, which pose a major risk to the lives, livelihoods and food security of the 64% of the rural population who depend on agriculture (GoB 2011; IPCC 2007). Scholar argued that farmers' capacity to adapt to the compounding influences of climate change, which can affect households' resources and resilience, is uncertain due to their poor socio-economic conditions (Wood et al. 2014; Lobell et al. 2008; IPCC 2007; Adger and Vincent

G. M. M. Alam · K. Alam
School of Commerce, Faculty of Business, Education, Law and Arts, University of Southern Queensland, Queensland, Australia

G. M. M. Alam · M. N. Khatun
Bangabandhu Sheikh Mujibur Rahman Agricultural University, Joydebpur, Bangladesh

S. Mushtaq
International Centre for Applied Climate Sciences, University of Southern Queensland, Queensland, Australia

W. Leal Filho (✉)
School of Science and the Environment, Manchester Metropolitan University, Chester Road, Manchester M15 6BH, UK
e-mail: g.m.monirul.alam@usq.edu.au; gmmonirul79@gmail.com

W. Leal Filho and J. Nalau (eds.), *Limits to Climate Change Adaptation*,
Climate Change Management, https://doi.org/10.1007/978-3-319-64599-5_2

2005). Therefore, adaptation measures are important to help the poor communities to cope with extreme weather conditions and associated climatic variations (Niles et al. 2015; Gandure et al. 2013; Rosenzweig et al. 2013; Adger et al. 2003).

In Bangladesh, the coastal and riverine households are the most susceptible to the impacts of climate-driven hazards, including riverbank erosion (Alam et al. 2017a, b; Alam 2016, 2017; GoB 2010) and recent models of hydrological impacts of climate change in different climatic zones have shown this to be true across Asia (Eregno et al. 2013). In particular, the hazard of riverbank erosion is a common problem in Bangladesh, which contributes to the loss of both physical and material endowments through loss of land, natural resources and employment opportunities of the riverine rural households and thus threatening their food security and livelihoods (Alam et al. 2017a, b). Due to climate change, they are also expected to face a projected increase in mean annual temperature, uncertainty in rainfall, and surges in disease, pest and weed pressure on crops and livestock which might have disastrous impact on their livelihood (Alam 2016).

In Bangladesh, twenty districts out of 64 in the country are prone to the riverbank erosion (CEGIS 2012; GoB 2010); while another study asserted that some parts of 50 districts in the country are subject to riverbank erosion (Elahi et al. 1991) (see the map in Fig. 2.1). Moreover, resource-poor households in the riverine areas are more prone to the impacts of frequent floods and waterlogging due to their proximity to the river, which also increases their vulnerability (Alam et al. 2016). About 8700 ha of homestead and farming land are lost to riverbank erosion, which displaces approximately 200,000 people annually (GoB 2010).

Despite increasing recognition of the need of adaptation to reduce such households' vulnerability, limited research, however, has so far been carried out in the domain of adaptation (discussed in section "Review of Literature"). Moreover, not all communities within the country are uniformly affected by climate change and hazards due to differential livelihood options and resources for adaptation (Alam 2016). Therefore, hazards-prone resource-poor household's strategies and barriers to adaptation which is not so far explored, is crucial to formulate and implement effective and sustainable adaptation policies in Bangladesh and elsewhere.

Moreover, in terms of barriers to adaptation in Bangladesh, the following ones are commonly seen: (i) cultural values which mean that people tend to remain in areas where extreme events occur (e.g., coastal and riverbank areas), (ii) lack of access to funds to implement adaptation measures, especially those related to infrastructure (iii) lack of access to technologies to adapt, and (iv) lack of a policy framework to implement adaptation measures in a sustainable way. Apart from the above, a major barrier is related to the fact that social systems make it difficult for people to reduce the adverse effects of climate on their livelihood and well-being, increasing their vulnerability. This study, using survey data from the most severe riverbank erosion-prone areas in Bangladesh, aims to provide information on adaptation strategies of the household and barriers to adaptation. The research questions to understand this are: (a) what adaptation strategies can the resource-poor households adopt to enhance their resilience?; and (b) what are the barriers to adaptation?

Fig. 2.1 The study areas The Nagarpur and ChauhaliUpazilas

The paper is organized as follows: a review of the literature is presented in section "Review of Literature"; "Methodology" section presents descriptions of the study areas, the data collection procedure and the analysis of the data; the results are presented and discussed in section "Results and Discussion"; and "Conclusions" section provides a summary and some policy guidelines.

Review of Literature

Globally, mitigation and adaptation are the two major policy responses to climate change. Adaptation has the potential to reduce the adverse impacts of climate change (IPCC 2001). IPCC defines adaptation as the adjustment in human or natural systems in response to climatic or environmental stimuli, which buffer harm or exploit beneficial opportunities (IPCC 2001). However, adaptation strategies vary from sector to sector, community to community and place to place (Malone and Brenkert 2008; Smit and Wandel 2006). Scholars argue that all adaptation is not good (Eriksen et al. 2011; Nyong et al. 2007). For example, the adaptation measures that deliver short-term gains and economic benefits can lead to increased vulnerability in the medium or long run (Jones and Boyd 2011). According to Smith et al. (2000), to fully understand adaptation, it is important to know three fundamentals of adaptation namely, adaptation to what, who adapt and how adaptation occurs?

There are few adaptation studies in Bangladesh that mainly focus on drought prone areas in Bangladesh (see, for example, Alam 2015; Alauddin and Sarker 2014; Sarker et al. 2013; Habiba et al. 2012). Few studies also focus on low lying and saline-prone areas (Rashid et al. 2014; Anik and Khan 2012). Though these studies provide important policy inputs, however, these might not be effective and applicable to other hazard-prone communities due to the heterogeneity of the impact of the various hazards and the socio-economic conditions of the households and therefore their responses vary (Alam 2016). This is particularly important for the most vulnerable riparian communities who are poorly resourced. There are studies on displacement and the socio-economic impacts of riverbank erosion in Bangladesh (see, for example, Alam 2016, 2017; Alam et al. 2017a, b; Ahmed 2015; Lein 2010; Hutton and Haque 2004; Makenro 2000; Elahi 1989; Zaman 1989; Rogge and Haque 1987; Greenberg 1986; Hossain 1984). However, there is a lack of in-depth empirical research on how the resource-poor hazard-prone households' response to climate change and hazards and what the factors are limiting their local adaptive responses. Scholars have argued that local level adaptation knowledge is a key to promoting the resilience of vulnerable communities (Alam et al. 2017a, b; Hiwasaki et al. 2014; Alexander et al. 2011; Green and Raygorodetsky 2010; Ellen 2007; Nyong et al. 2007). In terms of policymaking, farmers' local knowledge of adaptation strategies will have immense significance if they are supported by relevant government organizations, NGOs and research for the overall sustainability of the adaptation process in the country.

Methodology

Study Area

This study is based on the data collected from riverbank erosion-prone areas in Bangladesh as a case study. A multistage sampling technique was employed to collect the data. The sever riverbank erosion-affected district, upazila[1] and riverine villages were selected through consultation with the local experts. For each village, respondents were selected randomly. For the field survey, the Chauhaliupazila of the Sirajgonj district and the NagarpurUpazila of the Tangail district were selected (Fig. 2.1). The area is about 200 km north of Dhaka, the capital of Bangladesh. The Jamuna[2] River crosses the study area. The erosion rate of the river was around 5000 ha/year in the 1980s but only 2000 ha/year in recent years (CEGIS 2012). Data were collected from six riverine villages—Kashpukuria, Moradpur, Kairat, Datpur, Kashkawalia and Atapara.

Sampling, Questionnaire and Data Collection

Selection of the units of analysis is considered as the entry point in social science research. The unit of analysis influences greatly to the decision of research design, data collection and data analysis. In this study, the unit of analysis was households and for data collection, the household head (either male or female) was the survey participant. In Bangladesh, household heads have the power to exercise decision-making over household's resources and setting strategies (Alam 2016). First, a complete list of riverine households in the selected villages was collected from the Department of Agricultural Extension. In each village, 15% of the household heads were interviewed, which gives a sample size of 380 households for the study. For the cross-sectional household survey, 5% of the population is considered to be adequate (Bartlett et al. 2001). To ensure randomness in the sampling, a computer-generated random number table was applied to the list to select the 380 households.

Before data collection commenced, a structured survey questionnaire was tested with 20 respondents to ensure the adequacy of the information obtained and to avoid any ambiguity in the questions. Data were collected using face-to-face interviews between January and May 2014. The questionnaire sought information on the impact of climatic hazards, including riverbank erosion, on livelihoods and their response strategies. In the case of adaptation strategies, the respondents were asked about their range of practices. The respondents were also asked to mention the factors that limiting their successful adaptation.

Data Analysis

Descriptive analysis and Chi-square test were conducted to see whether there were differences between the farming groups in the adoption of adaptation strategies.

[1]Lower administrative unit of government; below district level but above village level.

[2]Bangladesh is composed of the floodplains and the deltas of three main rivers—the Padma (Ganges in India), the Jamuna and the Meghna (Brahmaputra in India).

Due to the small size of the land holdings, the households were categorized as large farm (12%) (>2.5 acres), medium farm (28%) (1.5–2.49 acres), small farm (33%) (1.49–0.5 acres) and landless (27%) (<0.5 acres) for a meaningful presentation of the results.

Results and Discussion

The results of the study are discussed in different phases. In the first phase, the socio-economic characteristics of the respondents are presented. The adaptation strategies of the households and barriers to adaptation are described later.

Socio-economic Characteristics

The information of households' socio-demographic and economic characteristics is very useful to get an insight into the profile of the study households and to formulate effective policy interventions. This information can be served as the delimitation of the study so that whatever findings or outcomes derived from this study can be described within the domain of this profile (Alam 2016). Socio-economic characteristics of the study households are presented below.

Half of the household heads belong to the age group of 46–60 years (Table 2.1). Average age of the household heads is around 45 years. Currently, the life expectancy at birth in Bangladesh is 70.3 years (UNESCO 2015). The majority of household heads is male (88%) as against women of 12%. The average family size of 5.21 is relatively large compared to the national average of 5.0 (BBS 2012). More than 46% of households had six members or more. The mean education level of the household was below primary level (3.17 years). More than 29% of respondents did not attend school. In Bangladesh, the estimated literacy rate was 61.5% in 2015 (UNESCO 2015). Majority of the household heads had education level between primary and secondary level. Only 9% had more than secondary education level.

Households' farm size is relatively low since all households had experienced loss of some of their land. More than 27% household belongs to the landless group. Household occupation groups are classified according to the main source of income (i.e., >50%) (Table 2.1). As expected, most of the households in the study areas depend on agriculture (71%) which is relatively higher than national statistics (BBS 2012). Service holders or affluent households usually live in nearby town or other places that are free from erosion problems.

Since most of the farmers depend on agriculture, therefore their income level is also low. More than 50% of households belong to the income level of Tk 61,000 to Tk 150,000 per year (US$ 1 = Tk 77). Road and transport communication is also inadequate in the areas. Farmers mainly use vans, bicycles, rickshaws, scooters, and tempo driven by small machines to market their products.

Table 2.1 Some selected socio-economic characteristic of the study households

Characteristics/variables	Number	Percentage
Age of HH head (mean: 45; range: 25–65)		
≤ 30 years	36	10
31–45 years	134	35
46–60 years	191	50
61–65 years	19	5
Gender of HH head		
Male	335	88
Female	45	12
HHs family members (mean: 5.21; range: 3–11)		
3	31	8
3–5	174	45
≤ 6 members	175	46
Education (mean: 3.17 years; range: 0–16)		
Illiterate	109	29
Primary (level 1–5)	137	36
Secondary (level 6–10)	104	27
Higher secondary (level 11–12)	21	6
<Higher secondary (level 12–16)	9	2
Employment status		
Agriculture	271	71
Business + Agriculture	75	20
Services + Agriculture	34	9
HHs yearly income (Tk) (mean: 35,000 Tk; Std. 38456)		
≥ 35,000	39	10
36,000–60,000	137	36
61,000–150,000	151	40
≤ 151,000	53	14
Farm category (average farm size: 0.56 acres)		
Large farm (>2.5 acres)	45	12
Medium farm household (1.5–2.49 acres)	107	28
Small farm household (1.49–0.5 acres)	127	33
Landless (<0.5 acres)	101	27

HH household

Households' Adaptive Strategies

The study revealed that all the households were responding to climate change and hazards through adopting a range of agricultural and non-agricultural adaptation strategies. An agricultural adjustment includes various techniques to boost up crop production such as adoption of new crop varieties, changing plantation time and irrigation techniques. By adopting agricultural adjustments, households can lessen the effects of riverbank hazard and other climate change issues on production and income. Non-agricultural adjustments, on the other hand, represent the practices that dampen the production and income ability such as sell of rest of land, livestock and poultry.

The study identified 15 farming and non-farming adaptation strategies, which were practiced by the respondents based on their long-term knowledge and perceptions of climate change (Table 2.2). Hazard-affected households response varies depending on their socio-economic conditions as well as the political and social settings. Most of the households adopted more than one strategy. Based on the respondents' main choice, the most common adaptation practices were changing plantation time, cultivation of pulses, cultivation of spices and oil seeds, homestead gardening, tree plantation and migration (Table 2.2).

Adaptation strategies were, however, shaped by farming category. A significant difference was found (χ^2-test, $p < 0.003$) in the adoption of adaptation strategies. We conducted a post-hoc analysis to see the location of the difference. The result indicates that non-agricultural adaptation was practiced mostly by small and landless farm households while agricultural adjustments were practiced mainly by large and medium farms. This indicates that wealthier farmers are in a better position to respond to the challenges posed by climate change and variability through adopting different strategies in agriculture whereas small and landless farmers have few choices.

Farmers were found to change their cultivation practices in response to the changing climate. For example, crop cultivation was found to be diversified in the study area. In the past, farmers rarely cultivated horticultural and cereal crops, and large parts of their farmland remained fallow in the dry season. Large and medium farmers (farmers with cultivatable land) were found to be adopting the HYV rice and wheat varieties as part of their response to the changing climatic conditions. Important changes is that they were cultivating spices and oil seeds in the newly formed char lands which had remained fallow due to the unavailability of crop varieties suitable for such land previously. In responding to the adverse effects of climate change, households were changing the planting times of their crops. Most of the land in the char areas and/or near to the river is subject to water logging and flooding during the rainy seasons. But the crops are now cultivated in a way that enables harvesting to be done before a hazard can arise. This adjustment evolved from long-term local knowledge and perceptions about the climate. Vegetable cultivation appeared to be the most common adaptation strategy in the study area. During field visit it was observed that small and landless farmers cultivated different types of short duration winter and summer vegetables.

The continuous loss of land through riverbank erosion is the main problem for the households. In order to ensure the sustainable use of the available land, households were practicing homestead gardening and tree plantation, particularly the small and landless farmers (Table 2.2). Homestead gardening provides a continuous supply of nutrients in the food chain and can be an important source of income. However, the small and the landless farmers have limited access to financial institutions and extension services. They were undertaking small

Table 2.2 Adaptation strategies of the households in the study area

Adaptive measure	Responses	Farm category			
		Large	Medium	Small	Landless
Agricultural adjustment					
Change planting time	8	x	x	x	
Cultivation of pulses	11	x	x	x	
Cultivation of wheat and other crops	4	x	x	x	
Tree plantation	6	x	x		
Cultivation of spices and oil seed	10	x	x	x	x
Cultivation of local Aman rice	5	x	x		
Cultivation of vegetables	6	x	x	x	x
Cultivation of HYV rice varieties (e.g., BRRI-28, 29)	8	x	x	x	x
Livestock rearing	7	x	x	x	x
Poultry rearing	5		x	x	x
Duck rearing	3			x	x
Homestead gardening	5			x	x
Non-agricultural adjustment					
Migration	12			x	x
Off-farm work (van, rickshaw, tampo driving)	7			x	x
Petty business	3			x	x

[a]According to main adaptation strategies although there were multiple options

businesses such as grocery shops, a tea stall and vendoring that require less capital. Many of them had taken up driving as their occupation in the face of diminishing employment in farming. Government organizations and NGOs can play an enabling role in improving their livelihoods by providing training and financial support.

Migration, both seasonal and permanent, was also found to be an important adaptation strategy, especially for the small and the landless farmers. Households with limited agricultural land used to migrate in search of alternative livelihoods for a few months. This temporary migration is very common in the study area, especially during the rainy seasons when there is limited scope for farming and non-farming employment. This local level knowledge of adaptation is crucial for policy makers to support and promote adaptation strategies, and to turn them into effective and sustainable action.

Limiting Factors to Adaptation

While the study households were found adopting a range of adaptation strategies, however, they reported some barriers that preventing them from successful adaptation. The respondents were asked to mention the limiting factors to adaptation and categorise these into 'main barriers' and 'barriers' only. They mentioned nine limiting factors for successful adaptation. The main barriers include lack of information about riverbank erosion and related climatic issues, knowledge of appropriate adaptation, appropriate crop varieties and credit facility (Table 2.3). They also mentioned other post-production related problems such as lack of storage facilities, marketing and transportation facilities, which is crucial for policy intervention towards improving their livelihood.

Understandably the barriers were felt heterogeneously among the farming groups. For example, the households who had relatively less land ownership were mentioned the lack of credit, own land and knowledge about appropriate adaptation as the main barriers to adaptation. There was highly significant ($p < 0.007$) lower average land size among the households who mentioned these as a main barrier compared to the rest of the respondents who did not mention these as a main barrier (independent sample *t* test). In case of credit, the large and medium farmers have had relatively better access to both government and NGOs credit facilities. The small and landless farmers in particular have hardly any access to government credit facilities, which appeared to be the main barriers for them (Table 2.3). The lack of institutional access and credit facilities can limit their ability to get the necessary resources and technologies they might want to adapt to the changing condition. Since the small and landless farmers have resource limitations therefore access to financial institution is crucial for them to promote adaptation.

Moreover, the large and medium farmers stated the lack of storage and marketing facilities as their main barriers that might prevent them from getting the right price of their products. The fact is that traders were not able to come in the area due to the poor road transportation system. Even the small and landless labours who are in need of seasonal job were not able to go to the cities easily due to the poor transportation system.

Table 2.3 Perceived barriers to adaptation measures

Barriers to adaptation	Response by farm category			
	Large	Medium	Small	Landless
Lack of information about riverbank erosion and related climatic issues	√√	√√	√√	√√
Lack of appropriate variety	√√	√√	√√	–
Lack of knowledge concerning appropriate adaptation	√	√	√√	√√
Lack of credit/money/saving	–	√	√√	√√
Lack of suitable land for cultivation	–	–	√√	√√
Lack of own land	–	–	√√	√√
Lack of storage facilities	√√	√√	–	–
Lack of marketing facilities	√√	√√	√√	–
Lack of transportation facilities	√	√	√	√

Where, √√ = main barriers, √ = barriers

Conclusions

Most of the farmers of Bangladesh depend on agriculture for their livelihood, which is most vulnerable to climate change. This chapter provides information of vulnerable households' adaptation strategies based on their farming status and barriers to adaptation in Bangladesh. The study reveals that farmers have adopted a range of adaptation strategies, which vary significantly among farming groups. The large and medium farmers have adopted mainly agricultural adjustments such as diversifying crops and tree plantation whereas the small and landless farmers mostly adopted non-agricultural adjustment such as driving and migration. They have also adopted the strategies of homestead gardening, and poultry and duck rearing towards improving their food security and livelihood.

Study also reveals some barriers to adaptation, which limit their successful adoption of adaptation strategies. The important barriers include lack of information about riverbank erosion and related climatic issues, knowledge of appropriate adaptation, appropriate crop varieties and credit facility. The lack of institutional access and credit facilities particularly for small and landless farmers can limit their ability to get the necessary resources and technologies they might want to adapt to the changing condition.

Scientists need to continue to develop crop varieties, high-value and flood-prone crops and technologies suitable to local conditions, especially in the emerging char lands, to accelerate the adaptation process. The NGOs should come forward to disseminate various information including successful adaptation among farmers and stimulate them to adopt with appropriate support such as credit and technical support. Development of improved communication, transportation and access to markets and services also vitally important to accelerate the effective and logical adoption of adaptation processes in these vulnerable areas. This will enhance the resilience of vulnerable households in riparian areas across Bangladesh.

References

Adger, W. N., Huq, S., Brown, K., Conway, D., & Hulme, M. (2003). Adaptation to climate change in the developing world. *Progress in Development Studies, 3*(3), 179–195.

Adger, W. N., & Vincent, K. (2005). Uncertainty in adaptive capacity. *Geoscience, 337*(4), 399–410.

Ahmed, I. (2015). *People of many rivers: Tales from the riverbanks*. Dhaka: The University Press Limited.

Alam, K. (2015). Farmers' adaptation to water scarcity in drought-prone environments: A case study of Rajshahi District Bangladesh. *Agricultural Water Management, 148,* 196–206.

Alam, G. M. M. (2016). An assessment of the livelihood vulnerability of the riverbank erosion hazard and its impact on food security for rural households in Bangladesh. Ph.D. thesis, School of Commerce, University of Southern Queensland, Australia.

Alam, G. M. M. (2017). Livelihood cycle and vulnerability of rural households to climate change and hazards in Bangladesh. *Environmental Management, 59*(5), 777–791.

Alam, G. M. M., Alam, K., & Shahbaz, M. (2016). Influence of institutional access and social capital on adaptation choices: Empirical evidence from vulnerable rural households in Bangladesh. *Ecological Economics, 130,* 243–251.

Alam, G. M. M., Alam, K., Shahbaz, M., & Clarke, M. L. (2017a). Drivers of vulnerability to climatic change in riparian char and river-bank households in Bangladesh: Implications for policy, livelihoods and social development. *Ecological Indicators, 72,* 23–32.

Alam, G. M. M., Alam, K., & Mushtaq, S. (2017b). Climate change perceptions and local adaptation strategies of hazard-prone rural households in Bangladesh. *Climate Risk Management.* doi:10.1016/j.crm.2017.06.006.

Alauddin, M., & Sarker, M. A. R. (2014). Climate change and farm-level adaptation decisions and strategies in drought-prone and groundwater-depleted areas of Bangladesh: An empirical investigation. *Ecological Economics, 106,* 204–213.

Alexander, C., Bynum, N., Johnson, E., King, U., Mustonen, T., & Neofotis, P. (2011). Linking indigenous and scientific knowledge of climate change. *BioScience, 61*(6), 477–484.

Anik, S. I., & Khan, M. A. S. A. (2012). Climate change adaptation through local knowledge in the north eastern region of Bangladesh. *Mitigation and Adaptation Strategies to Global Change, 17*(8), 889–896.

Bardsley, D. K., & Wiseman, N. D. (2012). Climate change vulnerability and social development for remote indigenous communities of South Australia. *Global Environmental Change, 22*(3), 713–723.

Bartlett, J. E., Kotrlik, J. W., & Higgins, C. C. (2001). Organizational research: Determining appropriate sample size in survey research. *Information Technology Learning and Performance, 19*(1), 43–50.

BBS. (2012). *Yearbook of agricultural statistics.* Dhaka: Bangladesh Bureau of Statistics.

CEGIS. (2012). *Prediction of river bank erosion along the Jamuna, the Ganges the Padma and the lower Meghna rivers in 2012.* Dhaka: Centre for Environment and Geographic Information Services.

Elahi, K. M. (1989). Population displacement due to riverbank of the Jamuna in Bangladesh. In J. Clarke et al. (Eds.), *Population and disasters.* Oxford: Basil Blackwell.

Elahi, K. M., Ahmed, K. S., & Malizuddin, M. (Eds.). (1991). *Riverbank erosion, flood and population displacement in Bangladesh.* Dhaka: Riverbank Erosion Study, Jahangirnagar University.

Ellen, R. (2007). *Modern crises and traditional strategies: Local ecological knowledge in Island Southeast Asia.* Oxford: Studies in Environmental Anthropology and Ethno Biology.

Eregno, F. E., Xu, C. H., & Kitterød, N. O. (2013). Modelling hydrological impacts of climate change in different climatic zones. *International Journal of Climate Change Strategies and Management, 5*(3), 344–365.

Eriksen, S., Aldunce, P., Bahinipati, C. S., Martins, R. D. A., Molefe, J. I., Nhemachena, C., et al. (2011). When not every response to climate change is a good one: Identifying principles for sustainable adaptation. *Climate and Development, 3*(1), 7–20.

Gandure, S., Walker, S., & Botha, J. J. (2013). Farmers' perceptions of adaptation to climate change and water stress in a South African rural Community. *Environment and Development, 5,* 39–53.

GoB. (2011). *The sixth five-year plan, 2011–2015.* Dhaka: Ministry of Planning, Government of the People's Republic of Bangladesh.

GoB (Government of Bangladesh). (2010). *Comprehensive disaster management programme, Phase II (2010–14).* Dhaka: Ministry of Food and Disaster.

Green, D., & Raygorodetsky, G. (2010). Indigenous knowledge of a changing climate. *Climatic Change, 100*(2), 239–242.

Greenberg, C. (1986). *The adaptation process of riverbank erosion displaces in an urban environment: A case study of squatters in Sirajgonj, Bangladesh.* Unpublished thesis, University of Manitoba, Canada.

Habiba, U., Shaw, R., & Takeuchi, Y. (2012). Farmer's perception and adaptation practices to cope with drought: Perspectives from North Western Bangladesh. *International Journal of Disaster Risk Reduction, 1,* 72–84.
Hiwasaki, L., Luna, E., Syamsidik, & Shaw, R. (2014). Process for integrating local and indigenous knowledge with science for hydro-meteorological disaster risk reduction and climate change adaptation in coastal and small island communities. *International Journal of Disaster Risk Reduction, 10,* 15–27.
Hossain, M. Z. (1984). *Riverbank erosion and population displacement: A case of Kazipur in Pabna.* Unpublished Ph.D. thesis, Jahangirnagar University, Dhaka.
Hutton, D., & Haque, E. (2004). Human vulnerability, dislocation and resettlement: Adaptation process of riverbank erosion-induced displacement in Bangladesh. *Disasters, 28*(1), 41–62.
IPCC. (2001). Climate change 2001: Synthesis report. In *Contribution of working group I, II and III to the third assessment report of the intergovernmental panel on climate change, Geneva.*
IPCC. (2007). Climate change 2007: Impacts, adaptation and vulnerability. In *Contribution of working group II to the fourth assessment report of the intergovernmental panel on climate change, intergovernmental panel on climate change.* Cambridge University Press, Cambridge.
IPCC. (2014). *Climate change 2014: Impacts, adaptation and vulnerability.* In *Part A: Global and sectoral aspects, contribution of working group II to the fifth assessment report of the intergovernmental panel on climate change.* Cambridge University Press, Cambridge.
Jones, L., & Boyd, E. (2011). Exploring social barriers to adaptation: Insights from Western Nepal. *Global Environmental Change, 21*(4), 1262–1274.
Lein, H. (2010). Hazards and force migration in Bangladesh. *Norwegian Journal of Geography, 54*(3), 122–127.
Lobell, B. D., Burke, M. B., Tebaldi, C., Mastrandream, M. D., Falcon, W. P., & Naylor, R. L. (2008). Prioritizing climate change adaptation needs for food security in 2030. *Science, 319* (5863), 607–610.
Makenro, M. B. (2000). *World disaster report 2001.* Geneva: International Federation of the Red Cross and Red Crescent Societies.
Malone, E. L., & Brenkert, A. L. (2008). Uncertainty in resilience to climate change in India and Indian states. *Climatic Change, 91*(3), 451–476.
McDowell, G., Ford, J., Lehner, B., Berrang-Ford, L., & Sherpa, A. (2013). Climate-related hydrological change and human vulnerability in remote mountain regions: A case study from Khumbu, Nepal. *Regional Environmental Change, 13*(2), 299–310.
Niles, M. T., Lubell, M., & Brown, M. (2015). How limiting factors drive agricultural adaptation to climate change. *Agriculture, Ecosystem and Environment, 200,* 178–185.
Nyong, A., Adesina, F., & Elasha, B. O. (2007). The value of indigenous knowledge in climate change mitigation and adaptation strategies in the African Sahel. *Mitigation and Adaptation Strategies for Global Change, 12*(5), 787–797.
Rashid, M. H., Afroz, S., Gaydon, D., Muttaleb, A., Poulton, P., Roth, C., et al. (2014). Climate change perception and adaptation options for agriculture in Southern Khulna of Bangladesh. *Applied Ecology and Environment Science, 2*(1), 25–31.
Rogge, J., & Haque, C. (1987). *Riverbank erosion hazard, rural population displacement, and institutional responses and policies in Bangladesh.* Paper presented at the annual meeting of the Association of American Geographers, Portland, Oregon.
Rosenzweig, C., Elliott, J., Deryng, D., Ruane, A. C., Muller, C., Arneth, A., et al. (2013). Assessing agricultural risks of climate change in the 21st century in a global gridded crop model intercomparison. *Proceedings of National Academy of Science, USA, 111*(9), 3268–3273.
Salick, J., Fang, Z., & Byg, A. (2009). Eastern Himalayan alpine plant ecology, Tibetan ethnobotany, and climate change. *Global Environmental Change, 19*(2), 147–155.
Sarker, M. A. R., Alam, K., & Gow, J. (2013). Assessing the determinants of rice farmers' adaptation to climate change in Bangladesh. *International Journal of Climate Strategies and Management, 5*(4), 382–403.

Smit, B., Burton, I., Klein, R., & Wandel, J. (2000). An anatomy of adaptation to climate change and variability. *Climatic Change, 45*(1), 223–251.

Smit, B., & Wandel, J. (2006). Adaptation, adaptive capacity and vulnerability. *Mitigation and Adaptation Strategies of Global Change, 16*(3), 282–292.

Thomas, D. S., & Twyman, C. (2005). Equity and justice in climate change adaptation amongst natural-resource-dependent societies. *Global Environmental Change, 15*(2), 115–124.

Tompkins, E. L., & Adger, W. N. (2004). Does adaptive management of natural resources enhance resilience to climate change? *Ecology and Society, 9*(2), 1–14.

UNESCO (The United Nations Educational, Scientific and Cultural Organization). (2015). *Literacy statistics metadata information table, institute for statistics, September 2015*. Geneva: UNESCO.

Wood, S. A., Jina, A. S., Jain, M., Kristjanson, P., & DeFries, R. S. (2014). Small holder farmer cropping decisions related to climate variability across multiple regions. *Global Environmental Change, 25,* 163–172.

Zaman, M. Q. (1989). The socioeconomic and political context of adjustment to riverbank erosion hazard and population resettlement in Bangladesh. *Human Organization, 48*(3), 196–205.

Chapter 3
Governance Limits to Adaptation in Cambodia's Health Sector

Daniel Gilfillan

Introduction

Around the world, the impacts of climate change are already observable in changing precipitation patterns and resulting changes in freshwater availability (WHO 2015). This will affect agricultural productivity, and there are significant implications for human health. There are direct impacts on food and water security, and follow on impacts such as the spread of infectious diseases changing (Costello et al. 2011); the implications for human health are particularly severe in developing countries (Lesnikowski et al. 2013). The purpose of this paper is firstly to explore existing governance structures and systems that limit adaptation in Cambodia's health sector. Secondly the chapter highlights both positive and negative aspects of the outlook for adaptation in Cambodia's health sector.

The rationale for undertaking this research is that, at its core, adapting to climate change is about people and communities. In this context, Lesnikowski et al. (2013) describe populations living in developing country contexts with weak governance systems and institutions as being disproportionately susceptible to climate-related health risks. This makes research into how developing country decision-makers choose to deal with these impacts a worthwhile endeavour.

The Intergovernmental Panel on Climate Change (IPCC) views adaptations in human systems as interventions designed to reduce damages associated with climate change impacts or, conversely, to maximize benefits (IPCC 2014). Resilience in the face of climate change relates to socio-ecological systems and their ability to cope with climate-related disturbances without losing their core identity (IPCC 2014). In the health sector, WHO (2008) recommends controlling climate sensitive health risks through national policies, plans and programs. Specific activities

D. Gilfillan (✉)
Fenner School of Environment and Society, The Australian
National University, Building 141 Linnaeus Way, Acton ACT 2601, Australia
e-mail: daniel.gilfillan@anu.edu.au

W. Leal Filho and J. Nalau (eds.), *Limits to Climate Change Adaptation*,
Climate Change Management, https://doi.org/10.1007/978-3-319-64599-5_3

include strengthening of human resources and health infrastructure, as well as strengthening surveillance and response systems. It is also important to consider the health impacts from other sectors such as water resource management, and health sector participation in preparation of national adaptation plans of action, and national communications required by the United Nations Framework Convention on Climate Change (UNFCCC).

Adaptation constraints are defined as those factors that make it more difficult for a system or society to adapt to climate change. In contrast, an adaptation limit is a constraint which is severe enough that adaptation can no longer occur (IPCC 2014). Adaptation limits can be either hard or soft. Hard limits are those, which cannot be overcome, such as the physiological capacity of organisms to adapt to a changing climate. In contrast, soft limits exist when it is conceivable that changing circumstances may allow what had previously been limits to be addressed as constraints (IPCC 2014). An example of a soft limit is where scarcity of local-scale modelling of climate change impacts results in decision-makers deferring decisions on climate change responses (see for example Morgan 2011).

Climate change adaptation requires both political action and social mobilisation (Huang et al. 2011), and so an understanding of climate change governance is fundamental to its implementation. Governance in this chapter is taken to be a society-wide process of decision-making (Graham et al. 2003). Governance structures include all elements in society that can influence policy decisions including government, civil society, business, and international organisations. For example, Picketts et al. (2014) argued that involving people at the local level is more likely to lead to appropriately targeted actions in that location.

This research about climate change adaptation governance in Cambodia's health sector is informed by two theories that explore networked systems. The Anglo-governance school has its roots in institutionalism and sees national governments as limited agents that must rely on guiding mechanisms and diplomacy to secure cooperation from other organisations within a broader network (Bevir 2009). Modern organisation theory values collaboration between individuals, systems and processes to achieve results (Hicks and Gullet 1975). By viewing adaptation through an organisational and governance lens, this research highlights adaptation limits in Cambodia's health sector associated with existing government and governance structures and processes. Some current transformations that may address these limits are noted.

Below is a review of relevant literature, followed by a description of the research methods used. Research results are then presented and discussed, highlighting two adaptation limits in Cambodia's health sector associated with governance systems.

Context

The literature review has four sections. First is an introduction to health and adaptation, followed by a review of the climate change adaptation and capacity development literature. Third is a review of the literature that discusses the coordination of climate change adaptation, and fourth is background information to the Cambodian context.

Overview of Health and Adaptation

In debates about climate change, the impacts on human health have received limited attention; for example in the lead up to the 15th Conference of the Parties to the UNFCCC in 2009 only 4 of 47 nations gave consideration to human health. (Costello et al. 2011). This is despite direct causal links between changes in climate and weather patterns and a number of key public health concerns, such as the distribution and prevalence of infectious diseases, including vector-borne and water-borne diseases, as well as increasing problems of water and food security (Lesnikowski et al. 2013; WHO 2015). For example, both drought conditions and flooding, which are expected to increase in countries like Cambodia, are linked to increases in the number of cases of water-borne diseases such as dysentery, typhoid and cholera (WHO 2015). To be effective, measures to adapt to these increasing risks will need to include the development of early warning capacity, through monitoring and surveillance, to control outbreaks, as well as building lay people's knowledge and awareness about the risks of water-borne diseases (WHO 2016). The changes to climate-related health impacts are expected to disproportionately affect those who already face high health risks. An example is a poor population already suffering from a high ill-health burden and thus sensitive to climate-related diseases, and that is living in a place with weak institutions and high levels of political inequality (Lesnikowski et al. 2013). This accurately describes the situation for many communities in developing countries such as Cambodia.

Huang et al. (2011) noted in their review of 10 years of global adaptation literature that specialised government sectors and uncoordinated policy goals inhibit responses to climate change. This can be further complicated because climate change adaptation is just one development consideration among many for developing countries (see Sosa-Rodriguez 2014). Effective adaptation requires making policy decisions, then resourcing and implementing them (Klein et al. 2014). There are a variety of governance mechanisms that can support both adaptation policy-making and implementation, such as clear mandates and inclusive decision-making. However, some authors note the lack of knowledge about effectiveness of different adaptation pathways (Huang et al. 2011), which can delay and disrupt policy decisions. Dany et al. (2015) break this lack of knowledge into levels of knowledge availability and understanding about adaptation, and discuss financial capacity and ability to coordinate.

Capacity Development

The United Nations Development Programme (UNDP) defined capacity as "the means to plan and achieve" (UNDP 2009, p. 5). Capacity development is "the

Table 3.1 Typology of capacity

Type of capacity	Source(s)
Financial and resourcing	Dany et al. (2015)
Human resources, skills and understanding	Willems and Baumert (2003), Bowen et al. (2015) and Dany et al. (2015)
Coordination, management and decision-making	Willems and Baumert (2003), Bowen et al. (2015) and Dany et al. (2015)
Context (regulations, systems and information)	Willems and Baumert (2003) and Dany et al. (2015)

process through which individuals, organizations and societies obtain, strengthen and maintain the capabilities to set and achieve their own development objectives over time" (UNDP 2009, p. 5). A review of the climate change adaptation literature highlights a number of types of capacity that may impact on adaptation responses, as presented in Table 3.1.

Capacity development is of particular importance in the health sector because of the complexity of impacts of climate change on human health (Huang et al. 2011). As Costello et al. (2011, p. 1878) argued, "[w]eak health governance and management structures in developing countries require long-term investment in human capacity". Abouzahr and Boerma (2005) provided detail, noting an international donor focus on urgent requirements for health-related data rather than on longer-term capacity development, resulting in developing world health systems that are unresponsive, fragmented and complicated. This exacerbates the challenges of direct health-related adaptation as well as possibilities to coordinate with other sectors such as agriculture and water management.

Coordination

Adaptation to climate change is a cross-cutting issue and requires cooperation at different levels and across scales. Carlsson-Kanyama et al. (2013) found that lack of action at higher government levels could constrain adaptation at more local levels, for example where local governments are not granted the authority to impose sanctions for non-compliance with adaptation-related policies. There is also a recognised need for coordination across sectors and between organisations (Costello et al. 2009). Human health, for example, can be affected by waste-water management, which may be privatised or supported by non-government organisations (NGOs) in some jurisdictions. The complex nature of interactions between different actors can be a barrier to climate change adaptation because objectives, planning timelines, and levels of authority and resource access vary significantly (Adger et al. 2009). Coordination is also important because even where long-term objectives are aligned, organisations and government departments may have conflicting short-term goals.

Cambodia

Cambodia's adaptation responses in the health sector are located within the broader national adaptation policy framework. The country ratified the UNFCCC in 1995 with the Ministry of Environment initially having responsibility for its implementation. In 2003 the Cambodian Climate Change Office (CCCO) was set up within the Ministry of Environment, primarily to develop Cambodia's National Communications to the UNFCCC (Bowen et al. 2013). Cambodia's National Climate Change Committee (NCCC), charged with preparing and coordinating government strategies, plans and programs on climate change issues, was established in 2006. The NCCC was chaired by the Prime Minister and included 19 government ministries in its membership. Its main role was to manage greenhouse gas emissions, but also had a mandate to oversee work on climate change vulnerability and adaptation (Solar et al. 2010). In 2009 the CCCO was upgraded to become the Climate Change Department (CCD). In May 2015 the National Council for Sustainable Development (NCSD) was established, further elevating the visible importance of sustainability issues. The NCSD secretariat now incorporates the CCD (interview data).

In 2006 the Ministry of Environment prepared Cambodia's National Adaptation Programme of Action (NAPA), which included human health as one of four key areas for adaptation action (RGC and MoE 2006). Apart from health, the three other priority areas identified in Cambodia's NAPA were agriculture, forestry and the coastal zone (RGC and MoE 2006). Subsequent to this, Cambodia's National Climate Change Action Plan for Public Health identified transmission of vector-borne diseases because of increasing precipitation, water- and food-borne diseases linked to poor sanitation, and food insecurity related to extreme weather events as its focus areas (Climate Change Technical Working Group for Health 2014). Cambodia's health sector also has a national disaster risk management plan, which prioritises increasing internal capacity to prepare for and respond to disasters, as well as capacity to assist with disaster prevention and recovery (MoH 2015).

Cambodia's ongoing decentralisation programs, which have been implemented over two decades (see Turner 2002), are an additional feature affecting adaptation. The country has six administrative levels of government, ranging from the national government to groups, which sit below the village level (Rushton 2008), and these are all affected by the decentralisation programs.

Methods

This qualitative research was based primarily on in-depth semi-structured interviews with senior staff from the Cambodian government, from NGOs and from international organisations. The government officials interviewed worked at the Ministry of Environment (MoE) and the Ministry of Health (MoH). Interviews were conducted with representatives from the World Health Organisation (WHO), the Asian Development Bank (ADB), the UNDP and the United Nations Environment Programme (UNEP). In order to explore the implications of national level climate change adaptation decisions and policies, interviews were conducted with senior staff from NGOs working locally and nationally, in both service delivery and research.

Thirteen interviews were conducted, six with government officials, four with personnel from international organisations and three with NGO personnel. All interviews were conducted in English, and varied in length from 29 min to 1 h, 45 min (average length 63 min). Interviews were transcribed, and interview transcripts were returned to respondents for them to check the interview record for accuracy. The researcher read each transcript several times, identifying themes evident within each interview. Following this, all data for themes common across transcripts were grouped together, initially using highlighter colours, and then by copying data into a file created for each theme. These themed files were then reviewed to identify sub-themes and used as the basis for the results section of the chapter. The themes and ideas generated were further substantiated through analysis of policy documents and existing research.

Limitations

There are three limitations associated with this study. First, while public health has clear links with other sectors such as agriculture and water management, an in-depth study of these links is beyond the scope of this paper. Second, some respondents did not have much time available as field research was conducted during the lead up to the 21st Conference of the Parties to the UNFCCC (COP 21), and this reduced the depth of these interviews. Finally, all but two respondents were male, despite an active search for female interviewees. The research focus was government policy and its implementation, so this should not significantly impact on research findings. It does highlight a male dominated arena, which raises questions about decision-making processes. For example, pregnant women and children have been identified by the world health organisation as being among the most vulnerable in Cambodia to food insecurity (WHO 2015), and food insecurity is directly linked to climate change impacts. In spite of this, Cambodian women appear to be largely absent from the policy-making arena.

Results

The results are divided between capacity-related and coordination-related responses. Interview data is cited using interviewee numbers. For example interviewee number seven appears as (#07).

Capacity

Capacity-related results are broken into four areas. These are: (1) Cambodia's education system, (2) government salaries, (3) under-resourcing, and (4) concerns about a lack of capacity development focus in both government and NGO initiatives to date.

Education

While not universal, the educated and intellectuals in Cambodia during the Pol Pot era were often targeted for execution, and educational resources were destroyed (Deth 2009), leaving Cambodia with an extremely low level of human capacity. In 2010, apart from Myanmar, Cambodia had the lowest spending on schooling of any of the ten member states of the Association of South East Asian Nations, and this spending as a percentage of the budget halved between 2000 and 2014 (Strangio 2014). The ability to achieve results in developing sustainable development pathways depends on the talent and human resources available, and this has been constrained in Cambodia because of the lack of government funding and attention paid to the education system. As described by a senior government bureaucrat, "human resources is always a constraint for us. We have young dynamic staff, but they don't have much experience. Especially with the kind of experience on research and policy formulation" (#08). According to the same respondent, current low human resource capacity in Cambodia can be linked to a history of the government placing a higher level of importance on achieving political stability than on developing the education system. This has been changing over the last 3 years (#08), which links to the observation from an employee of an international organisation that there is political pressure for reform, "they [the Cambodian ruling party] had 96 seats in the previous election, but now they got, I think, 65 or 68 seats, so they lost 20 seats. So, it's pressure them for reforms" (#09).

Salaries

Problems with human resource capacity do not just relate to levels of education and experience. A respondent from the NGO sector described low government salaries pushing health staff to work in private clinics in addition to their government job with the result that, "even if you build the capacity of people, there is no guarantee that they will perform their roles, if they are focussed on something else – other business going on in their lives" (#04). This view is supported in the literature: "Many clinical staff members have second jobs in the medical industry, operating private clinics or owning pharmacies. They often arrive late to work in the public health centres because of commitments to their private patients" (Rushton 2008, p. 78). The sentiment was also echoed by a respondent from an international organisation: "[t]he biggest challenges is overall weak and limited commitment of government officials" (#09). This respondent did note some recent improvements associated with increasing salaries and changing management methodologies.

Limited Financial Resources

One aspect of resourcing that respondents talked about was the impact that broader governance arrangements, including decentralisation, have had on the ability to adapt. While there are guidelines to ensure that sub-national policy-makers take climate change considerations into account when making decisions (#09), one senior government official noted that in "many sectors when decentralisation is implemented, I think they decentralise more responsibility to you, but not authority or resources" (#08). This idea was echoed in the words of an NGO worker, "[o]ne of the complaints that comes from them [the Provincial Committee for Disaster Management] is the fact that they have no money issued to them until the disaster actually happens" (#04). The reason for not decentralising authority or resources may be because the central government does not wish to hand over power to sub-national levels (#08). This respondent argued that full decentralisation in the health sector is still a long way off and said,

> I would expect that the national Ministry of Health would empower more in the health sector to the sub-national government because I have found that the health centre has a very important role in serving the public interest. But looking at their resources—I found very disappointed (#08).

Limited Capacity Building

In a resource-constrained environment where there is political pressure to reform, the government is likely to prioritise *visible* adaptations, so that community members are reminded on a daily basis what their government is doing for them. This is linked to infrastructure and other tangible responses being favoured. One government official, talking about climate change adaptation and health noted, "Cambodian people would like to see more infrastructure, like buildings, or facilities—car, vehicle, like that—and then they say, 'oh, it's a big reform.' But to me that's not reform that's materialise …." (#08). This view may tie to Strangio's (2014) argument of a culture within the government of 'gift-giving' for political purposes. Another respondent, similarly, noted that:

> They [the Cambodian government] do not really want to spend lot of resources into the capacity building. **They want to see more visible infrastructure, visible infrastructure that the people can see**. So something that people cannot touch, cannot see, they [the government] do not really want (#09) (emphasis added).

However the government culture of gifting material things may be based on inaccurate assumptions, with an interviewee from a Cambodian research institute reporting that, "a number of time I met a number of health staff in the rural area, what they report to me is that, 'we don't want the new and modern machine, what we just want is the common one that we know how to use it well'" (#12).

The NGO sector also has a tendency to prioritise visible results over capacity development because it facilitates reporting to funding agencies. Generally, NGOs have timelines that reflect a focus on results and can lead to management decisions that do not necessarily link with improving the abilities of local government to perform their tasks. One NGO worker noted that, "you look for a work-around in the sense that if the Provincial Committee for Disaster Management don't have the systems in place … then we try and look for someone else who has better systems in place" (#04). For ease of reporting it can be tempting to measure indicators like funds disbursed or number of workshops conducted, but longer-term outcomes of an adaptation project including degree of empowerment (#10), level of ownership (#10) and capacity development are much more difficult to capture (see UNDP 2009). Perhaps for these reasons, despite Cambodia having had over 20 years of international support, an employee of an international organisation noted that the implementation capacity at the Ministry of Health was still lacking, "[b]ut coming into implementation is very limited. Mainly Ministry of Health would not have enough resources including financial capacity and human resource capacity to translate what they have written on the paper [laws, regulations, decrees etc.…] into practicality" (#11).

Coordination

Coordination issues in Cambodia that are identified in the adaptation literature include unclear mandates and government agencies with overlapping responsibilities. For example, while the Ministry of Environment (MoE) is the overall lead agency for climate change in Cambodia, the "Ministry of Health [MoH] and the MoWRM [Ministry of Water Resources and Meteorology] are the central actors in the health and water networks, respectively, but do not have the requisite roles and responsibilities to drive the CCA [climate change adaptation] agenda" (Bowen et al. 2015, p. 5).

Coordination of climate change actions was viewed by respondents as important. A respondent from an international organisation talked about it in terms of the health sector, "… health is not limited to just the health sector, but achieving health outcomes through other sectors, especially the water and sanitation hygiene [WASH] sector, which is largely outside the Ministry of Health" (#07). In order to meet the need to facilitate coordination and information sharing between different departments, technical working groups (TWGs) have been established, however respondents raised a number of concerns about the TWGs, including concern about choice of membership:

> … while there is a climate change technical working group at the ministerial level … there are some key departments within the Ministry of Rural Development which are not included – the Department for Rural Water Supply and the Department for Rural Health Care are not included on that technical working group on climate change (#04).

Other concerns related to the operation of the TWGs included that, "… on paper there is a technical working group, sub-groups to coordinate the climate change, but they rarely meet. And they have names of senior officials, but when they come to meetings it is only junior representatives" (#09). Respondents also noted that TWGs require project-specific (i.e. donor) funding in order to meet; "[i]t [cross-ministry collaboration] works well, activity by activity—it's not something that would take place without the injection of project funds" (#07). Likewise, a senior official from the Ministry of Health noted that, "… right now it has just started [a donor funded health and adaptation project], so not yet any monitoring or evaluation, just start. So maybe later, two month later, maybe quarterly or something like that we [the TWG] will meet and report each other activity and maybe make a recommendation …" (#05). These responses, supported by observations from other interviewees (e.g. #02, #03, #06), point to a national government that will allow inter-ministerial coordination mechanisms such as TWGs to operate, but does not see them as a high priority.

At a higher level than TWGs are the national level committees and councils including the NCCC and the NCSD. A respondent from an international organisation noted that inter-ministerial coordination in Cambodia is generally challenging because of a vertical system of government and a culture of people protecting their 'turf'. According to this respondent, staff in the environment ministry have not followed this route, but rather "they've given opportunities to other ministries to benefit from climate finance …" (#01). Despite this positive picture, an employee of an international organisation observed that, "there is Ministry of Environment initially—they were the line ministry, but they don't have any power to coordinate. They don't have resources … to make sure that coordination exists" (#09).

Discussion

The discussion is divided between capacity-related and coordination-related issues.

Adaptation Limit One: Capacity Development

Poor human resource capacity in Cambodia is a product not just of the targeted killings during the Pol Pot era. Neither the Cambodian government nor the international aid sector have placed a high enough priority on individual capacity development in the health sector. Following the UN sponsored elections, the international aid sector saw Cambodia as a 'clean slate' on which they could build state systems and structures, failing to take into account the deep rivalries between political factions (Strangio 2014). Because these rivalries were not openly acknowledged and addressed they led to a government focus on the urgent problem of achieving political stability rather than on longer-term issues such as wage growth, or developing a high quality education sector (#08).

A 'living wage' should support a worker's basic standard of living, including consideration of supporting his/her family and paying for health care, education and recreation (Anker 2011). However, when governments set wages they do not just consider the needs of workers, because there are "possible negative effects of higher wages on employment and economic development" (Anker 2011, p. 1). Wages for local level health workers are not sufficient, and so these government workers are likely to be distracted from their government role by the need to supplement their income (#04). This distraction has a detrimental impact on service delivery, and forms an adaptation limit for local communities.

While the Cambodian government's loss of 20 of 90 seats in the 2013 election puts pressure on them to convince the public that it is looking after their interests (#09), this will not necessarily lead to a focus on issues such as wages or capacity development. The government has a history of gifting material items (Strangio 2014), possibly linked to a perception that Cambodian people are swayed by physical infrastructure and other tangible objects (#09). The political pressure to reform may therefore exacerbate the focus on infrastructure as a response to climate change.

At the same time, the international aid sector is results focussed, and there are difficulties in linking increased capacity to particular capacity development interventions (e.g. UNDP 2009). In terms of evaluating projects, one NGO worker observed that "it's statistical bean counting, we're just counting for the sake of counting" (#10). Counting the number of workshops that are run or the number of people who attend trainings does not provide information on capacity development, which:

> is essential to the success of any development enterprise. Without it, the integrity of development achievements can be compromised and progress can remain rootless and illusory, separated from the capacities that already exist and vulnerable to the increasingly severe and complex challenges facing the world (UNDP 2009, p. 9).

The difficulty of measuring changes in capacity is combined with a competitive results-oriented aid environment, which can easily lead to NGOs focussing their attention on easily measurable aspects of their work. While attending individual trainings does not necessarily lead to capacity development, Bowen et al. (2013, p. 282) noted that the capacity developed during the collaborative process of developing Cambodia's National Adaptation Programme of Action (NAPA) was "at least as important" an outcome as the NAPA itself.

Adaptation Limit Two: Coordination

Lack of coordination at the government level has limited Cambodia's adaptation options. The NCCC coordinated the work of 19 different ministries, and climate change technical working groups (TWGs) brought together technical personnel from government agencies for coordination of effort. However TWG effectiveness was constrained in four ways, which together formed an adaptation limit.

Firstly, the government did not prioritise TWGs. While senior officials' names appear on TWG paperwork, these officials often sent subordinates in their place when TWGs did meet (#09). Secondly, TWGs did not meet unless there was project (donor) funding allocated to them (#05, #07). Thirdly, relevant departments were not always invited to participate. For example the departments of rural water supply and rural health care were reportedly not included in the ministerial-level climate change TWG (#04). Finally, up until 2015 there was a separation of the technical and political aspects of climate change action (#08); the TWGs did not operate in tandem with the NCCC. While each of these as an individual constraint may have been overcome, in combination they place limits on effective coordination of climate change adaptation.

There is room to hope that some of these coordination issues are being addressed in a systemic manner, with the establishment of the National Council for Sustainable Development (NCSD) in May 2015. The NCSD had not yet met at the time of the field research for this chapter, however the model brings together both political level personnel and the technical personnel who are intended to be the 'bridges' between the ministries:

> … this council [the NCSD] also demand for each ministry appoint a technical, like a focal point. So it's kind of a technical person who will be able to assist the secretary of state from each ministry to have a kind of a full information at the council meeting (#08).

The NCSD is funded by the Cambodian government (#08), indicating a higher level of interest and ownership in the council than there has been in the climate change TWGs. This should lead to a higher likelihood that those who are designated on paper as members will be the people who attend meetings.

There are three reasons for caution in assessing the likely effectiveness of the NCSD. First, the environment ministry does not have the resources and power to ensure coordination with other ministries (#09), so the NCSD's success may partly depend on higher-level support. Second, the NCSD structure does not address the issue of which ministerial departments will be involved, which may result in relevant departments not having a voice on the council. Finally, the Cambodian government has a history of developing policies, but not following through with policy implementation (Strangio 2014; see also Rushton 2008), and successful linking of the technical and political elements will depend on strong government commitment. It remains to be seen how effectively the NCSD coordinates Cambodia's sustainable development including climate change adaptation in health. This is an area for future research.

Conclusion

This paper makes two contributions to the academic literature on adaptation limits. First, it highlights two governance limits that affect the ability of Cambodia's health sector to deal with the impacts of climate change. The first limit is a lack of focus on developing human capacity in Cambodia. This can be seen directly in the prioritisation of infrastructure responses over capacity development by the Cambodian government, and the international aid sector's prioritisation of reportable results. It can also be seen indirectly in the low wages for government employees, which have resulted in public health officials arriving at work late because of external commitments. Finally it can be seen in the low levels of funding for education compared with neighbouring countries. The second governance limit to adaptation is poor coordination across sectors. Despite international donors supporting the establishment of TWGs to link technical personnel across ministries, including for adaptation in Cambodia's health sector, the working groups are not prioritised by government and only meet when there is project-specific donor funding available. When the working groups do meet not all relevant government agencies are invited to participate; for example the departments of rural water supply and rural health care were not invited to participate in climate change TWGs. Finally, up until 2015 the political and technical discussions on climate change in Cambodia were disconnected.

The second contribution of this chapter is a discussion about Cambodia's outlook in coordinating adaptation, and developing capacity to adapt to climate change. The newly established NCSD is a potential solution to the coordination issues described above. However, the operation of the NCSD should be monitored for the involvement of government departments and agencies as well as for implementation of its mandate. There is reason for cautious optimism, however, because the design of the NCSD includes mechanisms to support coordination between technical and political levels as well as across ministries. In addition, as a government-funded entity the NCSD has the potential to make significant improvements to the coordination of responses to the impacts of climate change in Cambodia's health sector. Despite some changes in the education sector over the last 3 years, the outlook for capacity development is not as positive. While there is political pressure for the government to reform, the reported 'gift-giving' culture within government may encourage further investment in infrastructure and other hard adaptation responses, rather than investment in people. At the same time, issues such as providing a living wage for health workers appear to remain at odds with broader economic interests.

Acknowledgements The author acknowledges: the Fenner School of Environment and Society at The Australian National University and the Rotary Club of Hall for financial and material support to undertake fieldwork; the interviewees; A/Prof Jamie Pittock, Dr Liz Hanna and the reviewers for their constructive feedback on earlier drafts, as well as Daniel Ferris for editing and proof-reading. The author is responsible for any errors.

References

Abouzahr, C., & Boerma, T. (2005). Health information systems: the foundations of public health. *Bulletin of the World Health Organization, 83*(8), 578–583.

Adger, N. W., Dessai, S., Goulden, M., Hulme, M., Lorenzoni, I., Nelson, D. R., et al. (2009). Are there social limits to adaptation to climate change? *Climatic Change, 93*(3–4), 335–354.

Anker, R. (2011). *Estimating a living wage: A methodological review*. Geneva: International Labour Organization.

Bevir, M. (2009). The construction of governance. *Rivista italiana di scienza politica, 39*(3), 417–440.

Bowen, K. J., Miller, F., Dany, V., McMichael, A. J., & Friel, S. (2013). Enabling environments? Insights into the policy context for climate change and health adaptation decision-making in Cambodia. *Climate and Development, 5*(4), 277–287.

Bowen, K. J., Miller, F. P., Dany, V., & Graham, S. (2015). The relevance of a coproductive capacity framework to climate change adaptation: Investigating the health and water sectors in Cambodia. *Ecology and Society, 20*(1), 13.

Carlsson-Kanyama, A., Carlsen, H., & Dreborg, K.-H. (2013). Barriers in municipal climate change adaptation: Results from case studies using backcasting. *Futures, 49*, 9–21.

Climate Change Technical Working Group for Health. (2014). *National climate change action plan for public health*. Phnom Penh: Cambodian Ministry of Health.

Costello, A., Abbas, M., Allen, A., Ball, S., Bell, S., Bellamy, R., et al. (2009). Managing the health effects of climate change. *The Lancet, 373*(9676), 1693–1733.

Costello, A., Maslin, M., Montgomery, H., Johnson, A. M., & Ekins, P. (2011). Global health and climate change: Moving from denial and catastrophic fatalism to positive action. *Philosophical Transactions of the Royal Society of London A: Mathematical, Physical and Engineering Sciences, 369*(1942), 1866–1882.

Dany, V., Bowen, K. J., & Miller, F. (2015). Assessing the institutional capacity to adapt to climate change: A case study in the Cambodian health and water sectors. *Climate Policy, 15*(3), 388–409.

Deth, S. U. (2009). *The People's Republic of Kampuchea 1979–1989: A draconian savior?*. Athens, OH: University of Ohio.

Graham, J., Amos, B., & Plumptre, T. (2003). *Principles for good governance in the 21st century*. Ottawa: Institute on Governance.

Hicks, G. H., & Gullet, C. R. (1975). *Organizations: Theory and behaviour*. New York, NY: McGraw-Hill.

Huang, C., Vaneckova, P., Wang, X., FitzGerald, G., Guo, Y., & Tong, S. (2011). Constraints and barriers to public health adaptation to climate change: A review of the literature. *American Journal of Preventive Medicine, 40*(2), 183–190.

IPCC. (2014). *WGII AR5 glossary*. Intergovernmental panel on climate change. Online publication. https://www.ipcc.ch/pdf/assessment-report/ar5/wg2/drafts/fd/WGIIAR5-Glossary_FGD.pdf. Accessed December 29, 2015.

Klein, R. J. T., Midgley, G. F., Preston, B. L., Alam, M., Berkhout, F. G. H., Dow, K., et al. (2014). Adaptation opportunities, constraints, and limits. In C. B. Field, V. R. Barros, D. J. Dokken, K. J. Mach, M. D. Mastrandrea, T. E. Bilir, M. Chatterjee, K. L. Ebi, Y. O. Estrada, R. C. Genova, B. Girma, E. S. Kissel, A. N. Levy, S. MacCracken, P. R. Mastrandrea, & L. L. White (Eds.), *Climate change 2014: Impacts, adaptation, and vulnerability. Part A: Global and sectoral aspects. Contribution of working group II to the fifth assessment report of the intergovernmental panel on climate change* (pp. 899–943). Cambridge: Cambridge University Press.

Lesnikowski, A., Ford, J. D., Berrang-Ford, L., Barrera, M., Berry, P., Henderson, J., et al. (2013). National-level factors affecting planned, public adaptation to health impacts of climate change. *Global Environmental Change, 23*(5), 1153–1163.

MoH. (2015). *National strategic plan on disaster risk management for health 2015–2020*. Phnom Penh: Cambodian Ministry of Health.

Morgan, C. L. (2011). *Limits to adaptation: A review of limitations relevant to the project "Building Resilience to Climate Change-Coastal Southeast Asia"*. International Union for Conservation of Nature and Natural Resources (IUCN), Gland, Switzerland.

Picketts, I. M., Déry, S. J., & Curry, J. A. (2014). Incorporating climate change adaptation into local plans. *Journal of Environmental Planning and Management, 57*(7), 984–1002.

RGC and MoE. (2006). *National adaptation programme of action to climate change (NAPA)*. Phnom Penh: Royal Government of Cambodia and Ministry of Environment.

Rushton, B. (2008). *Consuming a participation policy: Cambodian health committees*. Sydney: University of Sydney.

Solar, R. W., Carson, T., & Srey, M. (2010). *The scoping assessment for national implementation in Cambodia—Summary*. Bangkok: Regional Climate Change Adaptation Knowledge Platform for Asia.

Sosa-Rodriguez, F. S. (2014). From federal to city mitigation and adaptation: Climate change policy in Mexico City. *Mitigation and Adaptation Strategies for Global Change, 19*(7), 969–996.

Strangio, S. (2014). *Hun Sen's Cambodia*. Chiang Mai: Yale University Press and Silkworm Books.

Turner, M. (2002). *Decentralization facilitation: A study of decentralization in Cambodia with specific reference to education*. Phnom Penh: Education Quality Improvement Project (EQIP), Ministry of Education, Youth and Sport.

UNDP. (2009). *Capacity development: A UNDP primer*. United Nations Development Programme, New York, USA. February 10, 2016.

WHO. (2008). *Protecting health from the effects of climate change: Resolution of the regional committee for the western Pacific*. Manila: World Health Organisation.

WHO. (2015). *Climate change and health in the western Pacific region: Synthesis of evidence, profiles of selected countries and policy direction*. Geneva: World Health Organization.

WHO. (2016). *Climate change adaptation to protect human health: Uzbekistan project profile*. World Health Organisation, Online publication. http://www.who.int/globalchange/projects/adaptation/en/index7.html. Accessed March 13, 2016.

Willems, S., & Baumert, K. (2003). *Institutional capacity and climate actions*. Paris: OECD Environment Directorate, International Energy Agency.

Chapter 4
Land-Based Strategic Model by Integrating Diverse Policies for Climate Change Adaptation in Nepal

Adish Khezri, Arbind Man Tuladhar and Jaap Zevenbergen

Introduction

Climate change is an important item on the agenda since the rate of greenhouse gas emission (GHG) is increasing globally. Despite a number of mitigation policies, anthropogenic influences have led to increasing GHG emission over the last 40 years, particularly between 2000 and 2010. Nowadays almost all sectors and regions aim to reduce the risk of climate change through adaptation, via place and context specific solutions which reduces climate vulnerability. Integration of adaptation into policy design is required as '*population size, economic activity, lifestyle, energy use, land use patterns, technology and climate policy*' are the main drivers of climate change. Institutional arrangements include adaptation in economic options (insurance), laws and regulations (land tenure security and property right), and national and government policies and programs (sub-national and local adaptation plans and community-based adaptation) (Dasgupta et al. 2014).

This paper focuses mainly on adaptation in national policies and programs in Nepal, a mountain country that faces the adverse effects of climate change. Effectiveness of adaptation depends on policies and interaction across all scales. Lack of co-ordination in climate policies has been shown to make farmers suffer in the case of West African farmers since there are different stakeholders taking adaptation action individually or not focusing precisely on adverse effects of climate change (Moumouni and Idrissou 2013). Therefore, adaptation requires co-ordination to manage these interactions, limit overlapping adaptation options and consider priority adaptation actions for different sectors. The Cancun

A. Khezri (✉) · A. M. Tuladhar · J. Zevenbergen
Faculty of Geo-Information Science and Earth Observation (ITC),
University of Twente, P.O. Box 217, 7500 AE Enschede, The Netherlands
e-mail: a.khezri@utwente.nl

J. Zevenbergen
e-mail: j.a.zevenbergen@utwente.nl

W. Leal Filho and J. Nalau (eds.), *Limits to Climate Change Adaptation*,
Climate Change Management, https://doi.org/10.1007/978-3-319-64599-5_4

Adaptation Framework includes institutions at the national level for '*strengthening and, where necessary, establishing and/or designation of national-level institutional arrangements*' (UNFCCC 2011). Literature reveals that only 6 out of 99 countries have an adaptation framework to address adaptation and this is often a single policy or separate laws. Nepal also has no adaptation framework (Nachmany et al. 2015).

Land is an inseparable part of human activities and natural resources plus it is a source of life and wealth (UN/ECE 1996). The purpose of this paper is to explores if the role of land is considered in adaptation plans and programs and it revealed that it is not. Therefore, this paper introduces a land-based strategic model for climate change adaptation services by integrating a variety of plans of action from the sectoral national policies related to land, environment, agriculture, forestry, meteorology and hydrology. Then, an institutional approach suggests towards the implementation of the land-based strategic model as part of land administration services which include geo-information to strengthen local and community institutions. As such, adaptation strategies at national and sub-national levels plus adaptation policies in different sectors have been reviewed in section "Adaptation Strategies". The methodology has been explained in section "Materials and Methods". Analysis of in-depth interview with the key experts at national level, a series of Nepal policy documents, in-depth interview with key experts at district and international levels, and focus group interviews have been presented in section "Analysis". The result of the analysis has been discussed in section "Towards strategies". SWOT analysis has been used to identify an adaptation intervention strategy in section "Analysis Using SWOT Technique". Finally, a land-based strategic model has been proposed to fill in the fragmented adaptation structure. Land Administration (LA) will be introduced as a tool for the implementation of this land-based model.

Adaptation Strategies

Adaptation is a series of strategies that responds to the needs of national and sub-national levels, sectors, communities and individuals facing climate change (Lim et al. 2005). Different strategies are introduced to tackle the impacts of climate change. They include local responses from farmers such as diversifying crops and livestock (Morton 2007), food preservation and storage, water harvesting, integrating farming activities with livestock or fisheries and cropping time alteration (Howden et al. 2007), micro-insurance mechanism for crops and livestock (Biener and Eling 2012), saving, loan and credit systems (Andersson and Gabrielsson 2012) and development of adaptation policy for different zones and resources (Pielke et al. 2007).

The IPCC report 2014, working group II, consists of part A: Global and sectoral aspects, and part B: Regional aspects, which shows the different impacts of climate change across the world. The components of climate change risk, including exposure, vulnerability and physical hazards (Burkett et al. 2014b) happen at different spatio-temporal scales because of various climate variables, resources, people's perceptions, infrastructure, ecology, environment and economy. Therefore, adaptation came into action on different scales and sectorial levels which are supported by their policies. Studies reveal that adaptation has positive effects on reducing poverty, improving livelihood and reducing vulnerability when supported by a policy (Stringer et al. 2009). UNFCCC, international organizations and NGOs support adaptation. Multiple guidelines, strategies, policies and documents encourage adaptation planning (USAID 2007; World Bank 2009; Dodman and IIED 2012).

National and Sub-national Adaptation Initiative

Countries take national initiatives to fill the gap between international negotiations and nations by developing adaptation plan through NAPAs and National Adaptation Plans (NAPs) (Mimura et al. 2014). Adaptation integrates adjusting to and coping with the current changes (Stringer et al. 2009). Therefore, it requires information about causes of risks and vulnerabilities that people and environment undergo (Füssel 2007).

The national level provides policies and plans, and performs networking and coordination between states, local levels, communities, civil societies and NGOs. It involves vulnerability reduction and capacity building for different states, cities, communities, and individuals. Laws and regulations plus government policies and programs are institutional instruments to accelerate adaptation (Noble et al. 2014).

Local authorities contribute to adaptation through their governance structure, utilizing local knowledge to adaptation planning and their roles in the implementation of community-based adaptation (Mimura et al. 2014).

Adaptation Policies in Different Sectors

Different opportunities for adaptation have been identified across various sectors including water resources, terrestrial eco-systems, marine and coastal systems, food security and food production systems, urban and rural areas. Key economic and service sectors include human health and security, livelihood and poverty (IPCC 2014). Countries take effort to develop different policies to manage their resources. This section reviews the key climate-sensitive sectors policies including agriculture, water, forest and land.

Agriculture policy encourages agricultural development and removes barriers to ensure food security. Macroeconomic, tax and market management are among existing agricultural policies (Demeke et al. 2014). Land is necessary for agriculture and to sustain rural livelihood. Land distribution and land titling as land-related agriculture policies improve farmers' access to land and tenure security. Climate-smart agriculture and agro-ecology are introduced to overcome food insecurity (Demeke et al. 2014). Agriculture adaptation to climate change includes crop management at farm level.

Water policy responses to urgent water management need. The major adaptation strategies to improve water management are:

- Investment in water infrastructure (irrigation and drainage systems),
- Monitoring existing water systems (water pollution control),
- Water infrastructure engineering (dams and pumps),
- Control water supply and demand (pricing and legislation),
- Utilizing new technology (water recycling).

Water is consumed at different scales and various sectors including industry, energy, food production and health. The water adaptation policies mentioned facilitate resolving water and land conflicts because they improve water security, which is access to healthy water in terms of quantity and quality (UN WATER 2011).

Forest management policy is a thematic element of Sustainable Forest Management (SFM) to manage all types of forest and to meet climate change challenges. REDD+ and voluntary carbon markets are introduced at national and sub-national levels as a part of mitigation strategy. Changing land use such as deforestation has direct and indirect effects on forest biodiversity, forest products and ecosystem services. Land use, land use change and forestry (LULUCF) plays an important role in climate change mitigation. Many land-related adaptation options including soil preparation, forest risk minimization of erosion by diversification of forest products and avoiding landscape fragmentation contribute to carbon sequestration (FAO 2011). Therefore, land use related activities should be integrated in forest management practices.

Land policy is a set of action plans to manage complex legal and socioeconomic activities (Smith et al. 2014). It is a tool for land use, land management, land tenure security and property rights (USAID 2008). Agriculture, Forestry and Other Land Use (AFOLUL) is an important sector contribute to climate change mitigation including energetic agriculture waste and forest monitoring (Shubert et al. 2009). Land use planning minimizes risks by hazard zoning, limit urban and rural development in disaster-prone areas and control population and infrastructure density (Bajracharya et al. 2011). Land use, land use change and forestry (LULUCF) has been recognized as part of a mitigation strategy in '*forest management, cropland management, grazing land management and revegetation*' (UNFCCC 2014) to decrease human-induced activities related to land.

In this paper we focus on an integration of these varied adaptation strategies that are land-based. However, there are also other adaptation strategies that are taking place by different ministries, NGOs and civil society groups that were not included in this paper. Furthermore, we focus on a land-based integration whereas other integration attempts can also be made (e.g. climate change adaptation and disaster risk reduction).

Materials and Methods

This paper focuses on Nepal since it ranked 7 in Climate Risk Index, 2014 (Sönke et al. 2015) as one of the most vulnerable countries to the impacts of climate change. Nepal has a variation of climatic conditions from sub-tropical to temperate to alpine mountains because of its specific rise of topography from some meters up to 4000 m. Maximum summer temperatures range from 28 to 40 °C and minimum winter temperatures from 7 to 23 °C. The monsoon is from mid-June to mid-September with about 80% of rainfall. Nepal's annual average rainfall is 1500–2500 mm with the maximum of 4500 mm. Landslides and floods are frequent natural hazards which damage people and their livelihood (UNDP 2009). Annually, it has the increase rate of 0.025% of GHG emissions and the annual average maximum temperature increase of 0.06 °C (GoN 2011a).

Adaptation is necessary since the best mitigation strategies do not stop further climate change impacts (Burkett et al. 2014a) in the short/medium term. Based on UNFCCC (2016a), adaptation consists of '*observation of climatic and non-climatic variables; assessment of climate impacts and vulnerability; planning; implementation; and monitoring and evaluation of adaptation actions*'. Each component of adaptation involves different stakeholders who play roles at different scales. 84% of the people in Nepal live in mountain rural areas and their major source of livelihood is agriculture (UNDP 2009). Effective adaptation provides opportunities for mountain communities and increase their resilience to tackle their needs and priorities.

The research method used here is the analysis of a series of qualitative interviews that were held with the key experts of Ministry of Land Reform and Management (MoLRM), Ministry of Agriculture and Development (MoAD), Ministry of Forest and Soil Conservation (MoFSC) and Ministry of Science and Technology (MoST) on their roles and responsibilities mainly related to adaptation. Next, Nepal's policy documents of CCP, NAPA, LAPA, NAP and LUP are selected to identify if they considered adaptation and in-land perspective. For validation of the policies and plans of action, the in-depth interviews with the regional key experts at ICIMOD are analyzed. Furthermore, the in-depth interviews with local key experts at district level and focus group interviews at community level are analyzed to explore if they are involved in any land-related adaptation program.

In-depth interviews had been conducted with two key experts at each ministry but one key expert at each district office and different thematic areas of ICIMOD. It is because the key expert at both district office and ICIMOD are involved in adaptation implementation but not adaptation policy development. Focus group interviews had been conducted with two mountain communities which were involved in adaptation implementation (two other remote communities were too difficult to reach). Different sets of questionnaires were designed for each in-depth interview and focus group interview related to climate change, land and livelihood, adaptation planning, policy and implementation. For this paper, we only analyzed the questions related to adaptation policy.

The ministries, policy documents and ICIMOD have been selected because of the important role they play in national adaptation policy to climate change and reduction of mountain people's vulnerability, respectively. Furthermore, other ministries, NGOs and civil society groups which are mainly involved in adaptation implementation have been left out. Even thought, they might have interesting outcomes in their works.

'*Computer-assisted NCT analysis*' was used as a method for qualitative data analysis in ATLAS.ti. NCT consists of three connected steps of Noticing Things (quotation), Collecting Things (coding) and Thinking about Things (finding patterns and relation in data). Interviews and documents have been coded in the light of adaptation to climate change. The methodology followed for each document is based on an analytical framework to search for a proposed adaptation option considering land. As such, an inductive research approach is taken, as the objectives and drivers of policy development, level of involvement and adaptation strategies have been identified. An approach was taken based on a grounded theory which emphasizes on how often a code is repeated (Kaplan and Maxwell 2005) in order to draw up the key services that each organization needs and can provide and that can be placed in the adaptation. Then the main findings of the analysis are discussed using the SWOT matrix to identify the gap on climate information and services. Finally, a land-based strategic model is proposed and LA has been introduced as an implementation tool that covers both the spatial and social aspects of land.

Analysis

This section explains the analysis of in-depth interviews with the key experts at ministry level, policy documents, in-depth interviews with the key experts at international level, district offices and focus group interviews at local level.

Analysis of in-Depth Interview with Key Experts at National Level

The Ministry of Land Reform and Management (MoLRM) developed a LUP, 2012 and is not involved in adaptation. However, the LUP, 2012 indirectly addresses the most frequent impacts of climate change on land. The ministry is able to support policy makers by producing land-related data from land measurement and cadastral data to mapping at different scales. For instance, land experts combine different data types including topographic maps, land type, crop type, soil type and soil capacity to provide land use map for the maximum benefit of farmers. There is no level of involvement with the adaptation program but the ministry is involved on a national, district, community and individual level because of its roles and responsibilities.

The Ministry of Agriculture Development (MoAD) developed NAP, 2004 and is not involved directly in adaptation. It conducted a Pilot Project for Climate Resilience (PPCR) for community poverty reduction. This ministry responds to agricultural development. It focuses on agro-business for communities, farmer training, distributing seed and fertilizer, financing small agriculture enterprises and micro-insurance schemes. The involvement of MoAD is the same as that of MoLRM.

The Ministry of Forest and Soil Conservation (MoFSC) is not involved in adaptation. The ministry contributes to the mitigation of climate change through community forest management. Conserving forest contributes to some extent to adaptation since the communities use forest products. The ministry is involved on a national, district and community level because of its roles and responsibilities in the mitigation of climate change.

The Ministry of Science, Technology (MoST) is involved in adaptation policy development such as NAPA, LAPA, low carbon strategy and climate change budget code. They allocate 80% of the budget for vulnerability reduction of local people. This budget mainly comes from mitigation such as REDD+. The ministry requested collaboration from different ministries, district and local levels for NAPA and LAPA development. The level of involvement with adaptation is limited to raising public awareness by organizing workshops, public discussions and seminars from time to time.

Analysis of Nepal Policy Documents

CCP, 2011

Objective To engage Nepal internationally in minimizing the effects of climate change and as an urgent need for the development of mountain people, livelihood and eco-system (GoN 2011a).

Driver of Development Impacts of climate change on mountainous areas, people, livelihood and ecosystem; changing rainfall and temperature patterns.

Level of Involvement International (UNFCCC) and national (ministry of environment).

Adaptation Strategies Preparation of NAPA and LAPA, adopting promotion of development of Clean Development Mechanism (CDM), renewable energy and raising public awareness.

NAPA, 2010

Objective To be used as a strategic tool for vulnerability assessment and prioritizing adaptation actions (Ministry of Environment 2010).

Driver of Development Changing rainfall and temperature patterns, poverty.

Level of Involvement National and Village Development Committee (VDC) levels.

Adaptation Strategies Climate change vulnerability assessment; identification of six climate-sensitive areas including:

- Agriculture and food security,
- Water resources and energy,
- Climate-induced disasters,
- Forest and biodiversity,
- Public health,
- Urban settlement and infrastructure.

Identification of adaptation priority activities including:

- Promoting community-based adaptation through integrated management of climate-sensitive sectors,
- Building and enhancing adaptive capacity of vulnerable communities through agricultural development,
- Community-based disaster management,
- GLOF monitoring and disaster risk reduction,
- Forest and ecosystem management,
- Adaptation to climate change in public health,
- Ecosystem management,
- Sustainable management of water resources and clean energy supply,
- Promoting climate smart urban settlement.

LAPA, 2011

Objective To fill the gap of adaptation implementation and support vulnerable groups (GoN 2011b).

Driver of Development Impacts of climate change on agriculture, forestry, public health, water and sanitation, watersheds and micro-finance.

Level of Involvement National, VDC, Municipality, District Development Committee (DDC), ward, community and household levels. LAPA has been developed and piloted in 10 Nepal districts.

Adaptation Strategies No adaptation strategies. LAPA as a bottom-up approach introduced different steps and many different core tools for '*integrating climate change resilience into local-to-national planning processes*' (GoN 2011b) but there is no core tool on how to implement an adaptation plan.

NAP, 2004

Objective To reduce poverty and ensure food security through agricultural development (GoN 2004).

Driver of Development Decreasing fertile agricultural land, lack of access of farmers to resources including skills, labor and commercial bases, lack of infrastructure for agricultural development (roads, industry, communication, marketing), food insecurity and poverty.

Level of Involvement National, villages and household levels.

Adaptive Strategies No adaptation strategies.

LUP, 2012

Objective To achieve land use planning and sharing land resources benefits to mountain people (GoN 2012a).

Driver of Development Unmanaged use of land, inequality in access to land, government and public land violation, land fragmentation, food insecurity and poverty.

Level of Involvement International (UN 1992), national, VDC, Municipality, DDC, ward, community and household levels.

Adaptation Strategies No adaptation strategies.

Analysis of In-depth Interview with Key Experts at International Level (ICIMOD)

This section presents analysis of in-depth interviews at ICIMOD conducted in different thematic areas. ICIMOD is a knowledge-sharing center based in Kathmandu with the aim of facilitating mountain people to adapt to change and sustainable mountain ecosystems.

Ecosystem Services is mainly responsible for support and implementation of REDD+ and it is not involved in adaptation.

Adaptation to Change keeps track of climate and non-climate changes. This thematic area facilitates ministry offices and DDCs in pilot adaptation activities by providing information about changes in the specific area.

Livelihood works closely with mountain communities to improve their adaptive capacity. This thematic area facilitates communities by alternative livelihood options such as stitching, planting fruits and vegetables, farm bees and collecting herbal medicine and aromatics.

Mountain Environment Regional Information System (MENRIS) supports both adaptation and mitigation indirectly by providing data and information to decision makers and district level.

Geospatial Solution supports adaptation by providing different thematic data such as forest type, forest fires, forest productivity and evapotranspiration. The thematic area integrates technology and adaptation since it provides more options to stakeholders of adaptation.

Analysis of In-Depth Interviews with Local Key Experts at District Level

District livestock office gives advice to the community on how to treat animals.
District agriculture development office distributes seeds and fertilizers and trains farmers in vegetable cultivation and small irrigation schemes.
District forest office is not involved in adaptation but mitigation through REDD+ and community forest management.
District survey office is not involved in adaptation.
District planning office raises community awareness through Citizen Awareness Centers (CACs) in all VDCs, supports vulnerable groups through social security programs, rural micro hydro-plan and facilitates the installation of hydropower and early warning systems.

Analysis of Focus Group Interview at the Community Level

Two focus groups interviews were conducted with the communities in Bocha and Cherikot in the Dolkha district. Dolkha district is selected because of frequent landslides and floods.

The Cherikot and Bocha communities are mainly involved in the mitigation of climate change through REDD+. They get fertilizer and seeds from the district agriculture office to speed up their agriculture products. District livestock office facilitates Bocha communities with animal treatment since livestock is a source of their livelihood.

Towards Strategies

From the policies mentioned in Adaptation Stategies it can be concluded that land plays a role in all socioeconomic and environmental activities which are climate-sensitive and are confronted with '*environmental, social, economic and political stressors*' (Olsson et al. 2014). Land is a basis for housing, agriculture, infrastructure, water, forest and livelihood among others. Annually 1.6 GtG carbon is emitted from both changes in land use (forest conversation to agricultural land) and land use (respiration of vegetation and organic materials). Land is a source of carbon removal/emission by its forest, cropland, wetlands, grassland and settlement categories (UNFCCC 2016b) and needs to be considered in each adaptation policy, strategy, plan and program.

The roles and responsibilities of MoLRM, MoAD, MoFSC and MoST have been analyzed with regard to land. The MoLRM does not participate in any adaptation plans and programs. However, it has the ability to produce any land-related data on demand for the whole of Nepal, having offices in all 75 districts. The MoAD is involved in a series of small-scale farmers' capacity building programs. The MoFSC engages communities in forest management as part of mitigation. The MoST developed a CCP, a NAPA and LAPAs to address adaptation and to fill in the fragmentation between different scales of national, district and community.

Nepal policy documents had been reviewed to explore the consideration of an adaptation program with land perspective in their plans. CCP is a law (Townshend et al. 2014), executive with a mitigation framework (LSE 2011) and does not mention which organization(s) should be involved in the process. NAPA was developed in six basic themes, introducing a list of priority adaptation options. LAPA, an executive law, has an adaptation framework that supports identification of climate-vulnerable areas and prioritizes community-adaptation action. NAP focuses mainly on its responsibilities in agriculture sector. LUP does not consider adaptation and only mentions classifying climate change prone-areas in forestry zones. None of these policy documents consider land-based services as an adaptation option.

Analysis of the in-depth interview with the key experts at regional level revealed that ICIMOD focuses on both adaptation and mitigation at regional level and do not consider specifically land. Analysis of the in-depth interviews with the key experts at district level and focus group interviews at community level revealed that there is no land-based adaptation foreseen for communities. The analysis of the in-depth interviews at international, national, district and community levels plus policy documents lead to the identification of the important factors including land, policy, coordination and implementation tools. Presence of these factors make communities adapt better to climate change.

Analysis Using SWOT Technique

SWOT analysis is a strategic tool to harmonize intervention strategies and to make improvement in complex strategic situations (Helms and Nixon 2010), such as adaptation to climate change. SWOT as a matrix is used to analyze the current situation of adaptation. It is influenced by internal factors in terms of strength and weakness of policies and ministries in carrying out adaptation activities, while external factors determine if adaptation can take advantages of opportunities optimizing institutional arrangements and adaptation (Table 4.1).

The SWOT analysis reveals 19 strategies which are required to improve adaptation from a land perspective. Land relates to structural and non-structural measures: infrastructure and policy (Charoenkalunyuta et al. 2011). '*Integrated management of agriculture, water, forest and biodiversity*' has been proposed by NAPA. Moreover, one of the concepts that LAPA purposes is '*integrated*

Table 4.1 SWOT Matrix of adaptation strategy in land perspective

Internal factors	**External factors**	
	Opportunity (O) 1. Clear mandates and roles/responsibilities of each organization 2. Existing climate change policies 3. Identification of climate-sensitive sectors 4. Prioritization of adaptation action 5. Availability of different tools to promote adaptation	*Threat (T)* 1. No appropriate policy for adaptation 2. Lack of coordination among ministries for adaptation 3. No tool for adaptation implementation 4. Services of one ministry not available for other ministries 5. Adoption of adaptation 6. LAPA only for 10 districts
Strength (S) 1. Ministerial understanding and awareness of the importance of adaptation 2. Ministerial provision of different services to citizens 3. Different scales participation 4. Different scales experiences with adaptation	*SO strategies* – Definition of roles and responsibilities of stakeholders – Availability of diverse services – Stakeholders' involvement in adaptation at all scales – Use of potential tools in adaptation	*ST strategies* – Development of new adaptation strategy – Start and improvement of coordination at different scales – Sharing of different services, knowledge and experiences
Weakness (W) 1. Overlapping adaptation action 2. Same adaptation options for different districts and communities 3. No idea of communities' expectations 4. No improvement in community resilience 5. No consideration of the role of land as a basis for adaptation	*WO strategies* – Defining appropriate adaptation policy – Introducing new adaptation strategy – Applying new concepts and tools to adaptation – More adaptation options	*WT strategies* – Utilizing land-based strategy – Improving coordination among adaptation stakeholders – Improving climate information and services – Finding potential tools to implement adaptation – Developing adaptation framework – Improving communities resilience

approaches between various sectors and sub-sectors' to inform different stakeholders on their sectoral programs (GoN 2012b). '*Coordinating climate change related activities at the local level*' and '*enhanced coordination at the technical level*' at national level has been recommended after analysis of climate change institutions of Nepal (CPEIR 2011). Coordination adds value to what needs to be done in adaptation.

Integrated Land-Based Strategy

People, waterbodies, agriculture and forest areas, urban and rural areas, mountain and all natural and man-made structures are connected or attached to land or a specific location. As mentioned earlier, all are affected directly or indirectly by the impacts of climate change. Therefore, an integrated land-based strategy is required in adaptation to climate change. Different policies which were analyzed and introduced in this paper are related to the land sector and linked to the impact of climate change. The strategic model is shown in Fig. 4.1. Please note that the letter 'S' stands for service.

Figure 4.1 shows how these institutions can be linked in providing services for mountain people building their resilience. Analyses of policy documents and in-depth interviews revealed that each ministry provides service(s) to the mountain people. The services are the result of their institutional roles and responsibilities and not directly because of adaptation.

The services considering land are fragmented in their contribution to adaptation. Furthermore, the policy documents mentioned adopt policies related to their institutional arrangements. These are the services which LA is able to provide. Table 4.2 lists these services with respect to land. Some of the services are not the responsibility of MoLRM but they can be integrated with land administration.

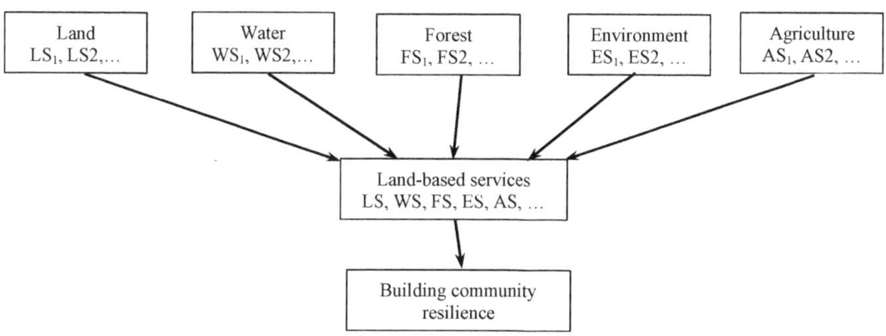

Fig. 4.1 Land-based strategic model

Table 4.2 Land information service(s) needed by policies

Policy document	Land information services
CCP NAPA LAPA	Administrative boundary level, vulnerable area, vulnerable group, hazard zone, natural disaster, land type, land management, land tax, land distribution, land registration, land measurement, land tenure, infrastructure, current adaptation and mitigation plan, land administration, carbon map, identification of land and water conflict areas
NAP	Administrative boundary level, soil type, land type, land consolidation, land use, land cover, hazard zone, natural disaster, infrastructure, identification of land and water conflict areas, land distribution, land registration, land measurement, land tenure

There is a need of implementing a land-based strategy to provide services to the mountain people. LA with its functionalities including land use, land tenure, land value and land development, is a tool. It has the ability to provide access to resources, use of resources and control over them in terms of information and services.

LA is country specific and can be adapted to support rules and regulations related to climate change policy and adaptation in any country. LA is able to collect, manage and disseminate land-related data mainly geo-information. Successful adaptation needs to consider the interest(s) of different stakeholders and resources available which are manageable through land administration.

Land governance, land policy, land management, land use planning, land tenure security, cadastral maps, carbon maps and transparency are a few of examples of land administration services in adaptation (FIG 2014).

Conclusion

Countries put effort into collaborating on an international level in adaptation to climate change. There are different adaptation strategies at national and sub-national levels because of the direct and indirect effects of climate change on different sectors. Institutional arrangements support adaptation by policy intervention to enable adaptation planning and implementation which is ad hoc and sporadic. Since land is a specific entity and source of natural, societal and environmental development, an integrated land-based strategic model is able to overcome these fragmented adaptation approach. The model integrates various services from different service providers to improve specifically mountain community resilience. Land administration as a tool supports the provision, integration, management and implementation of services from different sectors since it has the ability to engage multiple stakeholders at both policy and technical level across environmental, societal and economic sectors at different scales and time frames.

Acknowledgements This research paper was conducted at Faculty of Geo-Information Science and Earth Observation (ITC),University of Twente, and funded by Erasmus Mundus Action 2 Project SALAM, International Scholarship of the European Commission.

References

Andersson, E., & Gabrielsson, S. (2012). "Because of poverty, we had to come together": Collective action for improved food security in rural Kenya and Uganda. *International Journal of Agricultural Sustainability, 10,* 245–262. doi:10.1080/14735903.2012.666029.

Bajracharya, B., Childs, I., & Hastings, P. (2011). Climate change adaptation through land use planning and disaster management: Local government perspectives from Queensland Climate change adaptation through land use planning and disaster management: Local government perspectives from Queensland. In *17th Pacific Rim real estate society conference, climate change and property: its impact now later,* 16–19 January 2011, Gold Coast 1–16.

Biener, C., & Eling, M. (2012). Insurability in microinsurance markets: An analysis of problems and potential solutions. *The Geneva Papers on Risk and Insurance Issues and Practice, 37,* 77–107. doi:10.1057/gpp.2011.29.

Burkett, V. R., Suarez, A. G., Bindi, M., et al. (2014b). Global and sectoral aspects. Contribution of WGII to the fifth assessment report of the IPCC. In *Climate change 2014: Impacts, adaptation and vulnerability* (pp. 169–194).

Burkett, V. R., Suarez, A. G., Bindi, M., et al. (2014a). Point of departure. In C. B. Field, V. R. Barros, D. J. Dokken, et al. (Eds.), *Climate change 2014: Impacts, adaptation, and vulnerability. Part A: Global and sectoral Aspects. WGII of the fifth assessment report of IPCC.* Cambridge: Cambridge University Press.

Charoenkalunyuta, C. M., Tuladhar, A., & Zevenbergen, J. (2011). Community resilience in disaster prone areas based on land rights/ownerships. In *32nd Asian conference on remote sensing 2011,* ACRS 2011 (Vol. 1, pp. 556–571).

CPEIR. (2011). *Nepal climate public expenditure and institutional review (CPEIR).* Kathmandu: Government of Nepal, National Planning Commission with support from UNDP/UNEP/CDDE.

Dasgupta, P., Morton, J. F., Dodman, D., et al. (2014). Rural areas. In C. B. Field, V. R. Barros, D. J. Dokken, et al. (Eds.), *Climate change 2014: Impacts, adaptation, and vulnerability.* Cambridge: Cambridge University Press.

Demeke, M., Spinelli, A., Croce, S., et al. (2014). Food and agriculture policy decisions. Trends, emerging issues and policy alignments since the 2007/08 food security crisis.

Dodman, D., & IIED. (2012). *Cities and climate change initiative, tool series, developing local climate change plans, a guide for cities in developing countries.* United Nations Human Settlements Programme (UN-Habitat), Nairobi.

FAO. (2011). Climate change for forest policy-makers: An approach for integrating climate change into national forest programmes in support of sustainable forest management.

FIG. (2014). *The surveyor's role in monitoring, mitigating, and adapting to climate change.* University of Otago, Otago.

Füssel, H.-M. (2007). Vulnerability: A generally applicable conceptual framework for climate change research. *Global Environmental Change, 17,* 155–167. doi:10.1016/j.gloenvcha.2006.05.002.

GoN. (2004). *National agricultural policy 2004.* Kathmandu: Government of Nepal.

GoN. (2011a). Climate change policy, 2011.

GoN. (2011b). LAPA MANUAL: Local adaptation plans for action (LAPA).

GoN. (2012a). *national land use policy 2012.* Singhdurbar, Kathmandu: Ministry of Land Reform and Managment.

GoN. (2012b). Adaptation to climate change NAPA to LAPA.

Helms, M. M., & Nixon, J. (2010). Exploring SWOT analysis—Where are we now? A review of academic research from the last decade. *Journal of Strategy and Management, 3,* 215–251.

Howden, S. M., Soussana, J.-F., Tubiello, F. N., et al. (2007). Adapting agriculture to climate change. *Proceedings of National Academy of Sciences, 104,* 19691–19696. doi:10.1073/pnas.0701890104.

IPCC. (2014). Summary for policymakers. In C. B. Field, V. R. Barros, D. J. Dokken, et al. (Eds.), *Climate change 2014: Impacts, adaptation, and vulnerability. Part A: Global and sectoral aspects. WGII of the fifth assessment report of IPCC* (pp. 1–32). Cambridge: Cambridge University Press.

Kaplan, B., & Maxwell, J. A. (2005). *Qualitative research methods for evaluating computer information systems* (pp. 30–56). Berlin: Springer.

Lim, B., Burton, I., & Nations, D. P. U. (2005). *Adaptation policy frameworks for climate change: Developing strategies, policies and measures.* Cambridge: Cambridge University Press.

LSE. (2011). The London School of The London School of Economics and Political Science (LSE). http://www.lse.ac.uk/GranthamInstitute/law/climate-change-policy/. Accessed January 1, 2016.

Mimura, N., Pulwarty, R. S., Duc, D. M., et al. (2014). Adaptation planning and implementation. In C. B. Field, V. R. Barros, D. J. Dokken, et al. (Eds.), *Climate change 2014: Impacts, adaptation, and vulnerability. Part A: Global and sectoral aspects. WGII to the fifth assessment report of IPCC*. Cambridge: Cambridge University Press.
Ministry of Environment. (2010). *Government of Nepal national adaptation programme of action to climate change*. Kathmandu: Government of Nepal, Ministry of Environment.
Morton, J. F. (2007). The impact of climate change on smallholder and subsistence agriculture. *Proceedings of National Academy of Sciences USA, 104,* 19680–19685. doi:10.1073/pnas.0701855104.
Moumouni, I., & Idrissou, L. (2013). *Innovation systems for agriculture and climate in Benin: An inventory*. Greenwich: University of Greenwich.
Nachmany, M., Fankhauser, S., Davidová, J., et al. (2015). The 2015 Global Climate Legislation Study: A review of climate change legislation in 99 countries: Summary for policy-makers.
Noble, I. R., Huq, S., Anokhin, Y. A., et al. (2014). Adaptation needs and options. In C. B. Field, V. R. Barros, D. J. Dokken, et al. (Eds.), *Climate change 2014: Impacts, adaptation, and vulnerability. Part A: Global and sectoral aspects. WGII to the fifth assessment report of IPCC.* Cambridge: Cambridge University Press.
Olsson, L., Opondo, M., Tschakert, P., et al. (2014). Livelihoods and poverty. In C. B. Field, V. R. Barros, D. J. Dokken, et al. (Eds.), *Climate change 2014: Impacts, adaptation, and vulnerability. Part A: Global and sectoral aspects. WGII to the fifth assessment report of IPCC.* Cambridge: Cambridge University Press.
Pielke, R., Prins, G., Rayner, S., & Sarewitz, D. (2007). Climate change 2007: Lifting the taboo on adaptation. *Nature, 445,* 597–598. doi:10.1038/445597a.
Shubert, R., Schellnhuber, H. J., & Buchmann, N. (2009). *Future bioenergy and sustainable land use*. London: Earthscan.
Smith, P., Bustamante, M., Ahammad, H., et al. (2014). Agriculture, forestry and other land use (AFOLU). In O. Edenhofer, R. Pichs-Madruga, Y. Sokona, E. Farahani, S. Kadner, & K. Seyboth (Eds.), *Climate change 2014: Mitigation of climate change. WGIII of the fifth assessment report of IPCC* (pp. 811–922). Cambridge: Cambridge University Press.
Sönke, K., Eckstein, D., Dorsch, L., & Fischer, L. (2015). Global climate risk index 2016: Who suffers most from Extreme weather events? Weather-related loss events in 2014 and 1995 to 2014.
Stringer, L. C., Dyer, J. C., Reed, M. S., et al. (2009). Adaptations to climate change, drought and desertification: Local insights to enhance policy in southern Africa. *Environmental Science and Policy, 12,* 748–765. doi:10.1016/j.envsci.2009.04.002.
Townshend, T., Fankhauser, S., Aybar, R., et al (2014). The GLOBE climate legislation study: A review of climate change legislation in 66 countries.
UN. (1992). Report of the United Nations conference on environment and development: Rio de Janeiro, 3–14 June 1992: Vol. I–IV. United Nations, New York.
UN WATER. (2011). Climate change adaptation: The pivotal role of water (pp. 1–18).
UN/ECE. (1996). Land adminstration guidelines with special reference to countries in transition.
UNDP. (2009). Nepal country report.
UNFCCC. (2011). Cancun adaptation framework (pp. 1–31).
UNFCCC. (2014). LULUCF - Developments at past COP and SB sessions. https://unfccc.int/land_use_and_climate_change/lulucf/items/3063.php. Accessed January 1, 2016.
UNFCCC. (2016a). Elements of adaptation. http://unfccc.int/adaptation/items/7006.php#Observation. Accessed January 1, 2016.
UNFCCC. (2016b). *Handbook on the land use, land-use change and forestry (LULUCF) sector*. New York, NY: UNFCCC.
USAID. (2007). Climate change, property rights, and resource governance. Washington, DC.
USAID. (2008). Land tenure and property rights: Tools for transformational development.
World Bank. (2009). Development and climate change. Washington, DC: World Bank.

Author Biography

Adish Khezri is a Ph.D. candidate at Faculty of Geo-Information Science and Earth Observation (ITC), Twente University in the Netherlands. She is performing research on mountain community adaptation to climate change integrated with land administration.

Chapter 5
Climate Change and Migration in Bangladesh: Empirically Derived Lessons and Opportunities for Policy Makers and Practitioners

Johannes Luetz

Bangladesh: Demographic, Geographic and Climate Change Issues

In Bangladesh human movement is influenced by environmental and climatic change, together with traditional labour and seasonal migration patterns. To introduce the migration context in Bangladesh this section will commence with a discussion of the country's contextual backdrop regarding relevant demographic, geographic and environmental issues, including natural disasters.

Discounting city states Bangladesh has the highest population density in the world. With an average 1229 people living together on each available square kilometre of land, Bangladesh is home to more people than live in all of Russia combined (World Bank 2011a, pp. 344–345; Belt 2011, p. 64). Bangladesh is also among the poorest and most low-lying coastal countries in the world. According to the United Nations Development Programme (UNDP 2011, p. 126), Bangladesh is a Low Human Development country with approximately 150+ million citizens and a rank of 146 (out of 187 countries) on the Human Development Index (HDI). In terms of purchasing power parity (PPP), 49.6% of the population live below the international poverty line of US$1.25 per day (UNDP 2011, p. 144; attributed to World Bank 2011a, p. 346[1]), down from 57.8% in the year 2000 (ibid.[2]). In 2005, the most recent survey year available, 81.3% of the population lived on less than

---

[1]Data from 2005.
[2]See also http://hdr.undp.org/en/content/country-profiles.

J. Luetz (✉)
CHC Higher Education, Brisbane, Carindale, Australia
e-mail: j.luetz@unsw.edu.au

J. Luetz
University of New South Wales, Sydney, Australia

W. Leal Filho and J. Nalau (eds.), *Limits to Climate Change Adaptation*,
Climate Change Management, https://doi.org/10.1007/978-3-319-64599-5_5

US$2 per day (ibid.). Moreover, 57.8% of the population live in "multidimensional poverty", and 26.2% live in "severe poverty" (ibid.[3]) Although poor by these standards, Inman (2009) asserts that "the country has left behind its earlier reputation as an 'international basket case', a term once unkindly bestowed on it by a United States government official. Its economy is growing fast, the poverty rate is falling and the average life span has now stretched to 63 years" (Inman 2009, p. 18). In recent years improvements in human well-being have no doubt been made, for example "the fertility rate plunged from 6.6 births per woman in 1975 to 2.4 in 2009" (UNDP 2011, p. 9), and life expectancy at birth was most recently assessed in the Human Development Report at 68.9 years (UNDP 2011, p. 129), up from 55.2 years in 1980 (UNDESA 2011).

Since coastal areas abound with resources and trading opportunities, for millennia human populations all over the world have preferred to live "within 100 kilometres of coasts and near major rivers" (Small and Nicholls 2003; Small and Cohen 2004, cited in McGranahan et al. 2007a, p. 18). With a worldwide length of one million kilometres—or approximately 25 times the Earth's circumference—coastlines and their adjacent regions "are of extreme importance for humankind" (WBGU 2006, p. 40). Research by McGranahan et al. (2007a) assessed the vulnerabilities of coastal communities to risks resulting from climate change and defined a so-called Low Elevation Coastal Zone (LECZ) as "the contiguous area along the coast that is less than 10 metres above sea level" (p. 17). The research found that these LECZ constitute only 2% of the Earth's land mass but contain 10% of the world's population, and 13% of the world's urban population, thus having a higher rate of urbanisation than the rest of the world (ibid., p. 17; see also Brahic 2007). With more than 600 million people worldwide living in the zone, 75% of whom in Asia (ibid., p. 17), and "21% of the world's human population [living] less than 30 km from the sea" (Cohen et al. 1997; Gommes et al. 1998, cited in WBGU 2006, p. 40), the popularity of coastal regions is perhaps best evidenced by population growth rates of "twice the global average" (Bijlsma et al. 1996, cited in WBGU 2006, p. 40). Bangladesh exemplifies this trend: "[b]etween 1990 and 2000, the populations in the 0–10 m zones of Bangladesh … grew at more than twice the national population growth rate" (McGranahan et al. 2007b).

According to the United Nations Populations Fund (UNFPA), the coming two decades will see Asia's urban population increase from 1.36 billion in 2000 to 2.64 billion in 2030 (UNFPA 2007). UN Habitat predicts that 95% of the world's urban growth in the next two decades "will be absorbed by cities of the developing world, which are least equipped to deal with rapid urbanization" (UN Habitat 2006, p. 2). In short, much of the developing world is urbanising at an unprecedented rate, and it is doing so along, near or towards the coast. According to Small and Nicholls (2003), "[b]y the year 2030 approximately 50% of the world population could be living within 100 km of the coasts" (WBGU 2006, p. 40).

[3]"Population in severe poverty: Percentage of the population in severe multidimensional poverty—that is, those with a deprivation score of 50 percent or more." (UNDP 2011, p. 145).

The economic boom of countries in Asia[4] has been attributed in part to policies that promote coastal development and thereby encourage coast-ward human migration. In Bangladesh millions live in coastal proximity: 46% of the population of Bangladesh (more than 62 million) live within 10 m above sea level (McGranahan et al. 2007a, p. 26), and "approximately 50 million people live below five metres above sea level" (Maplecroft 2007b, p. 14). However, rapid urbanisation in coastal zones holds multiple risks. It exposes people to seaward hazards such as floods, cyclones, storm surges and sea level rise, while concurrently degrading sensitive ecosystems like mangrove forests and coral cover that traditionally protected these coastlines. At the same time coastal urbanisation is both rapid and often self-reinforcing: the more urbanised a coastal community becomes, the more it draws other would-be migrants into the suction of its magnetic field. The net effect can be an even further increase in the rate of coastal population growth and haphazard development practice in marginal environments exposed to seaward hazards like king tides, cyclonic surges, coastal erosion, land loss, etc. (McGranahan et al. 2007a, b; Nicholls and Small 2002).

Several big cities in South Asia (e.g., Dhaka, Chittagong, Khulna, Kolkota, Chennai, etc.) extend into the LECZ and face challenges which are exacerbated by the fast pace of informal or unplanned urbanisation (McGranahan et al. 2007a; Brahic 2007). During fieldwork media reports highlighted "unplanned development" and consequent problems as matters of growing concern (Shyamol 2011 p. 5; Ahsan 2011; The Independent 2011a, b). For example, Ahsan (2011)[5] mentions unplanned squatter settlements in Dhaka having resulted in virtually "the destruction of almost the entire natural sewage system of the city" (p. 1), and leaving it essentially "developed beyond reclamation" (p. 4).

Similarly, research by the World Bank identified Dhaka as

> the fastest growing mega-city in the world, with an estimated 300,000-400,000 new migrants, mostly poor, arriving to the city annually, [aggravating] large slums, poor housing, excessively high land prices, traffic congestion, water shortages, poor sanitation and drainage, irregular electric supply, unplanned construction, increasing air pollution and poor urban governance which results in growing problems of law and order. […] As migrants continue to arrive, they often end up in illegal settlements on precarious lands with major environmental concerns. The slums are located throughout the city with few services offered at high prices through middlemen, also called musclemen or *mastaans*, using illegal methods. […] Access to basic services such as water, sanitation, electricity, health and education by the poor is limited. Because most slums are not recognized as legal lands, the Government, NGOs and Donors generally do not provide services in these areas. As a result, a parallel structure has emerged with *mastaans* providing services for a high fee. (Baker 2007a, pp. xiii, xv-xvi).

[4]For example, "China's economic boom has been driven by policies that promote coastal development and which have encouraged one of the largest coastward migrations ever." (McGranahan et al. 2007b).

[5]Ahsan puts the blame on corrupt officials, namely "hawlka officers, attestation officers, circle officers, objection officers, and appeal officers… and settlement officers." (Ahsan 2011, p. 4).

In such precarious conditions it can be very difficult for newly arriving migrants to get themselves established. In Dhaka there is "only one public tap for every 500 slum dwellers" (UN Habitat 2003, cited in UNDP 2011, p. 48), and according to Belt (2011, p. 73), the percentage of urbanites with access to toilets has decreased from 59% in 1990 to 56% in 2010, with an overall coverage drop to 52% projected by 2030. Such mounting pressures can constrain migrants with no other options to source amenities outside of official government supply grids, often forcing them to pay even higher prices for basic service provision than wealthier urbanites. Baker (2007b) mentions "the poor pay 2 Taka for a bucket of water, whereas the wealthy can purchase 1000 litres through the official grid for 4.5 Taka" (cited in Luetz 2008a, Endnote 259, p. 116).

A number of studies have drawn attention to the possible consequences of slum settlement proliferation, which could exacerbate existing vulnerabilities.

According to UN Habitat,

> [n]early one billion people alive today—one in every six human beings—are slum dwellers, and that number is likely to double … Unprecedented urban growth in the face of increasing poverty and social inequality [will] increase … the number of people living in slums to about 2 billion by 2030. (UN Habitat 2007a, p. 1)

According to the former Vice President of the Asian Development Bank, Ursula Schaefer-Preuss,

> cities, especially in Asia, will be hardest hit when sea levels rise, with tens of millions of people likely to be forced from their homes … The world simply cannot sustain current urban development trends. (Schaefer-Preuss 2008, cited in Luetz 2008a, p. 90)

The eighth Secretary-General of the United Nations, Ban Ki-moon, synthesised the situation like this:

> [t]he vulnerability of cities is increasing due to climate change, which has accelerated extreme weather events and rising sea levels. At the same time, urban slums are expanding into areas vulnerable to floods, landslides, industrial pollution and other hazards. (UN Habitat 2007b, p. v)

According to the German Advisory Council on Global Change (WBGU),

> [c]limate change and urbanisation are dominant trends of global change. The interplay of the two trends in the major coastal cities of the developing world could cause an almost unmanageable situation, particularly if the arsenal of responses is limited by social, economic and institutional deficits. (WBGU 2006, p. 63)

And a press release by the United Nations University Institute for Environment and Human Security on the occasion of the UN Day for Disaster Reduction 2005 put it like this:

> [a]round the world vulnerability is on the increase due to the rapid development of megacities in coastal areas. […] Many cities are overwhelmed, incapable of handling with any degree of effectiveness the demands of a burgeoning number of people, many of whom take up shelter in flimsy shanties. Combine this trend with rising sea levels and the growing number and intensity of storms and it is the recipe for a disaster-in-waiting, with enormous potential to create waves of environment-driven migration. (UNU-EHS 2005, p. 3)

In summary, indications are that urban vulnerabilities are likely mounting, and that in view of Bangladesh's comparatively low rate of urbanisation current trends of rural-urban migration into Bangladesh's sprawling cities and megacities are likely to continue for years to come. This inference is made also on the basis of Bangladesh's rural-urban settlement ratio (71.6 versus 28.4%), which suggests that the continuing process of urbanisation, relative to other countries, is far from mature (World Bank 2011b). This possibility seems to make creative anticipatory urban planning, strategic preparations and overall migration preparedness increasingly more important. In short, the question how continued in-migration of additional millions of people into crowded cities like Dhaka can be accommodated and managed equitably and sustainably in the years ahead is a question that policy makers will need to address with care and foresight.

Aside from the vulnerabilities arising from demographic and urban development trends discussed above, Bangladesh is also susceptible to a number of natural disasters and related climate change processes, which are discussed next. These are already impacting on forced human migration in South Asia in a significant way and cannot be meaningfully discussed and understood in isolation of their context. Expressed in simple language, the cumulative impact of any natural disaster depends both on the environmental and socioeconomic context within which the disaster occurs, and the nature and force of the onslaught itself. In a country like Bangladesh where nearly 5% of the nation's citizens are affected annually by disasters (UNDP 2011, p. 152), and where more than 80% of the population fight for survival on less than two dollars a day (ibid., p. 144), the cumulative impact of disasters on human development can be significant. According to Ainun Nishat of the International Union for Conservation and Nature (IUCN) in Dhaka, Bangladesh is "nature's laboratory on disasters. […] We don't have volcanoes. But any other natural disaster you think of, we have it" (Inman 2009, p. 18). According to the Emergency Events Database (EM-DAT)[6] disaster definition,[7] the years 1990–2007 saw Bangladesh impacted by 164 natural disasters. Windstorms were both the most prevalent (50%) and most deadly disaster type (more than 143,000 people killed). Floods were less prevalent (28%) but appear to have exacted the most far-reaching

[6]The EM-DAT International Disaster Database, Université Catholique de Louvain, Brussels, Belgium [http://emdat.be] is a joint project of the Centre for Research on the Epidemiology of Disasters (CRED) and USAID's Office of Foreign Disaster Assistance (OFDA).

[7]"The Center for Research on the Epidemiologyof Disasters (CRED) defines a disaster as a 'situation or event, which overwhelms local capacity, necessitating a request to national or international level for external assistance; an unforeseen and often sudden event that causes great damage, destruction and human suffering.' For a disaster to be entered into the EM-DAT database, at least one of the following criteria must be fulfilled: 10 or more people reported killed; 100 people reported affected; declaration of a state of emergency; call for international assistance. Natural disasters which are entered into the database include the following disaster types (in alphabetical order): droughts, earthquakes, epidemics, extreme temperatures, floods, insect infestations, slides, volcanic eruptions, extreme high waves/surges, wild fires, and wind storms." (Luetz 2008a, p. 38; cf, http://emdat.be).

human impact with more than 113 million people[8] affected (Luetz 2008c, p. 2). The following discussion now briefly and consecutively introduces three disaster types which appear to be most immediately relevant for research into human migration in Bangladesh: (1) windstorms; (2) floods; and (3) erosion. These three disaster types are introduced here for three reasons. First, they are often working together in concert. Second, they are compounding pre-existing vulnerabilities. Third, they are poised to evolve significantly under climate change.

Climate Change and Windstorms in Bangladesh

In discussing windstorms it is helpful to note terminological and geographical appropriateness. According to the World Meteorological Organization,

> 'Hurricane', 'cyclone' and 'typhoon' are different terms for the same weather phenomenon which is accompanied by torrential rain and maximum sustained wind speeds (near centre) exceeding 119 kilometers per hour: In the western North Atlantic, central and eastern North Pacific, Caribbean Sea and Gulf of Mexico, such a weather phenomenon is called 'hurricanes'. In the western North Pacific, it is called 'typhoons'. In the Bay of Bengal and Arabian Sea, it is called 'cyclones'. In western South Pacific and southeast India Ocean, it is called 'severe tropical cyclones.' In the southwest India Ocean, it is called 'tropical cyclones.' (WMO n.d.)

A short study of the recent history of cyclones in the Bay of Bengal is pertinent. As reported by Indian novelist Amitav Ghosh (2008), it was a 19th Century British-Indian eccentric, Henry Piddington, who first coined the word "cyclone" in Calcutta (now Kolkata) in the 1840s to represent "whirling storms", an expression possibly derived from Greek "kukloma" (wheel, coil of a snake) and/or "kuklos" (circle) (McKean 2005, p. 421). Ghosh (2008) chronicles how Piddington, purportedly one of the earliest storm chasers to gain prominence, drew his inspiration from the British meteorologist William Reid. Growing enraptured with the circular phenomenon, which he compared to a "beautiful meteorite", Gosh reports that

> Piddington was among the earliest to recognise that a cyclone wreaks most of its damage not through wind but through water, by means of the devastating wave known as a 'storm surge.' In 1853, when the British colonial authorities were planning an elaborate new port on the outer edge of Bengal's mangrove forests, he issued an unambiguous warning: 'Everyone and everything must be prepared to see a day when, in the midst of the horrors of a hurricane, they will find a terrific mass of salt water rolling in…' But his warning was neglected and Port Canning was built, only to be obliterated by a cyclonic surge in 1867. […] The phenomenon of the storm surge has been extensively researched since Piddington's day, yet few public response systems have drawn the obvious lesson. To this day, the warnings that accompany a storm's approach typically say nothing about moving to high ground: Their prescription is usually to seek shelter indoors. As a result people tend to hunker down in the strongest structure within reach—only to find themselves trapped when the surge comes sweeping through. (Ghosh 2008, cited in Luetz 2008a, p. 96)

[8] This is a cumulative figure; multiple counts are possible if people were affected more than once.

On average Bangladesh is affected by "16 major cyclones per decade" (USDS 2012), and a number of studies suggest that continued warming of Indian Ocean surface waters could spawn even stronger cyclones in the future, a scenario which could be particularly difficult for coastal communities if increases in windstorm intensity are accompanied by rises in sea level (WBGU 2006, pp. 38–44; Sánchez-Arcilla and Jiménez 1997; IPCC 2007, pp. 13, 53; Emanuel 2005; Webster et al. 2005; Allison et al. 2009, p. 19). Some researchers have already noted that cyclones "in this region [Bay of Bengal] have gotten stronger in recent decades" (Elsner et al. 2008; Schiermeier 2008, cited in Inman 2009, p. 18), and impacts could be particularly detrimental for coastal communities in the Bay of Bengal.

According to the World Bank report Natural Disaster Hotspots: Case Studies, "millions of people have drowned due to storm surges around the world. […] However, most fatalities have occurred in Asia, and the major hotspot for fatalities due to storm surges is the Bay of Bengal" (Nicholls 2006, pp. 83, 94–95; Fig. 5.1).

There seems to be strong agreement among researchers that climate change will exacerbate storm surge risk: "by far the most certain aspect of climate change that will influence surge characteristics is global-mean sea-level rise. […] The overall conclusion is that the surge hazard will evolve significantly throughout the 21st century" (Nicholls 2006, pp. 89–90; linked to Church et al. 2001). The German Advisory Council on Global Change notes: "In most cases the most destructive results of sea level rise will [be] the increasing occurrences of storm surges" (WBGU 2006, p. 41). The warning should be taken seriously. With average ground levels in coastal areas as low-lying as "1.5 to 2 metres above sea level and [storm] surges reaching up to 6 metres" (IRIN 2007), and historical records indicating past cyclonic storm surge heights in the Bay of Bengal of up to 7, 9, 10 and 13.6 m (World Bank 2000, pp. 5, 15; Ghosh 2008; see also Nicholls 2006, p. 82), the incalculable and deadly risks posed by storm surges in coastal communities in the Bay of Bengal are considerable (Fig. 5.1).

> The main factors contributing to disastrous surges in the Bay of Bengal, especially Bangladesh, are: shallow coastal water, convergence of the Bay, high astronomical tides, densely populated low-lying islands and coastal areas, favourable cyclonic track, and the complex coastline including one of the world's largest river systems—the Ganges-Meghna-Brahmaputra. (Maplecroft 2007a, p. 15).

During fieldwork for this case study a number of respondents identified Cyclone Sidr (11–16 November 2007) as the primary trigger for their migration (e.g., Q09/Dest/Chittagong/20111201). A snapshot of this cyclone's overall damage illustrates the destructive force windstorms can unleash on low-lying coastal communities in the Bay of Bengal. According to the situation report released by the Bangladesh Disaster Management Information Center, Very Severe Cyclonic Storm Sidr killed 3292 people, injured 52,808, fully destroyed 563,877 households, and partially damaged 939,675. It affected 8,669,789 people, 2,000,848 families, and 30 of the 64 districts in Bangladesh. Moreover, "[c]rops on 596,516 acres of land were fully damaged while crops on 1,480,712 acres of land were partially damaged, […] 2400 educational institutes were fully damaged while 12,399 more were partially

South Asia	Location	Deaths
1970	Bangladesh	300,000-500,000
1737	India	300,000
1584	Bangladesh	200,000
1897	Bangladesh	175,000
1991	Bangladesh	138,000-140,000
1876	Bangladesh	100,000
1847	India	75,000
1854	India	50,000
1864	India	50,000
1833	India	33,000-50,000
1822	Bangladesh	40,000
1912	Bangladesh	40,000
1919	Bangladesh	40,000
1942	India	40,000
1839	India	20,000
1789 (uncertain)	India	20,000
1989	India	20,000
1965 (May 11)	Bangladesh	19,279
1965 (May 31)	Bangladesh	12,000
1963	Bangladesh	11,500
1961	Bangladesh	11,468
1985	Bangladesh	11,000
1876	Bangladesh	10,000
1971	India	10,000
1999	Orissa, India	10,000
Rest of the World	**Location**	**Deaths**
1881	China	300,000
1923	Japan	250,000
1694	Shanghai, China	100,000
1862	Zhujiang Delta, China	80,000
1724	Jiangsu Province, China	70,000
1922	Santao, Guangdong, China	60,000-70,000
1912	China	50,000
1780	Barbados, Martinique	20,000-22,000
1998	Honduras, Nicaragua (Hurricane Mitch)	10,000-17,000
1937	Hong Kong	11,000
1906	Hong Kong	10,000

Fig. 5.1 "In terms of fatalities, Bangladesh is presently the dominant storm surge hotspot globally" (Nicholls 2006, p. 102). The table depicts disaster deaths (> 10,000) associated with windstorms (since 1700), separated by region: South Asia, World; excerpted from Nicholls (2006, pp. 92–93; data sources for individual disaster events detailed on p. 93)

damaged, […] 1714 kilometres of roads were fully damaged, […] while 5409 more kilometres were partially damaged" (Ascension 2007, p. 4).

Throughout field research numerous interview participants variously indicated that they had lost family members to cyclonic storm surges (e.g., Q14/Dest/Chittagong/20111202).

Climate Change and Floods in Bangladesh

Being a part of the world's second largest delta system "comprising 100,000 km^2 of riverine flood plain and deltaic plain" (Sarker et al. 2011, p. 203), Bangladesh is naturally one of the most flood-prone regions in the world. During a typical year a quarter to a third of the low-lying country floods naturally during the monsoon season, and extreme floods can cover up to two thirds (Inman 2009, p. 18; Rowe 2004; UNDP 2007, p. 88). Figure 5.2 depicts floods in 2007 that saw more than 40% of the country flooded (UNITAR 2007).

Should Bangladesh's susceptibility to inundation be compounded by more intense tropical storms, more precipitation and rising sea levels as predicted to occur under climate change, the outlook could be rather grim (Maplecroft 2007a, b). A rise in sea level of one metre "would inundate 18% of land area, directly threatening 11% of the population. The impact on river levels from sea rises could affect over 70 million people" (Agrawala et al. 2003, cited in UNDP 2007, p. 100). Another study suggests that even under a 0.5 m rise in sea level projected by 2050, "Bangladesh is likely to lose about 11% of its land, affecting an estimated 15 million people" (Wheeler 2011, cited in UNDP 2011, p. 59). And a World Bank study suggests that "a one-metre rise would swallow about 15–20 per cent of Bangladesh's land area, where about 20 million people live today" (World Bank 2000, cited in Inman 2009, p. 19). Inman also predicts that "[m]onsoon rainfall is likely to increase and to fall in more intense bursts, making the annual floods broader, deeper and longer, and this could increase river erosion, too" (Inman 2009, p. 18). In The Copenhagen Diagnosis (Allison et al. 2009) eminent researchers further caution that "heavy precipitation rates may increase by 5–10% per °C of warming, similar to the rate of increase of atmospheric water vapor" (ibid., p. 19). In summary, increases in the prevalence and/or severity of future flooding events in Bangladesh do not appear to be implausible.

Climate Change and Erosion in Bangladesh

Criss-crossed by 230 rivers, Bangladesh's problems of river erosion are perennial. According to the Centre for Environment and Geographic Information Services (CEGIS), every year anywhere between 66,500 (BSS 2012) and 100,000 people (Shamsuddoha 2007) become homeless due to the effects of river erosion.

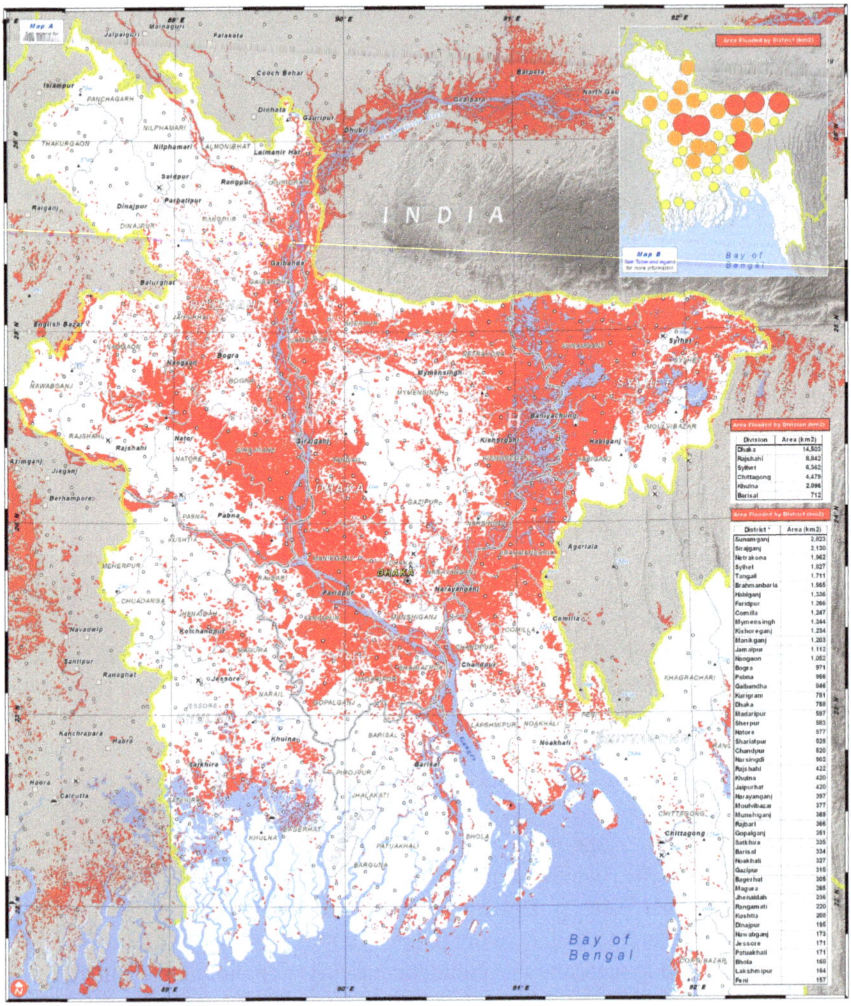

Fig. 5.2 Flood waters over Bangladesh during 2–5 August 2007 shown in *red* (analysis by UNITAR/UNOSAT 2007). The flood covered 42.41% of Bangladesh, left 848 people dead, affected 11.4 million, and caused US$100 million in economic damages (EM-DAT data, cited in Luetz 2008a, pp. 5, 7). Map quoted with permission from UNITAR/UNOSAT (2007): http://www.unitar.org/unosat/node/44/957

> According to UNICEF Emergency Specialist Mohammad Zulfikur Ali Khan, river erosion is causing countless poor farmers to become 'undone' and migrate to the cities …, adding further to the intense pressures of ramshackle slum settlements. With few saleable skills and having lost their farmland to the ocean or rivers, unschooled [migrants] have few options but to 'sell their physical labour and pull rickshaws'. (Luetz 2008a, p. 28; cf, 9)

There are some indications that river erosion may have been exacerbated by climate change related glacial melting in the Himalayas, as well as from rising sea levels along the coastal belt. Both issues are noted in the Fourth Assessment Report of the Intergovernmental Panel on Climate Change (IPCC): "[w]idespread mass losses from glaciers and reductions in snow cover over recent decades are projected to accelerate throughout the 21st century, … [eventually] changing seasonality of flows in regions supplied by meltwater from major mountain ranges (e.g. Hindu-Kush, Himalaya, Andes), where more than one-sixth of the world population currently lives" (IPCC 2007, p. 49). "Coastal areas, especially heavily populated megadelta regions in South, East and South-East Asia, will be at greatest risk due to increased flooding from the sea and, in some megadeltas, flooding from the rivers" (ibid., p. 50). Given that three of the region's major rivers—the Ganga, Yamuna and Brahamaputra—originate in the Himalayas, also known as the "water tower of Asia", South Asia may be particularly susceptible to short-term increases in glacial meltwater run-off. With approximately 5000 Indian Himalayan glaciers contributing 50–70% of their annual meltwater discharge into Himalayan river systems, South Asia appears to be quite vulnerable to changes in this long maintained and delicate equilibrium (Pearce 2012, pp. 8–9). According to the Wadia Institute of Himalayan Geology, in less than 50 years smaller glaciers measuring between one and five kilometres in length could be reduced to "small patches of ice" (Ghosh 2007). The IPCC synthesised that the current era of floods may be succeeded by an era of water stress: "[b]y the 2050s, freshwater availability in Central, South, East and South-East Asia, particularly in large river basins, is projected to decrease" (IPCC 2007, pp. 11, 50).

According to some researchers, coastal islands like Bhola were not previously as vulnerable to erosion:

> "[t]he erosion of Bhola only started in the 1960s. Before that the size was stable and the extent of erosion negligible. But from the mid-1960s the erosion began, and over the years the rate has accelerated. Huge quantities of meltwater from the Himalayas are crashing into coastal estuaries and are speeding up river erosion across Bangladesh through the Ganges and Brahmaputra. Additionally, large coastal areas are being gradually submerged due to global warming and rising sea levels. The collision between downward currents of freshwater and upward pressures of saltwater creates strong twirling and severe erosion. In short, Bangladesh is suffering from 'double erosion,' crumbling from the top down through increasing glacial meltwater run-off, as well as from the bottom up through rising sea levels"

(Shamsuddoha,[9] cited in Luetz 2008a, p. 27; cf, Terra Daily, 2005). According to Shamsuddoha the Island of Bhola has been reduced in size from an area of approximately 6400 km^2 in the 1960s to 3400 km^2 in 2004 (TerraDaily 2005). Once thriving population centres like Old Daulatkhan, Mirzakalu, Molongchara, Sarajgonj, Chowmohoni, Tazumiar have been permanently lost to the sea.

Shamsuddoha warns: "If this rate of erosion continues, the entire island of Bhola could be lost within the next 40 years" (Luetz 2008a, p. 28; cf Shamsuddoha and

[9]Face-to-face interview conducted by this researcher with Shamsuddoha, General Secretary Equity and Justice Working Group(EJWG), at the office of The Coast Trust NGO, Dhaka, 20 April 2008.

Chowdhury 2007, p. 23). A number of respondents interviewed for this case study on both the Island of Bhola and in Dhaka and Chittagong indicated that erosion was the unequivocal cause and primary trigger for their migration (e.g., Q08/Dest/Chittagong/20111201).

According to Satellite analysis prepared by CEGIS, in some places along the northeast coast of Bhola Island erosion claimed "six kilometres"[10] of land, and making thousands of coastal dwellers homeless.[11] It is not immediately clear, however, to what extent climate change may be implicated in or responsible for changes in river erosion in Bangladesh. Researcher Maminul Haque Sarker, a morphologist at CEGIS in Dhaka, explains that the interplay of multiple causal issues is complex and therefore difficult to disentangle as sediments settle along the coast and create new land which is protruding towards the sea. This process of land accretion is "adding nearly 20 square kilometres a year in the coastal areas" (Inman 2009, p. 19). During an expert interview for this case study at CEGIS in Dhaka,[12] Sarker discussed satellite based studies which show that at the present time erosion and accretion nearly cancel each other out, although accretion is presently seen as "the dominant process" (Q33/Exp/Dhaka/20121208). According to Sarker, "[t] here's a lot of accretion, and a lot of erosion, and they're almost in balance. [...] We are gaining land—but it's a net loss" (Inman 2009, p. 19). The reason why despite modest land gains Bangladesh is suffering an overall net land loss lies in the long time it takes for accreted land to be developed, averaging 12, 16 or 22 years depending on location (Sarker et al. 2011, p. 209). Before landless or displaced people can settle in areas where sedimentation has produced new land, elaborate and expensive embankments have to be built to offer these coastal lands the needed protection from tides and storm surges. In short, "coastal areas ... take decades to become productive" (Inman 2009, p. 19; cf CEGIS 2009; Sarker et al. 2011).

Given the dynamic nature of Bangladesh's river delta system, which has undergone remarkable morphological transformation during the last 250 years since modern mapping began (Sarker et al. 2011, p. 206),[13] the precise contribution of climate change to processes of erosion cannot be conclusively established.

According to Sarker, "there is a lot going on at the same time" (Q33/Exp/Dhaka/20111208), with multiple interrelated issues shaping and reshaping the delta through storms, floods, sea level rise, land subsidence, an earthquake in 1950, and climate change (Sarker et al. 2011). In this context it is difficult to project with any level of precision how future climate change may impact on present-day processes of erosion and accretion (Fig. 5.3). Notwithstanding these difficulties, Sarker et al. caution that "intense human impact and sea level rise may make the entire system more vulnerable to inundation and erosion" (ibid., p. 203).

---

[10]See https://youtu.be/PBJeelgnadU @ 21:42–24:15 min.

[11]See https://youtu.be/PBJeelgnadU @ 8:45–12:00 min.

[12]Expert interview with Dr. Maminul Haque Sarker at the Centre for Environment and Geographic Information Services (CEGIS), Dhaka, 8 December 2011.

[13]Sarker et al. (2011, p. 206) illustrate this through four maps showing four time periods.

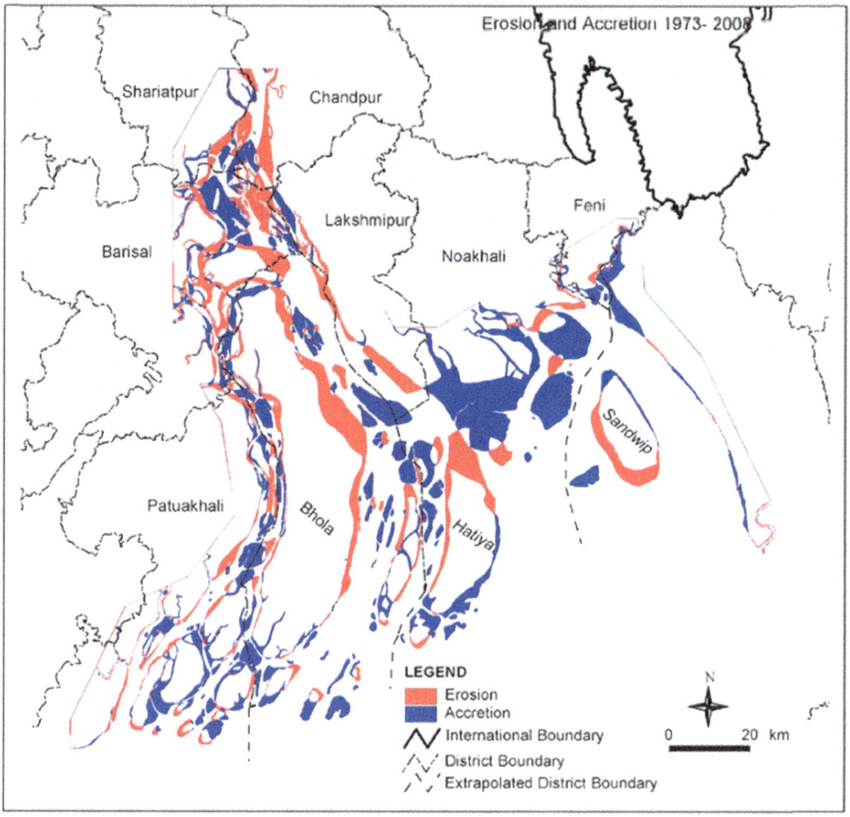

Fig. 5.3 Rate of erosion and accretion in the Meghna Estuary from 1973 to 2008. *Source* Map quoted with permission from Centre for Environment and Geographic Information Services (CEGIS 2009, p. 41)

What appears to be decidedly less ambiguous is the effect that the aforementioned disaster types are already having on human settlement through displacement.

Towards Synthesis: Disaster Homelessness and Preparedness

The precise contribution of climate change to migration is difficult to gauge, let alone quantify. However, "absence of evidence about a problem does not imply evidence of absence of a problem" (Myers and Kent 1995, p. 33). In fact, the gravest effect of climate disasters on human well-being in Bangladesh may very well be displacement or disaster induced homelessness.

The following account, excerpted from the United Nations Human Development Report *Fighting climate change: Human solidarity in a divided world*, illustrates the complex cause-and-effect relationship between natural disasters, climate change and human disaster homelessness:

> Flooding is a normal part of the ecology of Bangladesh. With climate change, 'abnormal' flooding is likely to become a standing feature of the future ecology. Experience following the flood event of 1998—dubbed the 'flood of the century'– highlights the danger that increased flooding will give rise to long term human development setbacks. The 1998 flood was an extreme event. [...] At its peak, the 1998 flood covered two-thirds of the country. Over 1000 people died and *30 million were made homeless*. Around 10 percent of the country's total rice crop was lost. With the duration of the flood preventing replanting, tens of millions of households faced a food security crisis. (UNDP 2007, p. 88; emphasis added)

Research by the World Bank found that 97.7% of people made homeless worldwide by natural disasters live in developing countries (Gilbert 2001, p. iii). The study found that "[b]y far the biggest causes are floods and windstorms, which together are responsible for 92.1% of all disaster homelessness" (Gilbert 2001, p. 1). Given that these disaster types abound in Bangladesh it seems unsurprising that the World Bank study identifies Bangladesh as the country with the highest prevalence of disaster homelessness in the world: between 1980 and 2000 more than 37.6 million Bangladeshis—or 29.9% of the population—were made homeless by disasters, over 13 million of them children (Gilbert 2001, pp. iii, 3; Luetz 2008b, p. 5).

According to Dilley, climate disasters appear to be particularly destructive in terms of the cumulative damage caused:

> [h]azards related to climate and weather affect more people ... than any other type of natural hazard by far. Over the past three decades disasters triggered by cyclones, droughts, and floods occurred 5 times as frequently, killed or affected 70 times as many people, and caused twice as much damage worldwide as did earthquakes and volcanoes, the two major geological hazards... [With] secondary impacts, such as landslides, epidemics, and pest infestations ... also considered, it becomes all the more evident that *climate and weather are primary concerns for natural disaster and risk management*. (Dilley 2000, p. 45; emphasis added)

A similar point is made by the United Nations Development Programme (UNDP):

> [o]n average around 262 million people were affected [by climate disasters] each year between 2000 and 2004, over 98 percent of them living in developing countries. [...] While over 98 percent of people affected by climate disasters live in developing countries,[14] economic impacts are skewed towards rich countries. The reason for this is that costs are assessed on the basis of property values and insured losses. (UNDP 2007, pp. 30, 77)

According to Atiq Rahman, executive director of the Bangladesh Centre for Advanced Studies (BCAS) in Dhaka and lead author of the Intergovernmental Panel on Climate Change (IPCC), climate change is already having enough of an impact in Bangladesh that it is understood to be partly responsible for human displacement: "I believe there are climate change refugees already" (Inman 2009, p. 20). Research by

[14]For details see "Underreporting climate disasters" (UNDP 2007, p. 77, Box 2.1, top of the page).

Rajan also warns that climate change could see millions of South Asians displaced by the end of this century: "Most of these people will be forced to leave their homes because of the sea level rise and drought associated with shrinking water supplies and monsoon variability. The bulk of them will come from Bangladesh as most of the parts of that country will be inundated" (Bhattasali 2008, para 9).

> Climate change is the biggest environmental threat faced by South Asia and may well be the biggest … challenge that the developing world will have to face in the coming decades. While the world has woken up to the threat of climate change, the true enormity of what this implies is still sinking in. […] About 125 million migrants, comprising about 75 million from Bangladesh and the remaining from densely populated coastal regions as well as other vulnerable parts of India could be rendered homeless by the end of this century (Rajan 2008a, p. 1).

According to Rajan, the number of South Asians displaced by climate change could be "10 times greater than the number of people who migrated during and after the partition of India in 1947" (Bhattasali 2008; Rajan 2008b; cf, Byravan and Rajan 2005a, b, 2006, 2008, 2009, 2010 for additional discourses and numerical predictions relating to possible scenarios of future forced migration in South Asia).

In summary, a combination of climate, demography, geography and topography has already made Bangladesh the world's number one hotspot for disaster homelessness (Gilbert 2001; Nicholls 2006). The amalgamation of "hazards and high vulnerability, particularly along the coasts in … Bangladesh, … results in very high risk to the population, relative to other areas of the world" (Maplecroft 2007a, p. 14). There is the chance that "climate change will almost certainly make these disasters worse, threatening to reverse the country's progress" (Inman 2009, p. 18; World Bank 2000, p. xi). In this context the combined effect on human migration emerges as an important policy concern.

During field research in Bangladesh for this case study some of the aforementioned scenarios involving numerical predictions of possible future human migrations were described by two Dhaka migration experts as "unhelpful" and "hype" (Q24 + Q25/Exp/Dhaka/20111205) because they tended to "scare people" and/or "lead to complete denial" (ibid.). Similarly, academics and staff of international organisations have called possible numerical forecasts of large numbers of future forced migrants "alarmist" or have cautioned against such warnings (McAdam and Saul 2010a, 2010b; Q24 + Q25/Exp/Dhaka/20111205; IOM 2010, p. xv, 31; Gemenne 2009).

While accurate prognoses of future migrations are inherently difficult, if not impossible to make, this researcher measuredly rejects the use of the term "alarmist" on the grounds that its use seems to insinuate exaggeration, a point corroborated by dictionary definitions of this word: "alarmist … someone who is considered to be exaggerating a danger and so causing needless worry or panic" (McKean 2005, p. 36). In light of the growing body of evidence linking climate change to the erosion of livelihoods the notion of "exaggeration" appears not only scientifically ill-informed but also runs counter to the premise of preparedness which seeks to pre-empt problems *before* they materialise beyond reasonable hopes of resolution. Therefore, the mere possibility (not probability or certainty) of humanitarian scale displacements and resultant human suffering is seen here to be reason enough to invoke a response of preparation, irrespective of whether or not large displacements will ultimately materialise. While gargantuan challenges can lead to torpidity, inaction or so-called "paralysis of analysis", the point bears repeating that the very notion of preparedness implies readiness before both need and certainty arise. As the United Nations has advocated regarding climate change adaptation: "Hoping and working—for the best while preparing for the worst, serves as a useful first principle for adaptation planning" (UNDP 2007, p. 198). This argument calls for a raft of "no regrets" policy approaches, which arise from this research and are set forth in "Policy Recommendations" Section of this paper.

Research Design, Methodology, and Data Collection and Analysis

Data for this case study were sourced in Bangladesh during fieldwork in November and December 2011, and are supported by observations made and conversations conducted during that time, as well as during months of preparation and follow-up. Field research was also aided by a preceding research visit to Bangladesh in April 2008, during which options for heightened community level disaster preparedness were explored, and data were gathered for the World Vision Asia Pacific Annual Disaster Report 2008 Planet Prepare (Luetz 2008a). This earlier research visit also enhanced familiarity with cultural, environmental, and socioeconomic issues, and laid the foundation for strategic partnerships in areas of translation/interpretation, logistics and overall research support. In light of this earlier research visit it also appeared sensible to return to some of the same sites and, where possible, interview some of the same community members again to try and gauge if/how conditions in respective localities and interview participant perspectives had evolved or changed. This proved particularly profitable in Dalalkandi, Tajumuddin, a coastal community on Bangladesh's biggest Island Bhola where two informants contacted in 2008 were revisited for additional commentaries on erosion and displacement processes, and ultimately confirmed that the places where interviews had occurred in 2008 (Luetz 2008a, pp. 26–28) had already "disappeared" due to coastal erosion (Q6/Exp/Migr/Orig/20111129).

Expressed in simple language this case study does not seek to argue for or against the causality nexus between migration and climate and/or environmental change. To the contrary, it conceptualises both present and future forced human movement, including that which may be related to climate change, as causing the least disruption and suffering, if dealt with *before* its manifestation and possible deterioration into a humanitarian scale crisis requiring ad hoc emergency intervention. To this end the research sought to assess the contributing role that climate change related problems and rapid onset disasters may have on human movement in the environmental context of densely settled megadeltaic regions.

As elaborated in Luetz (2017), a key objective was to conduct grassroots level field research across the whole spectrum of human migration or displacement, incorporating both communities of origin and destination. Moreover, there was a special emphasis on hearing from people personally affected by migration over and above intelligentsia primarily knowledgeable about them. To that end, semi-structured interviews were conducted in villages on the northeast coast of Bhola ("communities of origin"), Bangladesh's biggest island, where extreme levels of erosion on the order of "nearly six kilometres"[15] of coastline changes have been observed for a number of years (Q33/Exp/CEGIS-Dhaka/20111208). In Dalalkandi, Tajumuddin, community members consulted in 2008 (Luetz 2008a, pp. 26–28)[16] were revisited for additional key commentaries on erosion and community displacement processes (ibid.). In addition to "communities of origin" in Bhola, semi-structured interviews were also conducted in "communities of destination", namely in slums in Dhaka and Chittagong, Bangladesh's two biggest urban catchments, where in-migration continues on the order of 1000–2000 people per day (Baker 2007a, pp. xi and xiii; Muriel 2012). This comprehensive micro-level approach sought to contribute to a more holistic empirical understanding of displacement in Bangladesh, thereby aiming to raise anticipatory adaptation options for more equitable human migration based on preparedness informed policy and practice. A consolidated map detailing selected fieldwork sites is available at http://goo.gl/maps/byN0F.

Given that individual interviewees were typically encountered in groups, generally made up of the respondent and by-standing acquaintances who elaborated on issues discussed by volunteering their own personal experiences, the total number of people interviewed is far greater than the number of paper questionnaires. With a minimum of 14 such focus group discussions taking place and between 9 and 35 respondents participating in each conversation the total number of respondents in this field research is estimated by this researcher to be 289. This estimate is based on photos and voice recordings, including pertinent transcripts (see Appendix F in Luetz 2013).

An interview questionnaire tool was used to guide conversations into seven key areas of interest, of which this paper will discuss the following six: (1) interviewee demographics; (2) the migration situation generally; (3) migration push factors; (4) migration pull factors; (5) preferred migration destinations; (6) preferred migration management and/or solutions. The specific foci of these six sections are elaborated in Luetz (2013, pp. 78–80; 2017).

The questionnaire development mechanism broadly followed the five steps outlined by Radhakrishna (2007), and validity was ensured by accommodating feedback and input from the research, ethics, and pilot communities. Testing of the interview questionnaire occurred during a 17-day pilot study in 2010 on two atolls

[15]See https://youtu.be/PBJeelgnadU @ 22 min.

[16]Both respondents are depicted on photographs, Abdul Mannan (Luetz 2008a, pp. 26, 27–28), and Kaisor Ahamed (ibid., p. 28, bottom right).

in Bougainville/Papua New Guinea and led to the incorporation of simplifying features into the final questionnaire design.

With a total of 49 semi-structured interviews taking place, of which 48 were carried out on-site in a personal face-to-face manner, the aim of conducting at least 30 semi-structured interviews was slightly exceeded. Eight semi-structured interviews were held in Bhola Island ("communities of origin"), and 40 in the two urban conglomerates Chittagong (17) and Dhaka (23) ("communities of destination"). One 43-min key informant interview with a Member of Parliament was conducted via skype. Of all respondents queried, 96% had Bangladeshi nationality, 86% were Muslim (4% Hindu, 10% Christian), and 53% were female. Precise ethnic background was inconsistently provided by respondents[17] and is therefore inconclusive, albeit 92% of respondents[18] can be described succinctly by the catch-all ethnicity "Bengali".[19] Eleven key informant interviews (22% of the sample) were conducted with "experts", including a researcher/morphologist, local government officials, water resources and migration experts, development project officers and managers, and both local and international disaster management professionals with expertise in disaster risk reduction and community resilience. Of all interviews conducted, 38 were carried out in the country's lingua franca Bengali with the help of local guides and interpreters. The remainder was conducted in English. Respondent ages ranged from 18 to 82 years, with 37.5 years as the average age.[20]

Data analyses followed a mixed methods paradigm (Creswell and Plano Clark 2011), which comprised approaches informed by grounded theory (Charmaz 2006), comparing (Punch 2014; Tesch 1990), phenomenography (Marton 1981), and ethnomethodology (Silverman 1993).

Selected commentaries featured in this research follow a simple in-text referencing system that ensures respondent anonymity[21] while retaining pertinent interview details: upon specifying the questionnaire number (e.g. Q14) the in-text referencing system distinguishes between migrants (Migr), non-migrants (NMigr), host communities (Hst), and experts (Exp); communities of origin (Orig) and destination (Dest); and interview locations (e.g., Bhola) and interview dates in the format YYYYMMDD. This system makes it possible to perceive contextually significant interview nuances with ease while concurrently reducing footnotes or cross-references.

Methodological approaches are further elaborated in Luetz (2013, 2017).

---

[17]The question seemed confusing to both interpreters and interviewees alike.

[18]There were four expatriate experts.

[19]Two Bangladeshi experts identified as "Garo" and "Santal", otherwise the field for ethnicity was left blank.

[20]Five participants did not disclose their age, wherefore the average age is derived from a sample size of 44.

[21]Unless attribution by name was expressly agreed or requested by the respondent.

Results and Key Research Findings

From the research a number of key findings have been synthesised and consolidated under themed subheadings below. Selected key commentaries shed light on views, conditions, experiences, interpretations and value judgments pronounced by respondents in communities of out- and in-migration.

Environment Related Pressures Are Powerful Push Factors for Migration

Problems arising from "erosion/land loss" were consistently viewed (Fig. 5.4) as the most important environment-related migration push factors (ERPF). Individual respondents also mentioned additional ERPF not included in the questionnaire that in their view contribute to migration: "floods",[22] "forest related resource mismanagement; cutting down of trees, mangroves, etc.",[23] and "depletion of fish stocks due to over-fishing and a decline in water quality".[24]

While migration is typically triggered or enhanced by multiple factors, one question specifically prompted respondents to try and identify "the most important" issue involved. Responses to this question can be broadly grouped into three categories: (1) erosion; (2) windstorms; and (3) floods. Quantitative data are supported by a selection of key respondent commentaries, which provide a richer context and offer nuanced insights not otherwise captured.

The following two key commentaries reflect the experience of land loss and forced migration from Bhola Island (where progressive erosion has displaced thousands of people) to informal settlements in bourgeoning slums (where thousands of arriving migrants face harsh human adaptation challenges in urban catchments).

The first commentary is representative of islanders made homeless by erosion on the northeast coast of the Island of Bhola ("communities of origin"):

> The bridge[25] now leads to the open sea … of the estuary. It used to be the road to Tajumuddin but the erosion took it away. An entire Union[26] (formerly called Molongchora) was lost, 35 km^2 of land. This affected 100% of the people, about 40,000–45,000. The land loss occurred within the space of about five years. The government built a shelter for the displaced people to house them, but the water is already destroying the buildings and the

---

[22]Comment made in Chittagong.

[23]Comment made in Dhaka.

[24]Comment made in Dhaka.

[25]See https://youtu.be/PBJeelgnadU @ 8:45–12:00 min.

[26]Bangladesh is organised into seven Divisions, 64 Districts, hundreds of Sub-Districts or Upazilas, and thousands of Unions comprising tens of thousands of Villages. A completely disappeared Union implies that several villages were lost to erosion (Q02/Orig/Bhola/20111128).

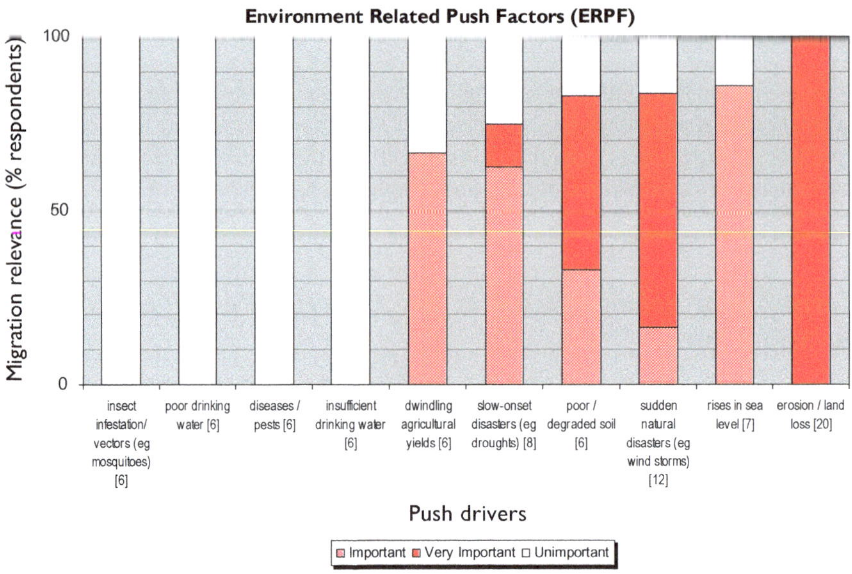

Fig. 5.4 Environment Related Push Factors (ERPF). Important and very important environment related pressures to migrate; Q = 49 FG = 289 N = [ ]

> people have abandoned them. Now people are hoping that concrete slabs can slow down the land loss… (Motasim Billa, Batamara Paksia, Borhanuddin; Q02/Orig/Bhola/20111128; Bengali verbatim citation available in Luetz 2013, Appendix F.1)

The second commentary reflects the experience of a migrant from Bhola Island in her new home in a Chittagong slum ("communities of destination"):

> Everybody in this slum is a migrant,[27] many from Barisal, most are here because of floods or cyclones. Rent in the slum is expensive. One room is 2200 Taka,[28] two rooms are 4000 Taka. I migrated from Bhola where a partial Union disappeared [due to erosion]. My village [of origin] is called Mohammadpur, Keramatgonj, Upazila Chorfashion, District Bhola. The main reason I migrated is erosion. (Khaleda Begum, Bondortila Ward 39, Chittagong City Corporation; Q08/Dest/Chittagong/20111201; Bengali verbatim citation available in Luetz 2013, Appendix F.2)

Respondents recurrently explained that the primary destructive impact of natural hazard events resulted from the knock-on effects of livelihood loss.

[27]See https://youtu.be/PBJeelgnadU @ 13:40–14:15 min.

[28]At the time of research equivalent in value to approximately US$29.

Livelihood Security

Environmental change—of which climate change is a part and to which it contributes—was repeatedly stated as the primary reason for migration. Several interviewees indicated that they had "lost everything" due to river erosion, coastal erosion or cyclonic devastation, and found themselves without land and/or means to rebuild their lives. Patience and tolerance levels naturally vary from respondent to respondent with some of those interviewed trying longer and harder to stay in the degraded, destroyed, salinated and often eroding regions before (eventually) concluding that to go is better than to stay. It is a clear finding of this research that there is a migration "tipping point" which is commonly shared across all strata of society at the precise point in time when livelihoods are compromised beyond reasonable hopes of recovery. When this point is reached, people migrate.

The following key commentary reflects the widely shared experience that people tend to remain in their communities of origin for as long as is feasible, even after being afflicted by slow or sudden onset disasters, but will feel compelled to move at the very moment (tipping point) when their livelihoods are compromised. While there was agreement among respondents that their places of origin were "better" (Q29/Dest/Dhaka/20111206) or their places of destination worse ("Dhaka is an unhealthy environment"; Q27/Dest/Dhaka/20111205), there seemed to be a broad consensus that the new life was overall preferable because "in Dhaka opportunities for income generation are a bit better" (Q26/Dest/Dhaka/20111205).

> Three years ago I came here with my husband and family, mother-in-law and father-in-law. Cyclone Sidr destroyed all household and other properties. At that time we had no income opportunity in that area and so we moved here. After Cyclone Sidr [15 November 2007] we stayed on for one more year, it was a struggle period, then we moved. We moved when there was nothing to earn, when we felt completely helpless [tipping point]. (Hanufa Sheik, Bondortila Ward 39, Chittagong City Corporation; Q09/Dest/Chittagong/20111201; Bengali verbatim citation available in Luetz 2013, Appendix F.2)

Livelihood security appears to emerge as the last line of defence beyond which uncontrolled migration may be unavoidable. The creation of sustainable livelihoods in predominantly rural areas therefore emerged as an important priority for migration management planning and preparedness informed climate change adaptation. Moreover, numerous interviewees[29] noted that their newfound urban life was preferable to their former rural subsistence on the grounds that income generation opportunities in urban slums are superior to those generally encountered in rural areas. Notwithstanding this qualified preference, a majority of interviewees added the caveat that they "liked" their rural areas and lives better and would return immediately if income generation was more readily available in those places (cf, Gray and Mueller 2012).

[29]See, e.g., https://youtu.be/PBJeelgnadU @ 20:23–21:00 min.

Urban Vulnerabilities and Perceived Poverty

There is robust empirical evidence that many migrants move from rural vulnerability to compounded urban vulnerability characterised by poor or absent amenities such as electricity, health care, water and waste management facilities. Prevailing conditions in many informal settlements have been labelled "subhuman" by researchers who unanimously agreed with the sentiment expressed by one expert that "if people knew about the conditions in the slums they might not go there" (Q23/Exp/Dhaka/20111205). According to McKean (2005, p. 1685), the term "subhuman" connotes "a lower order of being than the human [or] not worthy of a human being."

This view appears to be shared not only by researchers (e.g., Black et al. 2011), but also by many newly arriving slum dwellers who seemed genuinely surprised by the adverse realities of daily life encountered in informal slum settlements and were struggling severely for daily survival. Respondents queried indicated monthly household incomes in the range of 1500–4500 Taka (with 3000 BDT[30] given as the approximate average monthly figure). Rent in informal settlements averages around 1200–1400 BDT per month[31] for a space of approximately 2.5 m × 2.5 m. Some rental quarters were significantly smaller and recurrently characterised by moist or wet concrete floors with no furniture, often housing extended families with little children. In such conditions diseases can combine with other difficulties to make a bitter fate worse as the following testimony exemplifies:

> My old village is called Gattchua, Sandip Upazila and District. When I was five years old we migrated to Chittagong with my family. We left because of river erosion. I know this because my parents told me the story. A few years later, in 1991, my father, mother and older brother returned to Gattchua for a short visit. As they were returning [by boat] the cyclone hit the launch and it sank. All my family members drowned. I'm still grieving. My brother's wife's family lives with me now. I was 25 when it happened. I also lost three children (two sons, one daughter) to sickness (typhoid and pneumonia) caused by financial difficulties [in tears]. My deceased children were 2 1/2 and 1 1/2 years old when they died, the youngest died of pneumonia at 2 1/2 months of age. I survive as a maidservant. I earn 1500 to 2000 Taka[32] per month. My son is 26 years old and works as a rickshaw puller. (Sakina Begum, Noya Sohor Khulshi Chittagong, Q14/Dest/Chittagong/20111202; Bengali verbatim transcript available in Luetz 2013, Appendix F.6).

By far the most significant non-environment related push factor (NRPF) for migration appeared to be "unemployment". Tellingly, this result was matched or superseded only by "poverty", which respondents considered unanimously as "very important". Although "poverty" was not included in the questionnaire survey tool as a possible cause of migration, there was an overwhelming desire on the part of Bangladeshi respondents to see their challenges appreciated within their ever

[30]At the time of research 3,000 Taka were equivalent in value to approximately US$40.

[31]At the time of research equivalent in value to approximately US$16–18.

[32]At the time of research equivalent in value to approximately US$20–26.

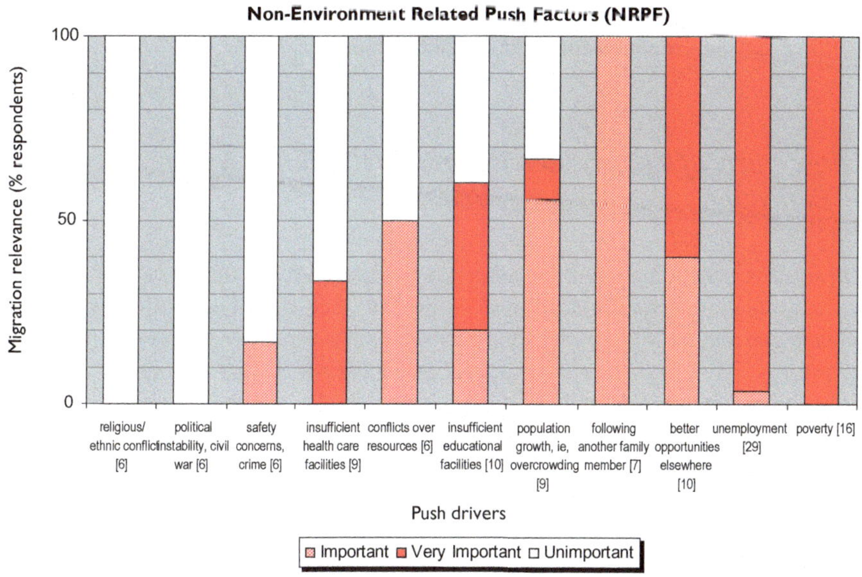

Fig. 5.5 Non-Environment Related Push Factors (NRPF). Important and very important non-environment related pressures to migrate; Q = 49 FG = 289 N = [ ]

present context of "poverty". Hence these data were subsequently integrated into the vertical-bar graph (Fig. 5.5).

Individual respondents also mentioned additional NRPF not included in the questionnaire that in their view also contribute to migration: "poverty",[33] "new experience",[34] and "liquidation of assets to treat kidney disease".[35]

While human movement is typically triggered or enhanced by multiple factors one question specifically asked respondents to try to identify "the most important" issue involved. Responses to this question can be broadly grouped into four socioeconomic categories: (1) lack of gainful employment (variously described as "lack of income generating opportunities", "better income opportunities elsewhere", "unemployment—we're completely landless", etc.); (2) poverty; (3) lack of education; and (4) disparate factors variously described as "new experience", "following another family member because family ties are strong", "liquidation of assets to treat disease", etc. The research also broadly confirmed that established corridors can reinforce migration, with respondents considering "following another family member" as "important" for their migration. Migrants also described that they often felt vulnerable in the slums with "no security"

[33]Throughout field research poverty was frequently raised as an important factor contributing to migration. In some instances it was declared as the primary underlying cause of the migration.

[34]Comment made in Chittagong.

[35]Comment made in Dhaka.

(Q09/Dest/Chittagong/20111201): "At any moment the landlord can evict us if he wants to develop the area for higher profit" (ibid.).

Large Parts of Society Are Eclipsed by Reliable Access to Education

There is overwhelming empirical evidence that a majority of slum dwelling children are not attending school but are instead helping their parents as full- or part-time income generators. Moreover, subhuman slum subsistence seems to be perpetuated by the pervasive lack of reliable access to education. A vast majority of migrant respondents interviewed indicated "primary school" or "no schooling" as their highest level of educational attainment (Fig. 5.6). The average number of school years completed by slum dwelling migrants is likely no more than an average primary school grade three or four. Naturally education raises climate change adaptation options, and inversely, lack of education limits them. There can be no doubt that the unequivocal remedy with which to tackle Bangladesh's present and possible future forced migration scenarios will invariably involve a heightened commitment to free and compulsory education for all. This can be safely understood and prioritised as "no regrets" good development practice which will pay dividends irrespective of which climate change scenario is ultimately realised (Starke and Mastny 2010, pp. 55–82). As one humanitarian expert put it: "educate those children in the slum, and you will break that vicious cycle of poverty in which they find themselves" (Q23/Exp/Dhaka/20111205). Moreover, education also represents sound preparedness practice in general and will both heighten community resilience and improve overall well-being and future prospects of children. Enhanced options for international migration in the future may also become more viable to a highly educated populace (Belt 2011, p. 18).

There is some empirical evidence that transitional education for working children can channel children into the formal education system. One so-called "Access Challenged Children" (ACC) educational programme in Kamalapur (Dhaka) is tailored specifically and exclusively to meet the needs of working and street children so as to open doors of opportunities to youth whose education could not otherwise be realised. The growth and popularity of this project underscores both the high demand and local community uptake of this approach, which appears well suited as an educational "bridge" for those without any other options. In light of migrant aspirations that highlighted education as the third-most significant pursuit (Fig. 5.8), it seems safe to surmise that such transitional education programmes should perhaps be more proactively explored, piloted, developed, funded and implemented.

The issue of education will be developed further in section "Discussion of Core Issues".

Interview location	Age	Sex	Years attended	Highest level of education
Orig/Bhola	70	m	0	No school
Orig/Bhola	82	m	0	No school
Orig/Bhola	18	m	4	Primary
Orig/Bhola	55	m	5	Primary
Orig/Bhola	32	m	9	Secondary*
Orig/Bhola	32	m	9	Secondary*
Dest/Chittagong	20	f	5	Primary
Dest/Chittagong	20	f	5	Primary
Dest/Chittagong	30	f	0	No school
Dest/Chittagong	45	f	9	Secondary
Dest/Chittagong	49	m	0	No school
Dest/Chittagong	60	m	2	Primary
Dest/Chittagong	35	f	0	No school
Dest/Chittagong	45	f	0	No school
Dest/Chittagong	25	m	0	No school, neighbour taught bit
Dest/Chittagong	40	f	5	Primary
Dest/Chittagong	28	f	5	Primary, unaffordable school fees
Dest/Chittagong	26	m	7	Secondary
Dest/Chittagong	50	f	0	No school
Dest/Dhaka	57	f	2	Primary
Dest/Dhaka	26	m	1	Primary
Dest/Dhaka	34	f	9	Secondary*
Dest/Dhaka	35	f	0	No school
Dest/Dhaka	35	f	0	No school
Dest/Dhaka	25	f	1	Primary
Dest/Dhaka	19	f	3	Primary
Dest/Dhaka	20	f	2	Primary
Dest/Dhaka	20	f	2	Primary
Dest/Dhaka	30	f	0	No school
Dest/Dhaka	28	f	0	No school
Dest/Dhaka	31	f	3	Primary
Dest/Dhaka	50	f	0	No school
Dest/Dhaka	20	f	2	Primary
Dest/Dhaka	34	f	10	Secondary School Cert (SSC)
Dest/Dhaka	35	f	3	Primary
Dest/Dhaka	30	f	0	No school
Dest/Dhaka	30	f	1	Primary
Average Age	**26**		**2.8**	**Average years of schooling**

(* precise grade unknown, grade 9 assumed)

Fig. 5.6 Years of schooling completed by 37 interviewed non-expert adult migrants

Strengthened Government Institutions and Interagency Coordination

There are strong indications that initiatives to formalise the informal sector are likely to bring about more equitable conditions in urban slums. There were suggestions that influence and power needs to be wrenched from *mastaans* ("musclemen", hoodlums) controlling local level availability of rental space, land, amenities, etc. and exacting exorbitant fees for the same. Apparently fearing reprisals, a number of respondents were visibly afraid to speak openly when asked to comment on the activities of *mastaans* in their area of residence. Strengthening the capacity of local level government institutions and better interagency coordination will likely lead to a more equitable balance of power and thereby result in better recognition of the needs of the poorest. A stronger state presence and stronger and accountable state institutions in urban slum areas also seem necessary if the power vacuum in "state-absent" areas is not to remain filled by *mastaans* in the future, and if the mounting needs for services desired by the urban poor are to be adequately met. One unidentified respondent affirmed that economic structures of many government agencies are poorly developed: "higher wages for the local police would make them less inclined to accept bribes… salary structures are very poor, so police officers collude with criminals to obtain 'top-ups', for example to pay for their children's education… officials need higher salaries" (Unidentified respondent; Tajumuddin, Bhola Island; 29 November 2011).

Research: Fertile Land Development

According to Sarker[36] and the Centre for Environment and Geographic Information Services (CEGIS) in Dhaka, land accretion presently outpaces erosion to produce a slight overall net increase of land (Dr. Maminul Haque Sarker, Morphologist; Q33/Exp/Dhaka/20111208; cf Fig. 5.3). However, given that the full development of newly accreted land typically takes multiple decades, on balance Bangladesh is losing slightly more fertile land through erosion than the country is gaining through sedimentation and resultant accretion.[37] At the same time it is not clear that processes of land gain observed over recent decades will necessarily continue into the future, given that these rates of accretion appear to have been supported by an earthquake which occurred last century:

> The Assam earthquake of 1950 caused huge landslides in the Himalayas, which discharged billions of cubic metres of sediment into the Brahmaputra River in India. The fine fraction

[36]See https://youtu.be/PBJeelgnadU @ 21:42–24:15 min.

[37]Char development can take several decades: "It takes possibly an average twenty to thirty years from the first deposits to the emergence of new land" (CDSP no date, p. 2).

> of this sediment (silt and clay) reached the estuary within a few years and until the 1970s and caused net accretion of about 1100 km^2. […] The coarser fraction of the earthquake-derived sediment moved through the river system more slowly, [reaching] the Meghna Estuary during the mid-1980s and again increased the net accretion. […] This high accretion rate will not be maintained (unless there is another event delivering huge amounts of sediment occurs), because the effect of the Assam earthquake has already diminished. (Sarker et al. 2011, pp. 215–216; cf, Verghese 1999; Sarker and Thorne 2006)

It is also not clear that land accretion trends can continue as climate change progresses with projected impacts on river discharge, coastal erosion, sea level rise and cyclonic activity. In fact, Sarker notes that under climate change sediment input could be compromised, "unless a method of planned and effective sediment injection into the polders is adopted" (Sarker et al. 2011, p. 216). More research into "vertical accretion by injecting sediment into polders" seems to be expedient to enhance prospects for "an effective adaptation strategy against sea level rise due to climate change" (Sarker et al. 2011, pp. 203, 216). Moreover, the Char Development and Settlement Project (CDSP) sponsored by the Government of Bangladesh, and supported by the Government of the Netherlands, is a promising land reclamation, reinforcement and resettlement project aimed at developing accreted land areas for the benefit of displaced populations. To date the multiagency[38] initiative has developed nearly 100,000 ha of accreted land for a population of 896,000 (CDSP no date, p. 3), with plans underway to expand operations:

> Demand for land to settle landless households and for other purposes is only expected to increase in the years to come … feasibility studies have already been carried out for prospective areas where char development programmes can be undertaken in future. [The Bangladesh Water Development Board (BWDB) is also due to] experiment with construction of cross-dams in a number of places… Such dams are expected to accelerate the rate of land accretion. (CDSP no date, p. 6)

In light of current demographic and environmental trends, this research has found adaptation programmes such as the Char Development and Settlement Project (CDSP) to be fertile grounds for further research and development commitments.

Migration Pull Factors: Migrant Aspirations and Preferred Destinations

By far the strongest aspiration voiced by respondents was gainful employment. The comparative levels of prioritisation respondents ascribed each category are

[38]Implementing agencies: Bangladesh Water Development Board (BWDB) [Ministry of Water Resources]; Local Government Engineering Department (LGED) [Ministry of Local Government, Rural Development and Cooperatives]; Department of Public Health Engineering (DPHE) [Ministry of Local Government, Rural Development and Cooperatives]; Ministry of Land (MoL); Department of Agriculture Extension (DAE) [Ministry of Agriculture]; Department of Forest (DoF) [Ministry of Environment and Forest].

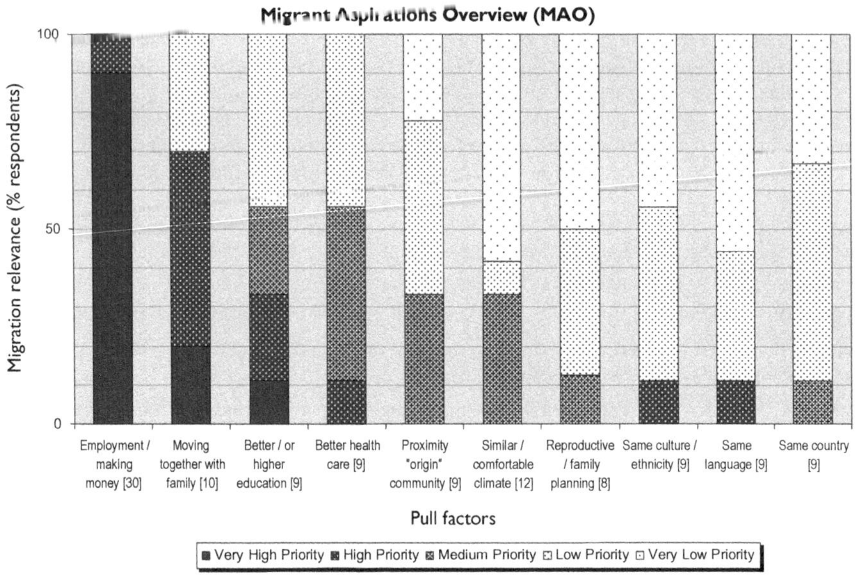

Fig. 5.7 Migrant Aspirations Overview (MAO): comparative levels of prioritisation as ascribed by respondents; Q = 49 FG = 289 N = [ ]

summarised by the Migrant Aspirations Overview (Fig. 5.7) and synthesised by the Migrant Aspirations Aggregate Index (Fig. 5.8).

While human movement is typically triggered or enhanced by multiple factors one question specifically asked respondents to identify "the most important" issue involved. Responses to this question were dominated by aspirations linked to gainful employment, which respondents variously described as "employment", "making money", "we came because my husband has work here", "my husband has work as a construction worker", "work—how can I continue my education if I have to feed my family?" Only four respondents indicated a different "most important" aspiration: "healthy environment" (two respondents), "moving together" (one respondent), and housing: "I would prefer to return, that's where my roots are, but there is no house for me to return to after Cyclone Sidr destroyed it in 2007" (one respondent).

Respondents also volunteered additional aspirations not included in the questionnaire: "Achieving a politically predictable future, political stability",[39] living in a "healthy environment, safety from fire and other risks",[40] and living in a "better environment" generally.[41]

[39]Comment made in Dhaka.

[40]Comment made in Bhorkut Slum, near Mohammadpur Beribadh embankment, Dhaka.

[41]Comment made in Staff Bosti Slum, West Agargaon, Dhaka.

Aspirations of forced migrant communities	Ascribed importance
Employment/making money	4.9
Moving together *with* the family	3.6
Better or higher education	3.0
Better health care	2.7
Proximity to the *origin* community	2.1
Same country	1.8
Same culture/ethnicity	1.8
Similar climate	1.75
Same language	1.7
Reproductive/family planning	1.6

Ascribed importance: weighted average calculated from assigned values: very high priority = 5 points; high priority = 4 points; medium priority = 3 points; low priority = 2 points; very low priority = 1 point.

Fig. 5.8 Migrant Aspirations Aggregate Index (MAAI)

Answers to questions about preferred migration destinations can be broadly grouped into two categories. The first category of answers is indicative of where people actually go/went. The second category of answers was offered in response to the hypothetical question where migrants would go "if there were no restrictions", whether political, financial, or otherwise. Answers in response to the actual migration destinations highlighted the following five issues (cf, Heissler 2013; Bélanger and Rahman 2013; Black et al. 2011; Dannecker 2013).

First, there is unplanned movement which occurs piecemeal and in step with progressive environmental degradation. This movement may be described as reactive rather than proactive. "About 1–2% of people migrate to Dhaka and pull rickshaws, the majority stay in the region and just keep on moving away from the eroding coast" (Q3/Orig/Bhola/20111129).

Second, answers confirmed that migrants choose destination cities based on their perception of prospects of gainful employment. "People go to Dhaka, Chittagong, Khulna or wherever there are many options for jobs, even nearby district towns" (Q20/Exp/Chittagong/20111203).

Third, choice of destination was more often than not guided less by what was perceived as desirable than what was perceived as practically affordable. "We came to this area [Babuls Colony Slum, Beltola, Chandranegor, Chittagong] because it is cheaper than elsewhere" (Q19/Dest/Chittagong/20111203). "I came here because the house rent is cheaper" (Q30/Dest/Dhaka/20111206).

Fourth, migration takes place along established migration corridors as migrants follow in the footsteps of other migrants, usually neighbours or relatives. "My husband decided to come to this area [Chittagong] because he knew a relative, and didn't know Bangladesh, or where else to go, … the relative was his only contact." (Q9/Dest/Chittagong/20111201; cf, Heissler 2013)

Fifth, settlement can be transient. "I would like to go back to my village—even with river erosion [problems]; if someone built a house [in a safe area], yes, we would go back" (Q17/Dest/Chittagong/20111202). "We're trying to save some money. If we can save enough we would like to go back" (Q31/Dest/Dhaka/20111207).

Answers in response to the hypothetical question highlighted severe restrictions on human movement and seemed to underscore that migrants do not often consider what they might want in an "ideal world" but rather what seems practically attainable under the prevailing circumstances (cf, Bélanger and Rahman 2013). "There are two or three cases where people have migrated to Saudi Arabia, Oman, Malaysia… But now Saudi Arabia gives no more visas. To go overseas young people need brokers. Because they are not economically solvent they sometimes borrow large amounts of money with 50% interest" (Q3/Q4/Orig/Bhola/20111129). "People pay the broker, or dalal, who collects money … Saudi Arabia, Abu Dhabi, Kuwait, Qatar, Singapore, Oman are easier and not so expensive" (Q6/Orig/Bhola/20111129; cf, Dannecker 2013, p. 45).

Unsurprisingly, the presence of Bengali people and/or culture was also a determining factor for choice of preferred migration destination. [People want to go] "to the same community, where the culture is already established" (Q1/Exp/Dhaka/20111123). However, rationality alone does not explain labour migration: "images and myths" also accompany the phenomenon, including perceived Muslim "Brother Country" affinities (Dannecker 2009, pp. 51, 54).

In short, overall the "premise of preference" was found to be an unfathomable starting position for poverty constrained forced migrants who have rarely been afforded the "luxury of choice".

Discussion of Core Issues

Everyday 1000–2000 new arrivals come to settle in Dhaka, the majority of this migration can be linked to income generation. According to some estimates Dhaka is presently the world's fastest-growing megacity (Baker 2007a, pp. xi, xiii; Muriel 2012). Unless there is a change in current trends Dhaka may also be on course to become the "world's third largest city" (ibid.). It is not clear whether initiatives by the national government to advance rural development and livelihood options can adequately reduce the present rate of rural-urban migration which outpaces the supply of urban land, housing and basic amenities and services.

This section offers a richer understanding of Bangladesh's bigger context within which erosion, floods, sea level rise, cyclones and other climate change related factors contribute to human movement. It will discuss three areas of interest, which appear particularly relevant to human migration in Bangladesh, namely the issues of (1) poverty, (2) education, and (3) urban planning.

Poverty Matters, First and Foremost

Throughout field research it was evident that poverty was a key underlying driver for human migration in Bangladesh, often working in concert with environmental issues to overwhelm human coping capacities. This was reflected in numerous interviewee commentaries[42] which broadly resemble this featured sample:

> I came 18 years ago due to poverty after floods destroyed my house in the village of Dhormopur, Post Office Laxmipur, Upazila and District Gaibandha in northern Bangladesh. My husband and I are day labourers breaking bricks in Chittagong. My husband earns 200–250 Taka[43] per day, I earn 150–200 Taka. [Our salary] depends on the volume [of broken bricks], my children are also brick breakers. The rent for our one room is 1000 Taka[44] per month. The pull to come to Chittagong was opportunities and employment, but the flood was the trigger. It's better here in the new place. In the old place there were no income generating opportunities, so it's better here (Josho Rani, Noya Sohor, Chittagong; Q13/Dest/Chittagong/20111202).

One humanitarian expert articulated what seemed to emerge during numerous interview situations, namely the impression that in many instances migrants were

> moving from one vulnerable situation [in rural areas] into what would seem … an almost more vulnerable situation … in the urban area, with limited security … and at the mercy of exploitative landlords, and very poor services in the city for slum dwellers. […] There is a real question mark … whether they're actually moving into something better. (Q22/Exp/Dhaka/20121204)

During the course of fieldwork one accompanying Bangladeshi research assistant, guide and translator succinctly synthesised the impact poverty is having on migrants in Bangladesh:

> People [we met during interviews] have moved for different reasons, [including] river erosion, Cyclone Sidr [2007], the Cyclone in 1991, income generation, … children's education, … but poverty is always implicated. While climate change cannot always be identified as a cause of migration, poverty is present in all cases. If you ask around in the slums you find that most people come [to the cities to alleviate poverty through] income generation. (Q7/Exp/Chittagong/20111201)

The severe hardships faced by the migrating poor in Bangladesh can also be deduced from a summary overview of two focus group discussions during which respondents indicated their monthly incomes and highest levels of schooling. What seemed to emerge is a conspicuous correlation between extreme levels of poverty and very low levels of educational attainment (Luetz 2013, Figs. 5.21 and 5.22, pp. 202–203).

---

[42]See https://youtu.be/PBJeelgnadU @ 18:19–19:05 min.

[43]At the time of research equivalent in value to approximately US$2.50–3.50.

[44]At the time of research equivalent in value to approximately US$13.

In sum, the two issues of poverty and educational attainment seem strongly interrelated and cannot be meaningfully discussed in isolation of each other. The next section now focuses more deeply on critical aspects related to education.

Education Matters

Young children working were easily and recurrently seen throughout field research. It appeared that numerous parent respondents could not afford to send their children to school and instead sent them to work and contribute to the pool of disposable household income so the family could make meagre ends meet (cf, ILO 2006).

During an interview in Bhola Island a former student of a school that had disappeared due to erosion six months prior to the fieldwork visit described how he had been forced to discontinue his education because of the need to work:[45]

> The school that was here was abandoned five or six months ago. It was in use 15–20 years; 250–300 students attended grades one to five. The name of the [disappeared] school is Dorichandpur Mustafijur Rahman Registered Primary School, Hazikandi, Post Office Hatshoshigonj, Chandpur Union, Tajumuddin Upazila. Now the [new] school is in someone's home inside the embankment. I only finished grade four but didn't attend school after that. Since then I have been working to support my family because of poverty. I can't go back to school. My father is the only breadwinner. His earnings are not enough. If I start studying again my family will be in crisis. If someone else helps my family I could resume my studies. Others are also in this same situation. [Even at 18 years of age] I would have to be readmitted to grade five. I would be prepared to live through the humiliation but can't afford to [financially]. (Q5/Orig/Bhola/20111129; Bengali verbatim transcript available in Luetz 2013, pp. 468–478, Appendix F.4[F.12] to F.5[F.22]).

Throughout interviews it seemed that schooling was repeatedly aborted not merely because of unaffordable opportunity costs associated with deferred earnings but also because of unaffordable incidentals associated with the daily provision of primary or secondary education. Among such costs which respondents regarded as unaffordable were monthly tuition fees of 40 Taka[46] and 50 Taka[47] for primary school enrolment per student per month,[48] annual exam fees on the order of 50–100 Taka[49] (Q5/Orig/Bhola/20111129), and incidentals to cover notebooks, pens, pencils, school bags, etc. (ibid.) which respondents repeatedly deemed prohibitive without the assistance of either international development organisations (Q19/Dest/Chittagong/20111203) or relevant help from the Bangladeshi government (Q29/Dest/Dhaka/20111206).

---

[45]See https://youtu.be/PBJeelgnadU @ 3:20 min.

[46]Kindergarten to grade two; (at the time of research equivalent in value to US$0.50).

[47]Grades three to five; (at the time of research equivalent in value to US$0.65).

[48]Alhaz Abul Kashem Chowdhury Primary School, Chandranagor East Nasirabad, Chittagong.

[49]At the time of research equivalent in value to approximately US$0.65–1.30.

Four impediments to uninterrupted schooling were recurrently mentioned:

First, in several areas there seemed to be an apparent lack of available schools. During one focus group interview in a Chittagong slum,[50] which included an estimated 40 children it emerged that not a single child present was attending school. Several mothers concurred: "in this place there is no school, it is too far away" (Q16/Dest/Chittagong/20111202). Of the 40 children who were not attending school, three had previously attended a school[51] but were no longer doing so "because the school was closed" (Q17/Dest/Chittagong/20111202). During interviews in another Chittagong slum[52] one teacher estimated that "there are 5000 school age children [aged 5–16 years] in this area; 50% do not attend school. This school is not sufficient for all children, we are four teachers, two male, two female, together we teach 500 students" (Q18/Dest/Chittagong/20111203).

Second, schools are frequently operating in two shifts (with children divided into two age groups). This practice has the tendency to limit the teaching exposure of children to either the morning (9 am–12 pm) or the afternoon (12 pm–3 pm) periods (Q5/Orig/Bhola/20111129; Q19/Dest/Chittagong/20111203).

Third, school buildings frequently serve dual or multipurpose functions, usually as schools, cyclone shelters, community and/or evacuation centres etc., wherefore education can be easily and lengthily interrupted, for example in the aftermath of disasters when children are routinely displaced from their classrooms. "During the rainy season people… move to the shelter house, … it's a school shelter, people stay there before, during and after the cyclone" (Q18/Dest/Chittagong/20111203). One humanitarian expert interviewee said: "Cyclone seasons are October to November and March to May. People take shelter in cyclone shelters, but there aren't enough, so people also take shelter in schools" (Q7/Exp/Chittagong/20111201).

Fourth and finally, some respondents also indicated safety concerns: "[my husband and I] have three children. None attends school. The school is far and there is no one to accompany my young daughters on the way to school" (Q31/Dest/Dhaka/20111207).

Throughout field research the low level of education was evidenced by demographic details disclosed as part of the semi-structured interview process. Of 37 non-expert adult migrant respondents (average age: 26 years) who indicated their highest level of educational attainment, 14 said that they had never attended school. Based on these 37 respondents' data, the average adult migrant life-long school attendance is 2.8 years (Fig. 5.6). This statistic broadly corresponds with UNDP data which estimates that the "[m]ean years of schooling" is 4.8 years (UNDP 2011, p. 129) and that the average "[p]upil-teacher ratio" is 45.8 (UNDP 2011, p. 160).

---

[50]Oli Sowdagor Colony, Hafiznogor, Bakulia Ward #18, Chittagong.

[51]See https://youtu.be/PBJeelgnadU @ 16:05–18:10 min.

[52]Gdi Moti Jhorna Lalkhan Bazar, Chittagong.

Most frequently school non-attendance was explained on the grounds of present-day livelihood pressures which were perceived to be so severe as to force the children to work and contribute family income as garbage collectors and/or recyclers, domestic workers, servants, street vendors, hotel boys, garage workers, burden bearers/carriers, couriers, etc. (Q32/Exp/Dhaka/20111207; WVB ACC 2011, pp. 1–2). UNICEF Bangladesh estimates that there are approximately 4.7 million working children (aged 5–14) in Bangladesh, and that "half of all child labourers do not attend school at all. [...] As a result, working children get stuck in low paying, low-skilled jobs, thereby perpetuating the cycle of poverty" (UNICEF BD 2010, p. 3).[53]

A model school, which accommodates both the felt needs of livelihoods and education has been established by a development organisation in Dhaka.[54] The school centre for Access Challenged Children (ACC) caters exclusively to the needs of working children and their parents. While parents of working children commonly maintain that their children are indispensable as income-generators (with respect to the needs of today), the NGO maintains that the children's education is indispensable for income generation (with respect to the needs of tomorrow). The ACC caters to both perspectives. A vibrant classroom with 100 + ACC students was encountered during the research visit (Q32/Exp/Dhaka/20111207). With 400 students enrolled (WVB ACC 2011) and additional hundreds of children waitlisted (Q32/Exp/Dhaka/20111207), community uptake in this Dhaka slum already appears to have been positive, and there are hopes that the prototype school model may be more widely replicated and mainstreamed (WVB ACC 2011; WVB KADP 2011). Incidentally, features of this prototype school closely resemble the Basic Education for Hard to Reach Urban Working Children (BEHTRUWC), which UNICEF supports in conjunction with the Government of Bangladesh (UNICEF BD 2010, pp. 4, 7).

In summary, if education raises options, lack of education forecloses them. According to recent interagency research, "[o]ver eight percent of 7–17 year-olds, almost three million [Bangladeshi] children in absolute terms, have never attended school. [...] Children's employment in Bangladesh appears incompatible with schooling, underscoring the importance of child labour as a barrier to achieving Education For All" (UCW 2011, p. ii; cf ILO 2006). In this context it seems unsurprising that education emerged as the third-most highly valued migrant aspiration mentioned by respondents during field research (see Figs. 5.7 and 5.8). Preparing for a better tomorrow (through education) while meeting the needs of today (through livelihoods) seems to emerge as the predominant challenge facing development organisations and policy makers today. One expert interviewee said:

[53] Attributed to ILO 2006.

[54] The Access Challenged Children education programme is operated by World Vision Bangladesh in conjunction with local Community Based Organisations (CBOs), which assist in child selection processes (WVB KADP 2011, p. 14).

> The two priorities [for migrants] are education and livelihoods. But livelihoods is more important than education for the destitute person because it feeds them today. Education is about tomorrow. And they don't have the option of thinking about tomorrow. And it's back to the old poverty trap again. This is where policy makers and NGOs need to push harder on the things that provide for today but not ignoring the future. (Q22/Exp/Dhaka/20121204)

Having discussed the interrelated issues of poverty and education, the next sub-section will now look at prospects for a policy posture of better urban planning, including planned migration.

Planning Matters

Throughout field research the importance of planning the urban demographic development of burgeoning cities was recurrently highlighted as an important priority for policy makers to focus on. With an additional 300,000 to 400,000 newly arriving migrants crowding into Dhaka each year (Baker 2007a, pp. xi, xiii), and thinly stretched social services struggling to meet rising demands, it is not hard to see why proactively planning the city's demographic evolution would be fundamental to more equitable migration management outcomes (Black et al. 2011; Kartiki 2011). Service supply shortages were an issue of concern for numerous respondents as the following commentary exemplifies:

> In this place there is no school, … no water supply, … no gas line, … no toilet. We only use a tube well with saline water, we only use firewood … one case costs 10 Taka.[55] There are only two toilets for 31 families, and one tube well for 31 families. (Q16/Dest/Chittagong/20111202)

This perspective is mirrored by Baker who reports that

> only 43 of the 1925 identified slums have a public toilet within 100 meters. […] Social services can be quite far for some slum dwellers exacerbating problems of access. Only seven percent of slums have a public health clinic and 26 percent have a government school (Baker 2007a, p. xiv). The largest slum, Korali Basti in Banani, with more than 12,000 households, does not have a single public toilet or health clinic. Problems of poor sanitation are made worse by the high population density in slums, which have considerable implications for transmission of communicable diseases and other negative externalities. (ibid., p. 7)

One humanitarian expert interviewee explained that dire slum conditions were a consequence of poor urban planning:

> The government has the responsibility but no interest in the slum areas. It guarantees security in residential areas where people have an identity and a permanent address, but not in the slums. People in the slums do not pay taxes, so the government takes little interest [in these] … floating people; the majority of people in the slums are floating people. You seldom see government officials conducting surveys, and if there are any, there is no

[55] At the time of research equivalent in value to $US0.13.

follow-up or action evident based on the survey findings or any reports. What is needed is urban planning… (Q32/Exp/Dhaka/20111207; emphasis original)

One Member of Parliament interviewed for this case study underscored the critical role of planned housing and urban governance:

The slums, in fact, are all over … because 'slum' doesn't have a definition, so you have pockets in many parts of the city, but these are the two [Bhashantek Bosti and Kuril Bosti] where you've got a lot of people who are housed, the density is probably 200 times greater than is otherwise the case in Bangladesh. […] Work … would really have to start with the slums that you already have in Dhaka city. I'm not saying that everyone has actually moved because of climate reasons. But it gives you a possible scenario of what might happen in the future when you have [more] people moving. *The slums are there because it's not planned. If you had planned migration, then you wouldn't actually have slums. So the slums are an indication of the fact that you have not been able to put in place a policy of planned migration.* (Saber Hossain Chowdhury, interviewed 23 November 2011; emphasis original)

Several humanitarian expert interviewees underscored the need for "better city planning"(Q20/Exp/Chittagong/20111203; Q24/Exp/Dhaka/20111205) and "decentralisation" (Q21/Exp/Chittagong/20111203) to alleviate congestion pressures of cities that were already "quite full" (ibid.).

Notwithstanding the plausible rationale for more effective planning, a fourth humanitarian expert was sceptical whether planned migration could ever work in practice:

I honestly doubt [migration] would ever work as a pre-planned exercise, because it requires the huge huge drive by policy makers and authorities to actually plan the process of relocation and prepare somewhere for these people to go which would have all sorts of implications for land, for housing, for livelihoods, for all sorts of things. And it would also require the commitment of the communities involved to be actively planning towards that, and they would need to know that there was something worth going towards before they agreed to actually leave their home, regardless of their vulnerabilities at home, and I'm convinced that people act in very rational ways, and they don't leave their homes unless they absolutely have to, so I honestly don't think it would work in practice. I don't think you'd ever get the commitment, nor specifically in Bangladesh would you get the policy makers to drive a process that would enable that to happen. (Q20/Exp/Dhaka/20111204)

However, according to Mohammed Maboud, an expert from Dhaka's North South University, prospects for planned migration could be enhanced by education:

… investing in educating Bangladeshis would not only help train professionals to work within the country but also make them desirable as immigrants to other countries—sort of a planned brain drain. Emigration could relieve some of the pressure … in the decades ahead. It's also a way to bolster the country's economy; remittances sent back by emigrants account for 11 percent of the country's GDP. 'If people can go abroad for employment, trade, or education and stay there for several years, many of them will stay.' […] By the time climate change hits hardest, the population of Bangladesh could be reduced by 8 to 20 million people—if the government makes out-migration a more urgent priority. (cited in Belt 2011, p. 81)

To conclude this discussion section it may be worth asking whether addressing matters of poverty, education and planning might not fittingly be regarded as "no regrets" adaptation policy prerogatives that appear to make sense irrespective of which climate change scenarios are ultimately realised, and would even seem to make sense in the complete absence of climate change. Given that related development priorities are coupled to relatively long lead times implies insistent institutional support and sustained policy maker prioritisation.

Policy Recommendations

As this research has shown, reasons for migration are numerous, interdependent and complex. In the view of this researcher it is not possible to disaggregate the intertwined mix of factors why people decide to move. A monocausal attribution of reasons why people in Bangladesh migrate therefore appears methodologically unsound, given that causes for migration are inclusive, not exclusive (Betts 2010, p. 378; see also Andersen 2002, p. 8). Throughout fieldwork the amalgamation of reasons why people had moved was as varied and diverse as the circumstances of the individual respondents involved.

While climate change cannot be readily isolated or identified in the mix of factors contributing to human migration, poverty always seemed to be rather unmistakably implicated. Hence there is far "more push than pull" (Q22-25/Exp/Dhaka/20111204-20111205)—"survival migration" is evident in an overwhelming proportion of people who are streaming to the cities in search of better fortunes (Betts 2010). Environmental problems generally compound existing vulnerabilities and can at times even be isolated as the primary "triggers" or tipping elements that push people "over the edge". The following analogy by a Bangladeshi government official aptly illustrates how climate change appears to be increasingly adding to the cumulative burden shouldered by millions of Bangladeshis:

> Let's say for example, one person is able to carry only 40 kg on his shoulders. That's his limit, and he's a poor man. Now on the top of that, I come and I give him one kilogram on top of that. So now the question will be: who is responsible for killing him? Is this the 40 kilograms he was already carrying on his head, or the one kilogram I have now put on the top of that? (Abal M Kamal Uddin, Comprehensive Disaster Management Programme (CDMP); cited in McAdam and Saul 2010a, p. 8)

This research into climate change and migration in Bangladesh highlights a number of lessons, opportunities and policy recommendations, which are presented in this section. As stated before, migration is viewed in this case study through the humanitarian lens of preparedness, which by definition is anticipatory in nature. Hence this research has sought to identify so-called "no regrets" policy approaches, which "make sense regardless of climate change projections" (Brunner and Lynch 2010, p. viii). The Intergovernmental Panel on Climate Change has defined "[n]o regrets options [as having] negative net costs, because they generate direct or indirect benefits that are large enough to offset the costs of implementing the options" (Markandya et al. 2001, pp. 474–475). For the purposes of this research, "no regrets policies" are conceptually perceived as policy priorities that are broadly beneficial irrespective of what particular scenarios are ultimately realised, and might even make sense in the complete absence of climate change.

While in terms of messaging this section could focus on any number of issues that would be worth addressing to ameliorate the conditions of forced migrants in Bangladesh, no other issues seem to hold as much preparedness promise as the areas of (1) livelihood security; (2) education; and (3) state planning. These three priorities are therefore ideationally proposed as the three legs of a solid policy tripod for more anticipatory climate change adaptation and migration management:

First, reduce poverty and promote livelihood security and resilience: Since poverty is the overwhelming reason why people are struggling to absorb the shocks of extreme weather events, disaster risk is best reduced through poverty reduction and the widespread mainstreaming of livelihood creation support mechanisms. It is a clear finding of this research that there is a migration "tipping point" which is commonly shared across all strata of society at the precise point in time when livelihoods are compromised beyond reasonable hopes of recovery. When this point is reached, people migrate. Livelihood security therefore appears to emerge as the last line of defence beyond which uncontrollable runaway migration becomes unavoidable. Moreover, livelihood security is also raised as a promising policy priority en route to sustainable development in predominantly rural areas where options for gainful employment are more constricted. Finally, and importantly, livelihood security is also poised to pay dividends with respect to bolstering the overall resilience of rural populations in the face of unabating natural disasters.

Second, promote free and compulsory education for all: There is overwhelming empirical evidence that a majority of slum dwelling children are not attending school but are instead helping their parents as full- or part-time income contributors. Yet at the same time it is quite clear that the aspired pathway out of subhuman slum subsistence lies in unintermittent education. A vast majority of migrant respondents interviewed indicated "primary school" or "no schooling" as their highest level of educational attainment. The average number of school years completed by slum dwelling migrants is likely to be no more than an average primary school grade three or four. Naturally education raises options, and inversely, lack of education limits them. There can be no doubt that the unequivocal remedy with which to tackle Bangladesh's present and possible future forced migration and climate adaptation scenarios will invariably involve a heightened commitment to education, free, uninterrupted and compulsory for all. This should be prioritised as "no regrets" good development practice, which will pay dividends irrespective of which climate change scenario is ultimately realised. Moreover, education also represents sound preparedness practice in general and will both heighten community resilience and improve the overall well-being and the future adaptation prospects of children. Enhanced options for future emigration may also become more viable to a highly educated populace. Finally, more transitional and informal education programmes in slums should be proactively explored, piloted, developed, funded and implemented.

Third, reduce urban vulnerabilities through targeted planning and service provision: There is overwhelming empirical evidence that many migrants move from rural vulnerability to increased urban vulnerability where problems are compounded by poor or lacking amenities such as electricity, health care, water and waste management facilities. Prevailing conditions in many informal settlements have been labelled "subhuman" by researchers who unanimously concurred with the sentiment expressed by one expert that "if people knew about the conditions in the slums they might not go there" (Q23/Exp/Dhaka/20111205). This view appears to be shared not only by researchers (e.g., Black et al. 2011), but also and importantly by newly arriving slum dwellers who seem genuinely surprised by the harsh and adverse realities of daily life encountered in informal slum settlements and struggle severely for daily survival. Mounting service supply pressures can also force migrants with no other options to source services outside of official government supply grids, thereby forcing many of them to pay even higher prices than more affluent urbanites. These urban vulnerabilities are best addressed through concerted and coordinated interagency planning, including the sustained mainstreaming of government institutions and services into slum areas where these are absent and where mounting congestion and service supply gaps could soon rise beyond any reasonable prospects for dignified human resettlement. Moreover, current trends of coastal people movements can mean that people are moving toward risk when they should be guided to migrate away from it. Political incentives that encourage urban construction projects in unsafe settings and/or political condonation of informal slum proliferation into vulnerable environments both need to be re-examined and checked to ensure that rising population densities are not occurring on marginal, hazard prone or non-climate proof lands. While interagency coordination is advised, the clear onus for planning inevitably falls on government.

Given the long lead times to implement and mainstream the above mentioned preparedness priorities, it is essential that the development of appropriate strategies is not needlessly delayed. To distil it further, and to conclude, the following three preparedness priorities and adaptation policy recommendations are advised:

(1) Create and safeguard livelihoods before these are compromised beyond reasonable hopes of recovery.
(2) Invest in education free, uninterrupted and compulsory for all before options are foreclosed and the hopes of a whole generation eclipsed.
(3) Plan anticipatory migration and mainstream services before slum conditions deteriorate beyond all reasonable prospects for human displacement with dignity, and importantly, before conditions degenerate beyond all hopes for remediation.

Concluding Synthesis

Large-scale human migration into sprawling cities is a fact of life for countless Bangladeshis. Climate change can be expected to catalyse this trend, making foresight a prudent policy posture of preparedness. The mix of factors contributing to human movement is difficult to dissect but there appears to be far more "push" than "pull", i.e., migrants are more typically forced by necessity rather than impelled by opportunity. Patience and tolerance levels naturally vary from respondent to respondent, with some of those interviewed trying longer and harder to stay in the degraded, destroyed, salinated and often eroding regions before (eventually) concluding that to go is better than to stay. It is a clear finding of this research that there is a migration "tipping point" which is commonly shared across all strata of society at the precise point in time when livelihoods are compromised beyond reasonable hopes of recovery. When this point is reached, people migrate. Livelihood security therefore appears to be the last line of defence beyond which uncontrolled migration may be unavoidable. The creation of sustainable livelihoods in predominantly rural areas is therefore identified in this research as an important area for policy maker attention. Although climate change is difficult to isolate as the definitive cause of human movement, it is impossible to dismiss as a core contributing agent. Moreover, lack of basic education is a key contributing factor limiting options for daily wage labourers and their respective families, constraining many to struggle for subsistence survival in subhuman conditions in urban slums where vulnerabilities and living conditions have been described as even harsher than the problems in rural communities of origin, which gave rise to the migrations in the first place. Poverty reduction, livelihood security, education, and proactive government planning are identified and recommended as essential priorities for adaptation policy and practice serving the cause of more equitable human migration in South Asia. In terms of future prospects there seemed to be agreement among some experts that more human movement mapping was profitable, and that non-governmental organisations had a key role to play to contribute to a more robust knowledge base through "evidence based research" (Q24+Q25/Exp/Dhaka/20111205). Although migrant respondents did not use the word "research" they recurrently appeared to express similar sentiments by requesting that more people learn about their situation and make it known to others, stating that the media was not generally available as a reliable mechanism to raise awareness because "journalists demand money" (Q5/Orig/Bhola/20111129).

Research Limitations

The research findings are subject to the following limitations. With no Bangla speaking proficiency on the part of this researcher and 78% of interviews requiring Bangla interpreters it is conceivable that information could have been either lost in translation or lack thereof. Given that interviews mostly took place in slum contexts, most respondents had little or no formal education, wherefore establishing comprehension through interpreters was often time consuming. This had a limiting effect on the size of the quantitative data sample. Despite the rich qualitative insights of this research it was ultimately impossible to compile a larger quantitative data sample given the budgetary constraints of time and money available for this fieldwork. Thus some findings are based in part on a comparatively small sample size and should be confirmed through other independent research. Finally, given this researcher's association with World Vision and an Australian public research university—although this did not feature explicitly in interviews—it is conceivable that the responses of some respondents may have been coloured by a possible hope to gain certain benefits or influence future research or development programming commitments.

Acknowledgements The author wishes to thank Kirsty Andersen for her copy editorial support, and Balaram Chandra Tapader for his research assistance in Bangladesh. Grateful acknowledgment for relevant Ph.D. research support is also made to John Merson, Eileen Pittaway, Richard Rumsey, Geoff Shepherd, and to the international development organisation World Vision. Finally, the author wishes to thank the Centre for Environment and Geographic Information Services (CEGIS), the United Nations Institute for Training and Research (UNITAR), and its Operational Satellite Applications Programme (UNOSAT).

References

Agrawala, S., Ota, T., Uddin Ahmed, A., Smith, J., & van Aalst, M. (2003). *Development and climate change in Bangladesh: Focus on coastal flooding and the Sundarbans*. Paris, France: Organisation for Economic Co-operation and Development.

Ahsan, M. (2011). 43 Dhaka canals grabbed: Task force hopes to reclaim only 13 canals. *New Age, IX*(211), pp. 1, 4.

Allison, I., Bindoff, N. L., Bindschadler, R., Cox, P. M., de Noblet, N., England, M. H., et al. (2009). *The copenhagen diagnosis: Updating the world on the latest climate science*. Sydney, Australia: Climate Change Research Centre, The University of New South Wales.

Andersen, L. E. (2002). *Rural–urban migration in Bolivia: Advantages and disadvantages*. Instituto De Investigaciones Socio-Economicas, Universidad Católica Boliviana, La Paz, Bolivia. http://www.childmigration.net/files/LAndersen.pdf. Accessed September 27, 2016.

Ascension, A. J. (2007). *Loss in the Wake of Sidr*. World Vision Bangladesh Communications, News and Features Packet 2007-522: 7 December 2007, Scribe 3.0, Input by Leeanne Grima, edited by Heidi Lenssen, input Date 5 December 2007, modification Date 8 December 2007, document on file with the researcher.

Baker, J. (2007a). *Dhaka: Improving living conditions for the urban poor. Bangladesh Development Series*, Paper No. 17, The World Bank, Dhaka, Bangladesh, India. http://go.worldbank.org/DBWC5G4ES0 and http://siteresources.worldbank.org/BANGLADESHEXTN/Resources/295759-1182963268987/dhakaurbanreport.pdf. Accessed May 1, 2008.

Baker, J. (2007b). *Dhaka: Improving living conditions for the urban poor. Bangladesh Development Series*, Paper No. 17, Interview, The World Bank, Dhaka, Bangladesh, India. http://siteresources.worldbank.org/BANGLADESHEXTN/Resources/295759-1182963268987/3930437-1182979937422/fullinterview.mp3. Accessed May 20, 2008.

Bélanger, D., & Rahman, M. (2013). Migration against all the odds: International labour migration of Bangladeshi women. *Current Sociology, 61*(3), 356–373.

Belt, D. (2011). The coming storm: The people of Bangladesh have much to teach us about how a crowded planet can best adapt to rising sea levels. For them, that future is now. *National Geographic*, May 2011, pp. 58–83.

Betts, A. (2010). Survival migration: A new protection framework. *Global Governance: A Review of Multilateralism and International Organizations, 16*(3), 361–382.

Bhattasali, A. (2008). S Asia in climate "crisis". A greenpeace report on climate change says that if greenhouse gas emissions grow at their present rate, South Asia could face a major human crisis. *BBC News*, Accessed 25 March 2008. http://news.bbc.co.uk/2/hi/south_asia/7313239.stm. Accessed March 27, 2008.

Bijlsma, L., Ehler, C. N., Klein, R. J. T., Kulshrestha, S. M., McLean, R. F., Mimura, N., et al. (1996). Coastal zones and small islands. In R. T. Watson, M. C. Zinyowera, & R. H. Moss (Eds.), *Impacts, adaptations and mitigation of climate change: Scientific-technical analysis* (pp. 289–324). Cambridge, UK: Cambridge University Press.

Black, R., Bennett, S. R. G., Thomas, S. M., & Beddington, J. R. (2011). Climate change: Migration as adaptation. *Nature, 478,* 447–449.

Brahic, C. (2007). Coastal living—A growing global threat. *New Scientist Environment*. http://environment.newscientist.com/channel/earth/dn11483-coastal-living–a-growing-global-threat-.html. Accessed January 8, 2008.

Brunner, R. D., & Lynch, A. H. (2010). *Adaptive governance and climate change*. Boston, MA, USA: American Meteorological Society.

BSS—Bangladesh Sangbad Sangstha. (2012). *River erosion may make 66,550 homeless this year.* http://www1.bssnews.net/newsDetails.php?cat=0&id=269479&date=2012-08-02. Accessed August 3, 2012.

Byravan, S., & Rajan, S. C. (2005a). Immigration could ease climate change impacts. A modest proposal to allow the big gas-emitters to take their share of responsibility. *Nature, 434,* 435.

Byravan, S., & Rajan, S. C. (2005b). Before the flood. *New York Times*, 9 May 2005. http://www.nytimes.com/2005/05/09/opinion/09byravan.html. Accessed July 22, 2011.

Byravan, S., & Rajan, S.C. (2006). Providing new homes for climate change exiles. *Climate Policy, 6,* 247–252. http://ssrn.com/abstract=950329. Accessed July 24, 2008.

Byravan, S., & Rajan, S. C. (2008). *The social impacts of climate change in South Asia*. http://ssrn.com/abstract=1129346. Accessed July 24, 2008.

Byravan, S., & Rajan, S. C. (2009). Warming up to immigrants: An option for the U.S. in climate policy. *Economic & Political Weekly, XLIV*(45), 19–23.

Byravan, S., & Rajan, S. C. (2010). The ethical implications of sea-level rise due to climate change. *Ethics & International Affairs, 24*(3), 239–260.

CDSP—Char Development and Settlement Project-III. (no date). *Government of Bangladesh, Government of the Netherlands*. Printed document obtained at the Center for Environmental and Geographic Information Services (CEGIS) on 8 December 2011, on file with the researcher.

CEGIS—Center for Environmental and Geographic Information Services. (2009). *Final report on monitoring planform developments in the eDP area using remote sensing*. Government of the People's Republic of Bangladesh, Ministry of Water Resources, Bangladesh Water Development Board, Dhaka, Bangladesh, India.

Charmaz, K. (2006). *Constructing grounded theory: A practical guide through qualitative analysis*. London, UK: Sage.

Church, J. A., Gregory, J. M., Huybrechts, P., Kuhn, M., Lambeck, K., Nhuan, M. T., et al. (2001). Changes in sea level. In J. T. Houghton, Y. Ding, D. J. Griggs, M. Noguer, P. J. van der Linden, X. Dai, K. Maskell, & C. A. Johnson (Eds.), *Climate change 2001: The scientific basis. Contribution of Working Group I to the third assessment report of the Intergovernmental Panel on Climate Change* (pp. 639–693). Cambridge, UK, and New York, NY, USA: Cambridge University Press.

Cohen, J. E., Small, C., Mellinger, A., Gallup, J., & Sachs, J. (1997). Estimates of coastal population. *Science, 278*(5341), 1209–1213.

Creswell, J. W., & Plano Clark, V. L. (2011). *Designing and conducting mixed methods research* (2nd ed.). London, UK: Sage.

Dannecker, P. (2013). Rationalities and images underlying labour migration from Bangladesh to Malaysia. *International Migration, 51*(1), 40–60.

Dilley, M. (2000). Climate, change, and disasters. In A. Kreimer, & M. Arnold (Eds.), *Managing disaster risk in emerging economies. Disaster risk management series no. 2* (pp. 45–50).

Washington DC, USA: The World Bank. http://go.worldbank.org/XHHDRBCL80. Accessed August 20, 2012.
Elsner, J. B., Kossin, J. P., & Jagger, T. H. (2008). The increasing intensity of the strongest tropical cyclones. *Nature, 455,* 92–95.
Emanuel, K. (2005). Increasing destructiveness of tropical cyclones over the past 30 years. *Nature, 436,* 686–688.
Gemenne, F. (2009). *Environmental changes and migration flows: Normative frameworks and policy responses*. Ph.D. Thesis, Institut d'Etudes Politiques de Paris and University of Liège.
Ghosh, A. (2008). Death comes ashore. *The New York Times*, 10 May 2008. http://www.nytimes.com/2008/05/10/opinion/10ghosh.html. Accessed August 20, 2012.
Ghosh, N. (2007). On thin ice. Gaumukh. Vanishing ice caps. When glaciers melt. *The Straits Times*, 23 June 2007, pp. S2–S3.
Gilbert, R. (2001). *Doing more for those made homeless by natural disasters. disaster management facility*. Disaster Risk Management Working Paper Series No. 1. Washington, DC, USA: World Bank. http://siteresources.worldbank.org/INTDISMGMT/Resources/housing.pdf. Accessed September 27, 2016.
Gommes, R., du Guerny, J., Nachtergaele, F., & Brinkman, R. (1998). *Potential impacts of sea-level rise on populations and agriculture*. FAO SD-Dimensions. http://www.fao.org/nr/climpag/pub/EIre0045_en.asp. Accessed September 27, 2016.
Gray, C. L., & Mueller, V. (2012). Natural disasters and population mobility in Bangladesh. *Proceedings of the National Academy of Sciences PNAS of the United States, 109*(16), 6000–6005.
Heissler, K. A. (2013). Rethinking 'trafficking' in children's migratory processes: The role of social networks in child labour migration in Bangladesh. *Children's Geographies, 11*(1), 89–101.
ILO—International Labour Organization. (2006). *Baseline survey on child domestic labour (CDL) in Bangladesh*. Dhaka: International Labour Office. ISBN 92-2-119282-2 & 978-92-2-119282-4. http://www.ilo.org/ipecinfo/product/download.do?type=document&id=4647. Accessed January 18, 2017.
Inman, M. (2009). Where warming hits hard. *Nature Reports Climate Change, 3,* 18–21.
IOM—International Organization for Migration. (2010). *Migration, climate change and environmental degradation: Definitional issues*. http://www.iom.int/jahia/Jahia/activities/by-theme/migration-climate-change-environmental-degradation/definitional-issues. Accessed June 4, 2010.
IPCC—Intergovernmental Panel on Climate Change. (2007). *Climate change 2007: Synthesis report. Contribution of Working Groups I, II and III to the fourth assessment report of the Intergovernmental Panel on Climate Change.* Geneva, Switzerland: Intergovernmental Panel on Climate Change. https://www.ipcc.ch/pdf/assessment-report/ar4/syr/ar4_syr_full_report.pdf . Accessed April 16, 2016.
IRIN. (2007). *Cyclone-damaged embankments urgently need repair*. 7 December 2007. http://www.irinnews.org/Report.aspx?ReportId=75735. Accessed January 7, 2008.
Kartiki, K. (2011). Climate change and migration: A case study from rural Bangladesh. *Gender & Development, 19*(1), 23–38.
Luetz, J. M. (2008a). *Planet prepare: Preparing coastal communities in Asia for future catastrophes, Asia Pacific Disaster Report*. World Vision International, Bangkok, Thailand. http://luetz.com/docs/planet-prepare.pdf. Accessed April 16, 2016.
Luetz, J. M. (2008b). *Bangladesh: Disaster monitor, 1*. World Vision Asia Pacific, Dhaka, Bangladesh, India. http://luetz.com/docs/dmfs1_bangladesh.pdf. Accessed March 21, 2013.
Luetz, J. M. (2008c). *Indonesia: Disaster monitor, 2*. World Vision Asia Pacific, Jakarta, Indonesia. http://luetz.com/docs/dmfs2_indonesia.pdf. Accessed March 21, 2013.
Luetz, J. M. (2013). *Climate Migration: Preparedness informed policy opportunities identified during field research in Bolivia, Bangladesh and Maldives*. Ph.D. Dissertation, University of New South Wales. http://handle.unsw.edu.au/1959.4/52944. Accessed May 31, 2016.

Luetz, J. M. (2017). Climate change and migration in the Maldives: Some lessons for policy makers. In W. Leal Filho (Ed.), *Climate change adaptation in Pacific countries: Fostering resilience and improving the quality of life*. Berlin: Springer.
Maplecroft. (2007a). *Maps: Issue report. Natural Disasters*. February 2007, Maplecroft.net, document on file with the researcher.
Maplecroft. (2007b). *Maps: Issue report. Climate Change*. February 2007, Maplecroft.net, document on file with the researcher.
Markandya, A., Halsnaes, K., Lanza, A., Matsuoka, Y., Maya, S., Pan, J., et al. (2001). Costing methodologies. In T. Banuri, T. Barker, I. Bashmakov, K. Blok, D. Bouille, R. Christ, O. Davidson, et al. (Eds.), *Climate change 2001: A report of Working Group III of the intergovernmental panel on climate change*. http://www.grida.no/climate/ipcc_tar/wg3/pdf/7.pdf. Accessed May 31, 2016.
Marton, F. (1981). Phenomenography: Describing conceptions of the world around us. *Instructional Science, 10*, 177–200.
McAdam, J., & Saul, B. (2010a). *Displacement with dignity: International law and policy responses to climate change migration and security in Bangladesh*, Legal Studies Research Paper, No. 10/113, The University of Sydney, Sydney, Australia. http://law.bepress.com/cgi/viewcontent.cgi?article=1258&context=unswwps-flrps10. Accessed September 27, 2016.
McAdam, J., & Saul, B. (2010b). An insecure climate for human security? Climate-induced displacement and international law. In A. Edwards & C. Ferstman (Eds.), *Human Security and non-citizens: Law, policy and international affairs* (pp. 357–403). Cambridge, UK: Cambridge University Press.
McGranahan, G., Balk, D., & Anderson, B. (2007a). The rising tide: Assessing the risks of climate change and human settlements in low elevation coastal zones. *Environment & Urbanization, 19*(1), 17–37.
McGranahan, G., Balk, D., & Anderson, B. (2007b). Climate change: Study maps those at greatest risk from cyclones and rising seas. International Institute for Environment and Development. http://www.iied.org/climate-change-study-maps-those-greatest-risk-cyclones-rising-seas. Accessed September 27, 2016.
McKean, E. (Ed.). (2005). *The new Oxford American dictionary* (2nd ed.). USA: Oxford University Press, New York.
Muriel, S. (2012). The fearless Ferrymen of Dhakas Buriganga River: Rush hour in the Bangladeshi capital sees thousands of Dhakas commuters boarding small wooden boats to cross the busy waters of the Buriganga River, one of the most dangerous waterways on earth, especially for the Ferrymen. *BBC News Magazine*, 25 August 2012. http://www.bbc.co.uk/news/magazine-19349949. Accessed August 25, 2012.
Myers, N., & Kent, J. (1995). *Environmental exodus: An emergent crisis in the global arena*. Washington, DC: Climate Institute.
Nicholls, R. J. (2006). Storm surges in coastal areas. In: M. Arnold, R. S. Chen, U. Deichmann, M. Dilley, A. L. Lerner-Lam, R. E. Pullen, et al. (Eds.), *Natural disaster hotspots: Case studies. Disaster risk management series no. 6*. The World Bank, Washington, DC, USA. http://siteresources.worldbank.org/INTDISMGMT/Resources/0821363328.pdf. Accessed June 27, 2008.
Nicholls, R. J., & Small, C. (2002). Improved estimates of coastal population and exposure to hazards released, Eos, transactions. *American Geophysical Union, 83*(28), 301. http://www.pik-potsdam.de/avec/peyresq2003/talks/0915/nicholls/background_literature/eos_coastal_population.pdf. Accessed June 30, 2010.
Pearce, F. (2012). Whose Water is it? China is taking control of Asia's great rivers at their source. *New Scientist, 214*(2862), 8–9.
Punch, K. F. (2014). *Introduction to social research: Quantitative & qualitative approaches* (3rd ed.). London, UK: Sage.
Radhakrishna, R. B. (2007). Tips for developing and testing questionnaires/instruments. *Journal of Extension (JOE), 45*(1). http://www.joe.org/joe/2007february/tt2.php. Accessed April 16, 2016.

Rajan, S. C. (2008a). *Blue alert. Climate migrants in South Asia: Estimates and solutions. A report by Greenpeace*. Accessed April 15, 2008. http://www.greenpeace.org/raw/content/india/blue-alert-report.pdf and 20 August 2012. http://www.greenpeace.at/uploads/media/blue-alert-report_web.pdf.

Rajan, S. C. (2008b). *Greenpeace report estimates climate impacted migration to be 10 times worse than partition*. Accessed April 15, 2008. http://www.greenpeace.org/india/greenpeace-report-estimates-cl# and 20 August 2012. http://www.greenpeace.org/india/en/news/greenpeace-report-estimates-cl/.

Rowe, H. (2004). *Bangladesh: Death toll from South Asian floods rises to 925, millions stranded*. Reliefweb, 26 July 2004. http://reliefweb.int/report/bangladesh/death-toll-south-asian-floods-rises-925-millions-stranded. Accessed August 13, 2012.

Sánchez-Arcilla, A., & Jiménez, J. A. (1997). Physical impacts of climate change on deltaic coastal systems (II): Driving terms. *Climatic Change, 35*(1), 95–118.

Sarker, M. H., Akter, J., Ferdous, M. R., & Noor, F. (2011). Sediment dispersal processes and management in coping with climate change in the Meghna Estuary, Bangladesh. In *Proceedings of the workshop held at Hyderabad, India, September 2009* (Vol. 349, pp. 203–217).

Sarker, M. H., & Thorne, C. R. (2006). Morphological response of the Brahmaputra–Padma–Lower Meghna river system to the Assam earthquake of 1950. In G. H. S. Smith, J. L. Best, C. S. Bristow, & G. E. Petts (Eds.), *Braided rivers: Process, deposits, ecology and management* (pp. 289–310). UK: Blackwell Publishing.

Schaefer-Preuss, U. (2008). Global warming: Cities can and must cut energy use. *The Straits Times*, 25 June 2008, p. 26.

Schiermeier, Q. (2008). Hurricanes are getting fiercer: Global warming blamed for growth in storm intensity. *Nature*. http://www.nature.com/news/2008/080903/full/news.2008.1079.html. Accessed August 17, 2012.

Shamsuddoha, M. (2007). *Climate change would intensify river erosion in Bangladesh*. Campaign Brief 6, Coast Trust fact sheet edited by Mohammad Shamsuddoha, hardcopy on file with the researcher.

Shamsuddoha, M., & Chowdhury, R. K. (2007). *Climate change impact and disaster vulnerabilities in the coastal areas of Bangladesh*. http://www.preventionweb.net/files/4032_DisasterBD.pdf. Accessed August 20, 2012.

Shyamol, N. (2011). Chittagong Hill cutting goes unabated: Political influence gives perpetrators immunity. ["Illegal Hill Cutting is Rampant in Chittagong and Adjoining Areas [and] the Prime Cause of the Landslide Here."], *The Independent*, 5 December 2011.

Silverman, D. (1993). *Interpreting qualitative data: Methods for analysing talk, text and interaction*. London, UK: Sage.

Small, C., & Cohen, J. (2004). Continental physiography, climate, and the global distribution of human population. *Current Anthropology, 45*(2), 269–277.

Small, C., & Nicholls, R. J. (2003). A global analysis of human settlement in coastal zones. *Journal of Coastal Research, 3*(3), 584–599.

Starke, L., & Mastny, L. (Eds.). (2010). *State of the world 2010—Transforming cultures from consumerism to sustainability*. London, UK: Earthscan.

TerraDaily. (2005). *Rising sea levels erode half Of Bangladeshs biggest island: Study, Dhaka, Bangladesh, India*. 15 June 2005. http://www.terradaily.com/news/climate-05zzx.html. Accessed April 10, 2008.

Tesch, R. (1990). *Qualitative research: Analysis types and software tools*. New York, NY, USA: Falmer Press.

The Independent. (2011a). Inhabitants to get civic amenities: Gazipur Pourasabha completes uplift works. *The Independent*, 6 December 2011, p. 8.

The Independent. (2011b). Long awaited dredging scheme starts in Ctg: CCC river renovation works to make instant impact. *The Independent*, 28 November, 2011, p. 5.

UCW—Understanding Children's Work. (2011). *Understanding children's work in Bangladesh.* Understanding Children's Work, Rome. http://www.ucw-project.org/attachment/Bangladesh_child_labour_report20111125_094656.pdf. Accessed September 22, 2012.

UNDESA—United Nations Department of Economic and Social Affairs. (2011). *2010 revision of world population prospects*. http://hdrstats.undp.org/en/indicators/69206.html. Accessed October 17, 2013.

UNDP—United Nations Development Programme. (2007). *Human development report 2007/2008. Fighting climate change: Human solidarity in a divided world.* New York, NY, USA: United Nations Development Programme. http://hdr.undp.org/sites/default/files/reports/268/hdr_20072008_en_complete.pdf. Accessed April 16, 2016.

UNDP—United Nations Development Programme. (2011). *Human development report 2011. Sustainability and equity: A better future for all.* New York, NY, USA: United Nations Development Programme. http://hdr.undp.org/sites/default/files/reports/271/hdr_2011_en_complete.pdf. Accessed September 27, 2016.

UNFPA—United Nations Populations Fund. (2007). *State of world population. Urbanizations second wave: A difference of scale*. http://www.unfpa.org/swp/2007/english/chapter_1/urbanization.html. Accessed March 29, 2008.

UN Habitat—United Nations Human Settlements Programme. (2003). *Water and sanitation in the world's cities: Local action for global goals*. London, England: Earthscan.

UN Habitat — United Nations Human Settlements Programme. (2006). *State of the world's cities 2006/2007*. Nairobi, Kenya.

UN Habitat—United Nations Human Settlements Programme. (2007a). *Slum dwellers to double by 2030: Millennium development goal could fall short*. UN-Habitat, Nairobi, Kenya. http://www.preventionweb.net/files/1713_463146759GC202120Slum20dwellers20to20double.pdf. Accessed August 4, 2008.

UN Habitat—United Nations Human Settlements Programme. (2007b). *Global report on human settlements 2007: Enhancing urban safety and security*. UN Habitat, Earthscan, London, England. http://www.unhabitat.org/pmss/getElectronicVersion.aspx?nr=2432&alt=1. Accessed August 4, 2008.

UNICEF Bangladesh. (2010). *Child labour in Bangladesh: Key statistics*. www.unicef.org/bangladesh/Child_labour.pdf. Accessed September 21, 2012.

UNITAR—United Nations Institute for Training and Research. (2007). *Map of flood water over Bangladesh*. UNOSAT. http://www.unitar.org/unosat/node/44/957. Accessed October 12, 2016.

UNU-EHS—United Nations University–Institute for Environment and Human Security. (2005). *As ranks of "environmental refugees" swell worldwide, calls grow for better definition, recognition, support*. http://icecap.us/images/uploads/50millionclimaterefugeesby2010.pdf. Accessed September 27, 2016.

USDS—U.S. Department of State. Bureau of Western Hemisphere Affairs. (2012). *Background note: Bangladesh*. http://www.state.gov/r/pa/ei/bgn/3452.htm. Accessed August 13, 2012.

Verghese, B. G. (1999). *Water of hope: From vision to reality in Himalaya-Ganga development cooperation*. Dhaka, Bangladesh: The University Press Limited.

Webster, P. J., Holland, G. J., Curry, J. A., & Chang, H.-R. (2005). Changes in tropical cyclone number, duration, and intensity in a warming environment. *Science, 309*(5742), 1844–1846.

WBGU—German Advisory Council on Global Change. (2006). *The future oceans: Warming up, rising high, turning sour. Special Report*. German Advisory Council on Global Change, Berlin, Germany. http://www.wbgu.de/fileadmin/templates/dateien/veroeffentlichungen/sondergutachten/sn2006/wbgu_sn2006_en.pdf. Accessed September 27, 2016.

Wheeler, D. (2011). *Quantifying vulnerability to climate change: implications for adaptation assistance*. Working Paper 240, Center for Global Development, Washington, DC, USA. www.cgdev.org/content/publications/detail/1424759 and http://www.cgdev.org/files/1424759_file_Wheeler_Quantifying_Vulnerability_FINAL.pdf. Accessed October 12, 2016.

WMO—World Meteorological Organization. (no date). *FAQs—Tropical cyclones: 1. What is the difference between 'hurricane', 'cyclone' and 'typhoon'?* http://public.wmo.int/en/About-us/FAQs/faqs-tropical-cyclones. Accessed October 12, 2016.

World Bank. (2000). *Bangladesh: Climate change and sustainable development*. Washington, DC, USA: World Bank. http://www-wds.worldbank.org/external/default/WDSContentServer/WDSP/IB/2001/04/13/000094946_01033105302920/Rendered/PDF/multi0page.pdf. Accessed August 17, 2012.

World Bank. (2011a). *World development report 2011: Conflict, security, and development*. Washington, DC, USA: The World Bank. http://siteresources.worldbank.org/INTWDRS/Resources/WDR2011_Full_Text.pdf. Accessed October 12, 2016

World Bank. (2011b). World Development Indicators (WDI) & Global Development Finance (GDF), World Data Bank. Selected data and metadata exported 7 March 2013 from http://data.worldbank.org/country/bangladesh and http://databank.worldbank.org/ddp/home.do.

WVB ACC—World Vision Bangladesh Access Challenged Children. (2011). *Program: Kamalapur ADP*. World Vision, Bangladesh, India, draft background paper on file with the researcher.

WVB KADP—World Vision Bangladesh Kamalapur Area Development Programme. (2011). *Programme Management Report*. World Vision, Bangladesh, India, unpublished reporting document on file with the researcher.

Author Biography

Dr Johannes M Luetz is Senior Lecturer, Postgraduate Coordinator and Research Chair at CHC Higher Education in Brisbane, Australia. Prior to joining CHC he was a Lecturer at the University of New South Wales (UNSW) in Sydney in the School of Social Sciences, where he also earned his Ph.D. in Environmental Policy and Management. He has worked extensively with World Vision International on research projects raising awareness of the growing effects of climate change on poor and vulnerable communities in Asia, Africa and Latin America. Dr Luetz holds the academic title of Adjunct Lecturer School of Social Sciences, Faculty of Arts and Social Sciences, UNSW Sydney.

Part II
Limits to Climate Change Adaptation in Africa

Chapter 6
Limits to Climate Change Adaptation in Zimbabwe: Insights, Experiences and Lessons

Nelson Chanza

Introduction

Adaptation is increasingly becoming a sought after concept in climate policy, research and practice. This is owing to growing acknowledgement of the inadequacy in mitigation of greenhouse gases (GHGs) to curb dangerous anthropogenic interference with the global climate system. The concept of adaptation in climate science and practice is generally understood as involving a set of interventions targeted at climate stimuli. As such, the success of these measures can be influenced by numerous factors that are both climatic and non-climatic. Definitions of climate change adaptation abound. This preponderance can be attributable to increasing scholarly appetite to examine the concept of climate change adaptation. As such, it is important to specify the definition as relating to response mechanisms of adjustment to the new conditions created by change and variability in the climate system (Adger et al. 2005; Nyong et al. 2007; Bizikova 2012; IPCC 2014). This deployment means that adaptation responses occur in both human systems and ecosystems (IPCC 2014). For the purposes of covering the broad sectors that are affected by climate change in Zimbabwe, this study explores adaptation limits in both socio-economic sectors and ecological systems.

Zimbabwe is a highly vulnerable country to the threats posed by climate change. Given that the economy is primarily agro-based with the majority of the peasantry populated in rural areas and dependant on climate-sensitive livelihoods such as arable farming, livestock rearing and various ecosystem services, the threat of climate change is indeed a reality. Already, limits to climate change adaptation exist, where the country fails to adequately minimise the negative effects of climate change. Much of the scholarly attention to climate change in the country has

N. Chanza (✉)
Department of Geography, Bindura University of Science Education, Bindura, Zimbabwe
e-mail: nchanza@gmail.com

W. Leal Filho and J. Nalau (eds.), *Limits to Climate Change Adaptation*, Climate Change Management, https://doi.org/10.1007/978-3-319-64599-5_6

focused on impact identification (Hartmann et al. 2002, Chamaille-Jammes et al. 2007; Aguilar et al. 2009), adaptation (Gumbo 2006; GoZ/UNDP/GEF 2009; Chagutah 2010; Gukurume 2013) and potential mitigation practices (Brown et al. 2012; Chanza and de Wit 2015). Most of these studies, however, are based on fragmented collection of empirical evidence across the country's climate sensitive sectors such as agriculture, water, human settlements, health, energy and biodiversity (Chanza 2014). Furthermore, there are wide knowledge gaps in informing appropriate mitigation and adaptation policy interventions (AAS 2013). The impacts of climate change in Zimbabwe are mainly evident in the increase in extreme weather events such as floods, drought, high temperatures and violent storms and lightning, and desiccation of vleis, streams, rivers and other surface waters, including the recently observed low water levels in Lake Kariba, which threatens hydro-electricity power (HEP) generation. The socio-economic implications are already unfolding in rampant electricity cuts to households, institutions, farmers and industry. The cumulative effects are seen in low agricultural productivity and poor economic performance.

Elsewhere, the Fifth Assessment Report (AR5) of the Intergovernmental Panel on Climate Change (IPCC) has reported evidence on limits to adaptation that many countries face (Klein et al. 2014). The primary objective of this report is to examine the limits and barriers that are evident in the adaptation system in Zimbabwe. Such an investigation can serve to deepen current understanding of the process of adaptation in developing countries in order to contribute to policy development and decision making process by communities facing climate risks. It is intended that the climate change management community would use these insights, evidence and experiences drawn from Zimbabwe to guide formulation of robust adaptation strategies that are critical in building resilient communities in the face of climate change. Specifically, the paper draws from evidence in such key sectors as agriculture, water, human health and settlements, energy, industry and biodiversity to exposit that there are already signs of limits to adaptation.

In documenting the evidence, insights and experiences with climate change phenomena, the report reviewed literature covering climate change adaptation, which is mainly drawn from studies done in Zimbabwe. It uses a sector-based assessment of adaptation efforts in those areas that already show noticeable impacts of climate change to demonstrate paucity in adaptive capacities. Media search on reports of climate change was also conducted to gather more information on sectoral impacts and adaptation gaps in the country. Acknowledging that area-specific accounts of adaptation practices could have shown pockets of successful adaptation interventions across the country's landscape, this study maintains that the sectoral-based analysis is the best approach in assessing the strengths and limitations of adaptation interventions in Zimbabwe. This stance, as argued here, not only gives a more reliable illumination of the status of national adaptation strategies, but also enables generalisations to be made about the efficacy of adaptive measures in place. The report starts by a conceptualisation and scholarly treatment of key concepts to show what constitutes limits in climate change adaptation. It then presents the vulnerability to and impacts of climate change for each identified sector. This is intended to contextualise the analysis of adaptation challenges that these sectors face. With reference to climate change adaptation therefore, research that reviews the major factors underpinning successful adaptation is highly timely given that much empirical evidence is pointing towards challenges on the same (Chanza 2014, 2015).

Limits to Climate Change Adaptation: A Conceptualisation

Mitigation and adaptation jointly feature as critical concepts to curtail climate change impacts (IPCC 2014). As a concept, adaptation has a diverse and conflicting scholarly treatment in climate science, policy and practice. As such, competing definitions abound. The commonly used definition is that given by the IPCC (2007, p. 869) as: "adjustment in natural or human systems in response to actual or expected climatic stimuli or their effects, which moderates harm or exploits beneficial opportunities." Moser and Ekstrom (2010) deviated from this definition arguing that adaptation must not consider climate change alone but may be initiated or undertaken in a set-up of non-climatic windows of opportunity. From the perspective of analysing barriers and limits to adaptation that this chapter pursues, the view of Moser and Ekstrom deserves a closer attention. Harnessing such opportunities available in the adaptation system can be useful in building adaptive capacities, thereby minimising limits to adaptation. However, the growing failures of mitigation (Wilbanks 2005; Dow et al. 2013; UNEP 2013; WMO 2014) means immense challenges for adaptation. The scope of mitigation is on GHGs and sinks while adaptation focuses on those sectors and activities that are sensitive to climate impacts (Wilbanks 2005; Bizikova 2012). Given that mitigation and adaptation reinforces each other (Wilbanks 2005; IPCC 2014), most scholars agree on the need to prevent the occurrence of climate change (mitigation) in order to minimise pressure on adaptation, particularly in the developing world facing challenges of successfully adapting to climate destabilisations (for example, Adger et al. 2003; Chagutah 2010; Dow et al. 2013; Islam et al. 2013). Wilbanks (2005), for instance, opines that a highly adaptive society depends on the level of GHG stabilisation through mitigation. This suggests that adaptation is difficult or even a futile if mitigation fails to minimise the magnitude of the costs to be handled. Already, signs of adaptation failure are evident in both developed (Dow et al. 2013; Klein et al. 2014) and developing countries (Mertz et al. 2009; Islam et al. 2013; Amjath-Babu et al. 2016). In the latter case, limits to climate change adaptation are more pronounced (Klein et al. 2014) owing to complex obstacles to successful adaptation.

In praxis, adaptation occurs at multiple spatial and temporal scales (Adger et al. 2005; Klein et al. 2014). Accordingly, limits and barriers can be found across the spatial and temporal spectrum of adaptation practice (Islam et al. 2013). In consistency with scholarly treatment, this chapter refers to limits as constraints, barriers, gaps or deficiencies that lead to poor adaptation (Moser and Ekstrom 2010; Dow et al. 2013; Islam et al. 2013; Barnett et al. 2015). In this deployment, it can be understood that the adaptation system has factors that make it difficult to implement plans or devise actions to minimise climatic risks. Some scholars argue for the need to isolate barriers from limits, arguing that the former can be overcome with sufficient political will, social support, resources and effort (Moser and Ekstrom 2010; Islam et al. 2013), albeit treating them jointly in their writings. Islam et al (2013) agree with Moser and Ekstrom (2010) that limits are insurmountable while barriers are mutable or malleable. Inarguably, failure to surmount the barriers will lead to the manifestation of complex limits that causes maladaptation, particularly in the developing world. This study shares similar views that the difference between limits and barriers is rather blurred and prefers to treat them jointly. To IPCC (2012), these impediments mainly include resource limitations, institutional characteristics that stymie actions, or lack of integrity in ecological systems. In a study to determine limits and barriers to adaptation to climate change in Bangladesh coastal fishing communities, Islam et al. (2013) identify four forms of barriers as: natural (mainly related to the occurrence of the climatic event itself), technological (of radio signal offshore, inaccurate cyclone forecast, lack of safety equipment and navigational instruments, and poor quality boats and engines), economic (related to low incomes and lack of access to credit), social (lack of education, skills and livelihood alternatives) and institutional (support mechanisms and law and regulation enforcement). Clearly, the propensity and combined interaction of such factors in countries like Zimbabwe cannot be overlooked. Chanza (2014) reinforces earlier observations by McGuigan et al. (2002), Thornton et al. (2006), Mertz et al. (2009), and Hertel and Rosch (2010) that widespread poverty erodes capacities for successful adaptation. As shall be seen later in this discussion, there is a preponderance of evidence on the relationship between poverty and maladaptation in the Zimbabwean context. The next sections present evidence on the sensitivity of Zimbabwe's socio-economic and ecological systems and discuss the limits and barriers to adapting to climate change.

Limits to Adaptation in the Agriculture Sector

While some aspects of climate change such as longer growing seasons and warmer temperatures may bring benefits in other countries (IPCC 2014), in Zimbabwe, however, there are already signs of adverse impacts. These impacts have great negative implications on agricultural productivity, including reduced crop and animal productivity (IPCC 2014). Given the intrinsic relationship between food production level and climate, agriculture becomes one of Zimbabwe's most climate sensitive sectors. Accordingly, it is important to understand the vulnerability of the agricultural sector to climate change before examining the adequacy and effectiveness of interventions in place. Changes in mean rainfall and temperature as well as the increase in extreme events will affect agriculture (FAO 2008). In its much publicised report, Turn Down the Heat: Why a 4 °C Warmer World Must be Avoided, World Bank (2013) warns that extremes in temperature is likely to affect yields of rice, wheat, maize and other important crops. This development will adversely affect food security. Already, Zimbabwe has started to experience high temperature anomalies with negative consequences on national and household food security. The agro-ecological delimitation, shown in Fig. 6.1, is a function of a mixture of three main variables; namely, effective rainfall, temperature and soil quality, which in turn influence agricultural potential. Rainfall patterns and crop production progressively deteriorate as one moves from Region 1 to Region 5. Driven by a concern that climate change might have influenced the country's agro-ecological regions, a study by Mugandani et al. (2012) confirmed significant shifts in natural regions, which is indicative of climate change and variability. The scholars concluded that Zimbabwe's highly productive agro-ecological regions (Region 2 and 3) are shrinking in size at the expense of drier regions (Region 4 and 5). This projected drying trend affects the agricultural sector as cropping is mostly rain-fed resulting in reduced crop yields, low livestock productivity and poor pastures, which ultimately affects food security and livelihood options.

Given that the economy is primarily agro-based with the majority of the peasantry populated in rural areas and dependant on climate-sensitive livelihoods such as arable farming, livestock rearing and various ecosystem services, climate change is indeed a threat and limits to adaptation cannot be ruled out. With the assistance of various development partners, the Government of Zimbabwe has not been complacent in responding to drought posed by climate change. Current adaptation practices in the agricultural sector include focus on irrigation farming, use of

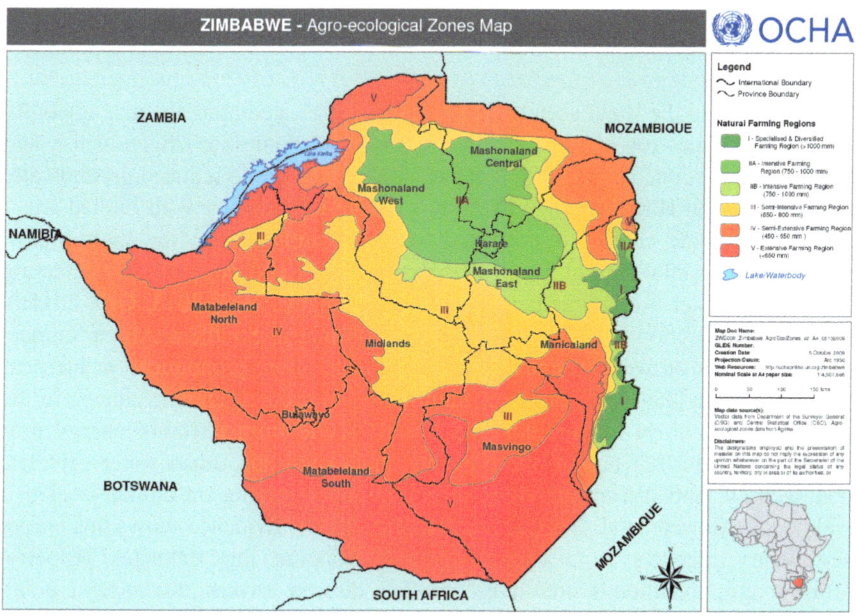

Fig. 6.1 Zimbabwe's agro-ecological zones (*Source* www.reliefweb.int)

drought tolerant crops such as small grains, conservation agriculture, and an assortment of indigenous based adaptation, among others (Mubaya 2010; Dube and Phiri 2013; Chanza 2014). However, uptake and adoption of such measures and practices tends to be influenced by access to information, magnitude of climatic event, institutional support and societal or individual behaviour (Patt and Gwata 2002). Similarly, in a study that looked at the possibility of transforming from rain-fed agriculture to irrigation farming in some sub-Saharan countries including Zimbabwe, Amjath-Babu et al. (2016) observe complex interaction of such factors as groundwater and energy availability, cost and availability of labour, access to market, infrastructural needs, farm conditions and natural resources stocks, climate, population density, as well as political variables. In order to address such adaptation gaps in communities facing climatic risks, scholars like Nyong et al. (2007), Berkes (2009), Turner and Clifton (2009), Orlove et al. (2010), Roos et al. (2010), Speranza et al. (2010), Nakashima et al. (2012) and Egeru (2012) call upon the need to use indigenous knowledge. However, in Zimbabwe, local knowledge still falls short of matching the magnitude and severity of risks potentiated by climate change (Chanza 2014). A case in point is the 2015/16 El-NINO induced drought where more than 30% of the country's 13 million people were in dire need of food aid, an event which necessitated the government to declare it a national disaster (Herald 10/02/16). The current mode of adaptation strategy seems to orient largely towards food aid and handouts in the event of climate-induced crop failure. Inarguably, this trajectory is not sustainable and is evidence of limits to successful adaptation.

Adaptation Interventions and Limits in the Water Sector

With reference to the water sector, climate change is exacerbating the unavailability of water resources for industrial, power generation, domestic, irrigation use and livestock production. This is evident from supply shortages to reduced groundwater levels (Davis and Hirji 2014). Zimbabwe is a semi-arid country with limited water resources. Climate change is predicted to significantly impact the hydrological cycle altering rainfall, runoff, and evapotranspiration and consequently affecting dynamics of water use and demand patterns (Kusangaya et al. 2013; GoZ 2015a). The earlier projection given by the IPCC (2007) that in Southern Africa, climate change's effects on water resources are likely to be more pronounced in the near future is quite evident in Zimbabwe. Summarily, impacts of climate change in the water sector manifest in droughts and floods. Aside from affecting energy production and industrial output, these events affect rural and urban water supplies, food security, and disrupt activities of communities relying on climate-sensitive livelihoods (Mawere et al. 2013; Davis and Hirji 2014). Evidence shows that before the country recovers from cyclone-induced flooding that damages property, infrastructure, livelihoods and lives, another disaster strikes. Kusangaya et al. (2013) observe that the cumulative effects of intense storms, heavy rainfall events, floods and droughts are threatening human and animal life, and causing social and economic hardships. Zimbabwe also lacks capacity to adequately deal with epidemics related to water such as cholera, typhoid, and other water-borne diseases.

In order to assess the adaptation practices for responding to water challenges in the country, this review concentrates on Zimbabwe's two largest cities: Harare (the capital) and Bulawayo (the second largest city). The country's seven catchment areas namely, Mazowe, Manyame, Save, Runde, Sanyati, Gwayi and Mzingwane are facing serious desiccation that is attributable to low rainfall amounts and high temperatures. As a result, more than 12,000 boreholes are now dry and dysfunctional owing to recurrent droughts and poor maintenance (Herald 01/02/16). Harare, which gets its water supply from Manyame catchment, is home to more than 16% of the national population. Already, the city faces acute water cuts, partly due to drying up, aggravated by the pollution of its major reservoir, Lake Chivero. Similarly, Bulawayo, located in the driest region of the country with inadequate supplies of surface and underground water has been experiencing serious water supply challenges over the past 100 years. The city's five reservoirs that were designed to supply it with water are all located in Mzingwane catchment, whose water yield continues to diminish owing to increased desiccation associated with climate change (Moyo et al. 2005; Baker 2012; Chirisa and Bandauko 2015). Of late, however, severe drying coupled with water pollution has forced government to decommission its three dams, leaving only two dams, Inyakuni and Insiza to supply the 'thirsty' city (Baker 2012).

In order to adapt to chronic water shortages of these cities, government, with support from development partners, has come up with several strategies and projects. These include: construction of new reservoirs, drilling of boreholes, water rationing, water governance and water-use education. Sadly, however, not much has been done with dam projects owing to lack of funds. The proposed Kunzvi Dam Project meant to address the water supply challenges in Harare is yet to commence, decades since the idea was mooted. Similarly, the National Matabeleland Zambezi Water Project, which intends to divert water from the Zambezi River about 400 km north of Bulawayo, has been in the pipeline for over 100 years. The government has been battling to secure US$800 million, which the project requires to construct a reservoir downstream of the confluence of Gwayi and Shangani rivers and to subsequently connect this to the City of Bulawayo using a 150 km pipeline (Baker 2012). Although boreholes have been drilled in many parts of Bulawayo and Harare, the sustainability of borehole yield in an increasingly dry climate is also questionable. Furthermore, this is just a short-term intervention given that boreholes are difficult to regulate and maintain. In Bulawayo, the unpopular policy mooted by the Bulawayo City Council in 2012, which compelled all households connected to the municipal sewer system to flush their toilets simultaneously twice each week in response to the water crisis did not help in abating the situation. The authorities argued that this was necessary to keep sewer pipes from being blocked and damaged. Clearly, limits to successful adaptation are evident here.

Overall, the adaptation toolbox in the water sector includes water harvesting techniques, water rationing and water recycling. Apparently, water harvesting techniques are still at a very small scale of adoption. Such interventions have been driven largely by non-governmental organisations such as Practical Action, Zvishavane Water Projects, Dabane Trust and Mvuramanzi Trust. Water recycling projects in such cities as Harare have been hampered by limited funds, human capital deficiencies and obsolete infrastructure. Capacity to adapt to climate change in this regard has also been constrained by rapid urbanisation and economic challenges that the country faces (Chirisa and Bandauko 2015). It can be concluded here that the effects of climate change on the water resources and the adaptation requirements ramify across other development sectors such as agriculture, human settlements, health, industry and energy. For instance, decline in water availability is also likely to have an adverse effect on hydro-electric power (HEP) generation.

Limits to Climate Change Adaptation in Human Health and Settlements

The climate science community attributes the increase in the likelihood of occurrence and strength of extreme weather and climate events to climate change (IPCC 2012, 2014). These events pose adaptation challenges in the health and human settlements sectors. The irregularities in the distribution of temperature and rainfall discussed in the preceding sections pose serious implications on the settlement sector. Overall, climate related hazards reported in the country include tropical cyclones, floods and flash flooding, drought and intra-season dry spells, thunderstorms and lightning, hailstorms, heavy rains, ground frost and heat waves (GoZ 2015a). This section presents the vulnerability of the health and human settlements sector to climate change. It then shows that current efforts to minimise risks of exposure are inadequate owing to widespread poverty and policy deficiencies among a set of barriers and limits.

Low-lying areas are vulnerable to climate change-induced flooding. Most of these places are concentrated northwards toward the Zambezi River Valley and southwards toward the Save River Valley and Limpopo River Valley owing to marked decrease in altitude. Of late, an increase in the frequency of floods and tropical cyclones has been linked to climate change (Unganai and Murwira 2010; GoZ 2015a). These events pose serious risks to the communities through loss of crops, infrastructure damages, injuries and exposure to diseases, loss of assets, homes and lives. It is on record that such disasters are related to climate change (GoZ 2015a). Currently, the regulatory framework for addressing disaster risks associated with climate change can be best described as weak and fragmented. For example, the available policies and laws governing settlement location, design and monitoring of building standards and quality are not in synchrony with the realities of climate change. This situation tends to be compounded by the combined effects of poverty and limited political will for law enforcement and budgetary support. As a result, people end up settling or erecting structures in hazard prone areas such as riverbanks, floodplains, wetlands and sloppy areas.

The major health threats due to climate change are mainly related to changing patterns of disease, water and food insecurity, vulnerable shelter and human settlements, extreme climatic events, and population growth and migration. The inundation of low-lying areas poses health problems due to the spread and harbouring of vectors that cause diseases like cholera, typhoid, malaria and meningitis. For example, in a study of climate change on epidemiology in Zimbabwe, Hartmann et al. (2002), using sixteen climate change scenarios, reveal that the geographical distribution of malaria could change, with previously unsuitable areas becoming suitable for transmission as the *Plasmodium* vector takes advantage of the new conditions. The water supply challenges in major cities cited earlier is a typical example of the susceptibility of urban settlements to climate change-induced drought. Already health problems associated with poor access to portable water are manifesting through cholera and typhoid outbreaks in some parts of Harare.

Given the predicted increase in frequency and intensity of extreme events such as floods, storms and droughts (GoZ 2015a), Zimbabwe faces immense challenges for mitigating climatic impacts on its vulnerable population. It is understood that the most affected people in the country are the majority rural communities with limited adaptive capacities (Chanza 2014). This is because these events are unfolding in a context where realities of other non-climatic stressors have already brought untold suffering to the people (Chirisa and Chanza 2009; Chagutah 2010), exposing the majority of the poor to extreme climatic events and disasters. Most scholars agree that the level of exposure to climatic threats is a function of community poverty, which in turn influences adaptive capacity (Patt and Gwata 2002; Brockhaus et al. 2013).

Limits to Climate Change Adaptation in the Energy and Industry Sector

The energy sector features prominently in climate change discourse in Zimbabwe. This is mainly both to its sensitivity to climate change and the ramification of the effects of energy crisis on the industry sector. Aside from this, the energy sector constitutes a largest share of the country's total GHG emissions (GoZ 2012). With this in mind, I have decided to treat these two sectors jointly in this section. An examination of the adaptation status in the energy and industry sectors serves to understand the impact that climate change has on the country's economy. Zimbabwe's current development goals are largely informed by its blueprint document, the Zimbabwe Agenda for Sustainable Socio-Economic Transformation (ZimASSET). Already, energy supply challenges are cited as a hindrance for propelling development in the country (GoZ 2013).

About 53% of the country energy supply is generated from HEP from Lake Kariba. Overall, Zimbabwe's power stations are currently performing at below generation design capacity owing to lack of adequate financing for operations, high transmission losses, obsolete and poor maintenance of equipment (Mtisi and Prowse 2012).

Of late, the problem has been compounded by low water levels in Lake Kariba, which is related to climate change. A study of hydro-electric power generation conducted in the Zambezi Basin, taken in conjunction with projections of future runoff, indicates that hydropower generation would be negatively affected by climate change (Nicol and Kaur 2009). This situation has far reaching implications on the industry sector. Among a deluge of other operational factors, energy crisis is also cited as a major deterrent to the anticipated productivity in the industrial sector. During peak periods, Zimbabwe has a power generation shortfall of between 600 to 1600 megawatts (MW). Given the global call for cutting carbon emissions, the country's propensity to switch to clean energy sources such as HEP, solar energy and biofuels is largely constrained by lack of funds (GoZ 2015b). In trying to cope with energy supply challenges, communities, largely in the rural areas, extract forest biomass for cooking and heating. The agricultural sector also uses fuelwood for curing tobacco. These energy sources are not sustainable as the extraction of biomass leads to the removal of GHG sinks and contributes to carbon emissions. These are clear examples of adaptation failure. On a different note, households, companies and institutions are investing in solar energy projects in the energy mix. Although concepts such as green energy and energy efficient buildings are also being promoted, wide adoption of such strategies tends to be limited by financial and technical constraints (GoZ 2015a).

Climate Change Threats on Biodiversity and Limits to Adaptation

About 67% of Zimbabweans live in the rural areas (ZIMSTAT 2013) on natural resources dependent livelihoods. They derive food, medicines, food, water, medicine, energy sources, building and craft materials, as well as spiritual, cultural and aesthetic services from biological diversity and functions provided by ecosystems (GoZ 2014). However, the climate-sensitiveness of these ecosystems means that these livelihoods are at threat. This makes climate change a key issue in Zimbabwe. Both terrestrial and aquatic ecological systems are showing evidence of sensitivity to change and variability in the climate system. In its Second National Communication to the United Nations Framework Convention on Climate Change (UNFCCC), Zimbabwe reported that plant diversity is projected to decline under both worst case and best case scenarios owing to climate change (GoZ 2012). Projected decreases in precipitation aggravated by increases in temperature given earlier by Unganai (1996), is reported to lead to corresponding decrease in net primary productivity across most parts of the country. Only the eastern highlands and the central regions of the country, which receive high rainfall seem to have the greatest adaptive capacity since the vegetation found there is less sensitive to climate variability and change. Table 6.1 summarises the significance of different types of terrestrial ecosystems in biodiversity conservation and the various threats they face from both climatic and non-climatic factors. The synthesis given in this

Table 6.1 Key threats to successful adaptation in Zimbabwe's terrestrial ecosystems (*Source* GoZ 2014)

Terrestrial ecosystem	Importance to biodiversity conservation	Adaptation barrier/threat
Montane forest grassland mosaic	Headwaters of four key rivers; part of the Afromontane centre of endemism; high species richness; important bird area; biodiversity conservation hotspot; important for ecosystem-based adaptation to climate change	Invasive alien species such as *Lantana camara* and wattle; artisanal mining, harvesting of non-timber forest products (NTFPs), closure of tea company
Kalahali *Acacia baikiaea* woodland	Various species diversity	Wildlife poaching; wild fires; climate variability; wood poaching; elephant population; human encroachment; gazetted forests under threat; invasive plant species; coal mining; Zambezi water channel through Hwange
Southern Africa bushveld	Dominated by mega herbivores	Conversion to agriculture; urban expansion; pollution; mining; unsustainable harvesting; bush encroachment by invasive species
Southern miombo Woodland (drier Zambezian miombo)	Centre of plant diversity; high mineral composition; 500 bird species with six confined to the region; high reptile endemism with 30 species exclusively found in the eco-region; four reptile species strictly endemic	Wildlife and firewood poaching; conversion to agriculture; deforestation to tobacco curing; commercialized harvesting of NTFPs; urban expansion; mining
Zambezian *baikiaea* woodland	Dominant tree species is the *Baikiaea plurijuga* (Zambezian teak) which is endemic to the ecoregion; Important bird area	Timber logging; recurrent wild fires; conversion to agriculture; climate change
Zambezian and mopane woodland	Important area for mammalian diversity; rich in mega fauna	Mining; conversion to agriculture; large elephant populations; invasive alien species; wildlife and timber poaching

table points to some challenges towards effective adaptation in this sector. It can be seen here that the main problem affecting biodiversity conservation and management relates to urbanisation and urban development, population pressure, poaching, poverty, deforestation, mining activities as well as the proliferation of invasive species under extreme climatic conditions. Furthermore, with the predicted drier climate in the country, fires are expected to increase in intensity (GoZ 2012, 2014). Already, the combined effects of climate change and poor enforcement of environmental laws are causing uncontrolled veldt fires. This results in serious losses on forests and grasslands and will diminish the ability of ecosystems to provide goods and services to society and wildlife.

In addition, over 80% of the country's landscape falls in drier regions where livestock ranging is the only possible livelihoods option (GoZ 2012). As earlier specified from the studies done by Mugandani et al. (2012), the increasing aridity associated with these zones suggests adaptation challenges for farmers whose livelihoods are largely dependent on rangelands, including communities and businesses relying on wildlife-based enterprises. Evidently, desertification associated with degraded rangelands and erratic rainfall pose serious adaptation challenges to communities in dry regions of the country. The negative impact on the livelihoods of farmers, particularly small-scale, resource-poor communal farmers is likely to worsen the poverty situation. This is an evident sign of limits to adaptation.

Similarly, the aquatic ecosystems are not immune to the effects of climate change. Zimbabwe has various water bodies and wetlands in the form of flowing rivers, dams, dambos, vleis and pans, which are widely distributed across the country. These freshwater ecosystems are important centres for aquatic biodiversity as they provide a unique range of habitats, regulation of water flows and flooding. They also support unique bird species, water and grazing for livestock and wildlife; and crop cultivation (GoZ 2014). However, as climatic conditions worsen, these services are likely to be affected by recurrent droughts, temperature extremes and high evapotranspiration rates and low rainfall inputs (GoZ 2012). The National Biodiversity Strategy Action Plan (NBSAP) lists the following threats on aquatic ecosystems: dam construction, development of irrigated agriculture, overgrazing, gully erosion and siltation, infrastructural development, pollution from urban activities, mining activities, overfishing and invasive alien species (GoZ 2014). Evidently, this shows that limits to adaptation in the biodiversity sector are both climatic and non-climatic.

Discussion and Policy Directions

The preceding sectoral-based analysis has shown that Zimbabwe has not been complacent in adopting strategies and practices for responding to climatic stimuli. In its Second National Communication to the UNFCCC, Zimbabwe singles out adaptation as a priority to cushion its vulnerable socio-economic development sectors. On record, a number of sector-specific adaptation options identified have been stymied by paucity of adaptation funding (Chanza et al. 2015). It has been argued here that both barriers and limits characterise the adaptation terrain, whose complexity tends to be aggravated by knowledge gaps and uncertainties in the future climate system.

In the agriculture sector, the need to promote irrigation infrastructure and adoption of rainwater harvesting technologies to mitigate drought and water scarcity has been severely hampered by lack of funds. Similarly, dam construction projects in the water sector are largely trailing in synchrony with the financial gaps as they take long to be completed. Bemoaning the poor utilisation of the borehole infrastructure in the country, the Minister of Environment, Water and Climate,

Mrs Muchinguri-Kashiri acknowledges that much still has to be down to speed up adaptation efforts in the water sector (New Zimbabwe 25/01/16). According to the 2015 National Climate Change Response Strategy (NCCRS) document, the challenge for the country is in developing adaptation and mitigation strategies that address complex impacts of the phenomenon (GoZ 2015a). With particular reference to developing countries such as Zimbabwe, variability and change in the climate system poses an additional burden to the poor where extreme climatic events act as a threat multiplier with negative outcomes for livelihoods. Clearly, limits to climate change adaption abound.

In the human settlement sector, vulnerability is worsened by poor coordination of disaster risk management (DRM) initiatives, low levels of public participation, lack of a culture safety among the general public and limited or absence of resources to erect disaster-resilient infrastructure. The available infrastructure is largely based on outdated historical climatic conditions, whose design fails to accommodate changes in the climate system (GoZ 2015a). Such infrastructure cannot withstand the current and projected extreme climatic events. Accordingly, the design of climate proofed infrastructure and settlements, as specified in the NCCRS, is an urgent necessity. Notwithstanding the incorporation of decentralised frameworks for the coordination of DRM interventions at multi-spatial levels in the proposed legislation (GoZ 2012), the country has been facing governance and coordination challenges that are likely to stymie such initiatives (Chirisa and Chanza 2009). For Zimbabwe to address the barriers and limits specified in this discussion, it has to swiftly adopt strategies for mainstreaming climate change into DRM. Such mechanisms are specified in the current Sendai Framework of Disaster Risk Reduction 2015–2030 as: multi-stakeholder involvement in understanding disaster risk; strengthening disaster risk governance to manage disaster risk; investing in disaster risk reduction; enhancing disaster preparedness for effective response, and to 'Build Back Better' in recovery, rehabilitation and reconstruction (UN 2015). In addition, the public health delivery system is limited to cope with present and potential injuries and health problems associated with disasters. Given that vulnerabilities to climate change tend to be location specific, it is necessary for the country's adaptation strategies to reflect decentralised and community-initiatives such as leveraging on indigenous knowledge and skills. In this view, the limits to adaptation can be minimised through, for instance, the prioritisation of urban and rural local authorities to deal with climate risks in their localities. However, for them to be effective, they need to be supported with adequate financial resources. Currently, most of these local authorities lack financial capital to maintain infrastructure meant to reduce disaster risks owing to limited funding from the central government.

Overall, there is need to strengthen the country's management information systems for comprehensive climate change adaptation and disaster mitigation. In this consideration, the generation and provision of reliable and appropriate climate information will be a key component in the climate change adaptation menu. In order to address adaptation limits at multiple scales, climate information should be available and usable at national, provincial, district and community levels.

Such interventions need to be supported by robust public awareness programmes on climate risks and disasters through education and training. Central in these strategies is the institutional strengthening of key departments such as the Meteorological Services Department (MSD) and the Department of Civil Protection (DCP). The former would need to give early warning systems (EWS) to minimise climate impacts resulting from drought and floods while the latter should be capacitated to coordinate climatological and forensic analyses of atmospheric and geological hazards, respectively.

Emerging from this analysis is that climate change imposes serious impacts across different sectors including agriculture, water, energy, human settlements and biodiversity. The biggest challenge which communities and authorities are grappling with is adaption to climate change impacts in all these sectors. In the context of Zimbabwe, agriculture, water and biodiversity are the most affected sectors. Adaptation limits include those related to inadequate financing mechanisms, weak institutional frameworks and poor climate change policies. The impacts of climate change in Zimbabwe are cross-sectoral, for instance, climate change affects water, energy, food security, biodiversity in a much more interrelated manner. As such, there is need to adopt a systems approach in climate change adaptation planning and policy formulation. There is also need to have cross-sector collaborations for new technologies to adequately respond to climate change impacts.

Conclusion

In Zimbabwe, the context of climate change adaptation is characterised by low adaptive capacities that tend to be aggravated by widespread poverty; limited financial resources for upscaling adaptation efforts, extension services and information sharing; climate uncertainties and knowledge gaps; policy and institutional gaps for coordinating a paraphernalia of adaptation practices; low technology uptake compounded by resource deficiencies and the magnitude and severity of climatic events themselves. Cumulatively, these factors constitute barriers and limits to successful adaptation. As a result, the country is seriously exposed to climatic risks. In order to address these constraints, Zimbabwe needs to speed up climate policy development and to avail funding for adaptation interventions. However, it has been argued here that the frequency and severity of climate events will not match with the national capacities to curtail them. Of particular emphasis is that current efforts to mitigate climatic risks and build adaptive capacities are currently constrained by economic challenges, a context in which the country is still recovering from the past economic melt-down. Accordingly, unless the country's adaptation efforts are complemented by funding, skills and technological support from the global community, the looming maladaptation reminiscent of adaptation limits will remain a key feature. From a global adaptation policy discourse, it is necessary to invigorate the paradox of climate change in such countries as Zimbabwe, whose contribution to GHG emissions is very insignificant, albeit suffering the greatest brunt of the phenomenon. Nevertheless, this should not be construed as an excuse for the country's failure to put in place robust adaptation interventions to climate-proof its highly sensitive socio-economic development sectors and ecological systems.

Future prospects for successful adaptation should be marked by a mixture of strategies and speedy implementation of planned interventions. In the agricultural sector, adaptive strategies should swiftly shift from the current passive orientation towards proactive ones. The country will minimise its adaptation limits by speeding up up-scaling of irrigation farming, supported by off-farm livelihoods diversification practices to boost household incomes and access to food. Responses in the water sector could be enhanced by mobilising domestic and international funds to speed up completion of dams to improve water availability and access. This should also be backed by a culture of water recycling, water governance and water-use education. In order to marginalise the vulnerabilities in the human settlement and health sectors, government needs to strengthen its EWS backed by early actions in drought and flood management. The current trajectory towards low carbon economy in the industry and energy sectors needs to be promoted. The success of this initiative, for example, could be supported through education programmes for behaviour change by energy users. Lastly, enforcement of environmental management laws should be strengthened to ensure that ecosystem-based benefits of prudent biodiversity management are realised.

References

Adger, W. N., Arnell, N. W., & Tompkins, E. L. (2005). Successful adaptation to climate change across scales. *Global Environmental Change, 15,* 77–86.

Adger, W. N., Huq, S., Brown, K., Conway, D., & Hulme, M. (2003). Adaptation to climate change in the developing world. *Progress in Development Studies, 3,* 179–195.

African Academy of Sciences (AAS). (2013). *DFID, ACU and AAS Launch Climate Impacts Research Capacity Building Programme in Africa,* Online, Available: http://www.aasciences.org/index.php/projects/circle. Accessed February 25, 2016.

Aguilar, E., Aziz Barry, A., Brunet, M., Ekang, L., Fernandes, A., Massoukina, M., et al. (2009). Changes in temperature and precipitation extremes in western central Africa, Guinea Conakry, and Zimbabwe, 1955–2006. *Journal of Geophysical Research, 114,* D02115.

Amjath-Babu, T. S., Krupnik, T. J., Kaechele, H., Aravindakshan, S., & Sietz, D. (2016). Transitioning to groundwater irrigated intensified agriculture in Sub-Saharan Africa: An indicator based assessment. *Agricultural Water Management, 168,* 125–135.

Baker, J. (2012) *Water supply challenges in Bulawayo, Zimbabwe. GIS in water resources,* Online, Available: www.ce.utexas.edu/prof/maidment/giswr2012/TermPaper/Baker.pdf. Accessed February 1, 2016.

Barnett, J., Evans, L. S., Gross, C., Kiem, A. S., Kingsford, R. T., Palutikof, J. P., et al. (2015). From barriers to limits to climate change adaptation: path dependency and the speed of change. *Ecology and Society, 20*(3), 5. doi:10.5751/ES-07698-200305.

Berkes, F. (2009). Indigenous ways of knowing and the study of environmental change. *Journal of the Royal Society of New Zealand, 39*(4), 151–156.

Bizikova, L. (2012) *Climate change adaptation, mitigation and sustainable development: opportunities for integrating research into policies. Linking climate models to policy and decision-making*. Paper 9: 215–228, Online, Available: http://projects.upei.ca/climate/files/2012/10/Book-1_Paper-9.pdf. Accessed February 2, 2016.
Brockhaus, M., Djoudi, H., & Locatelli, B. (2013). Envisioning the future and learning from the past: Adapting to a changing environment in northern Mali. *Environmental Science & Policy, 25,* 94–106.
Brown, D., Rance Chanakira, R., Chatiza, K., Dhliwayo, M., Dodman, D., Masiiwa, M., Muchadenyika, D., Prisca Mugabe, P., & Zvigadza, S. (2012). Climate change impacts, vulnerability and adaptation in Zimbabwe. *IIED Climate Change Working Paper No. 3*, October 2012, International Institute for Environment and Development (IIED), London.
Chagutah, T. (2010). *Climate change vulnerability and adaptation preparedness in Southern Africa: Zimbabwe Country Report*. Cape Town: Heinrich BöllStiftung Southern Africa.
Chamaillé-Jammes, S., Fritz, H., & Murindagomo, F. (2007). Detecting climate changes of concern in highly variable environments: quantile regressions reveal that droughts worsen in Hwange National Park, Zimbabwe. *Journal of Arid Environments, 71,* 321–326.
Chanza, N. (2014) *Indigenous knowledge and climate change: Insights from Muzarabani, Zimbabwe*, A Ph.D. Thesis, Nelson Mandela Metropolitan University: Port Elizabeth.
Chanza, N. (2015). Indigenous-based adaptation: An imperative for sustainable climate change strategies for Africa. In M. Mawere & S. Awuah-Nyamekye (Eds.), *Harnessing cultural capital for sustainability: A pan Africanist perspective* (pp. 85–134). Bamenda: Langaa Publishing House.
Chanza, N., Chirisa, I., Zhakata, W., & Makura, E. S. M. S. (2015). *The consideration of ethics and justice issues in the formulation of climate change policies in Zimbabwe*. National Climate Justice Project, Online, Available: http://blogs.law.widener.edu/ncj/category/zimbabwe/#sthash.NyIJ7sfR.dpbs. Accessed February 20, 2016.
Chanza, N., & de Wit, A. (2015). Rediscovering indigenous climate knowledge for better responses to climate change: Insights from Muzarabani. *International Journal of Climate Change: Impacts and Responses, 6,* 19–35.
Chirisa, I., & Bandauko, E. (2015). African cities and the water-food-climate-energy nexus: An agenda for sustainability and resilience at a local level. *Urban Forum*. doi:10.1007/s12132-015-9256-6.
Chirisa, I., & Chanza, N. (2009). How will climate change transform African local governance?—Assessing the role of civic engagement. *Journal of Public Administration and Policy Research, 1*(2), 035–046.
Davis, R., & Hirji, R. (2014) *Climate change and water resources planning, development and management in Zimbabwe*, An Issues Paper, 93731, World Bank/Government of Zimbabwe, Harare.
Dow, K., Berkhout, F., Preston, B. L., Klein, R. J. T., Midgley, G., & Shaw, R. M. (2013). Limits to adaptation, *Nature Climate Change*, 3, online, Available: www.nature.com/natureclimatechange. Accessed February 20, 2016.
Dube, T., & Phiri, K. (2013). Rural livelihoods under stress: The impact of climate change on livelihoods in South Western Zimbabwe. *American International Journal of Contemporary Research, 3*(5), 11–25.
Egeru, A. (2012). Role of indigenous knowledge in climate change adaptation: A case study of Tese Sub-Region, Eastern Uganda. *Indian Journal of Traditional Knowledge, 11*(2), 217–224.
FAO. (2008). *Climate change adaptation and mitigation in the food and agricultural sector*. Technical Background Document from the Expert Consultation Held on 5–7 March 2008, Online, Available: ftp://fao.org/docrep/fao/meeting/013/ai782e.pdf. Accessed Januray 26, 2016.
Government of Zimbabwe (GoZ). (2012). *Zimbabwe's Second National Communication to the United Nations Framework Convention on Climate Change*, Ministry of Environment and Natural Resources Management, Harare.

Government of Zimbabwe (GoZ). (2013). *Zimbabwe Agenda for Sustainable Socio-economic Transformation (ZIMASSET): October 2013 to December 2015*, Government of Zimbabwe, Harare.

Government of Zimbabwe (GoZ). (2014). *National Biodiversity Strategy and Action Plan*, Ministry of Environment, Water and Climate, Harare.

Government of Zimbabwe (GoZ). (2015a). *Zimbabwe National Climate Change Response Strategy*, Ministry of Environment, Water and Climate, Harare.

Government of Zimbabwe (GoZ). (2015b). *Zimbabwe's intended nationally determined contribution (INDC). Submitted to the United Nations Framework Convention on Climate Change (UNFCCC)*, Online, Available: www4.unfccc.int/submissions/INDC/Submission%20Pages/submissions.aspx. Accessed February 1, 2016.

GoZ/UNDP/GEF. (2009). *Coping with Drought: Vulnerability and Adaptation to Climate Change, a Focus on Chiredzi District, Zimbabwe. Synthesis Report*. Environmental Management Agency, Harare.

Gukurume, S. (2013). Climate change, variability and sustainable agriculture in Zimbabwe's rural communities. *Russian Journal of Agricultural and Socio-Economic Sciences, 2*(14), 89–100.

Gumbo, D. (2006) *Working together to respond to climate change. Annex I expert group seminar in conjunction with the OECD Global Forum on Sustainable Development. Zimbabwe: Country case study on domestic policy frameworks for adaptation in the water sector*. March 28, 2006, Online, Available: http://www.oecd.org/env/cc/36318866.pdf. Accessed February 08, 2016.

Hartmann, J., Ebi, K., McConnell, J., Chan, N., & Weyant, J. P. (2002). Climate suitability: For stable malaria transmission in Zimbabwe under different climate change scenarios. *Global Change and Human Health, 3*, 42–54.

Herald, 01 February 2016, 12,000 Boreholes run dry—Minister. Zimpapers, Harare.

Herald, 10 February 2016, Government seeks $1.5 billion for food. Zimpapers, Harare.

Hertel, T. W., & Rosch, S. D. (2010). Climate change, agriculture, and poverty. *Applied Economic Perspectives and Policy, 32*(3), 355–385.

IPCC. (2007). In M. L. Parry, O. F. Canziani, J. P. Palutikof, P. J. van der Linden, & C. E. Hanson (Eds.), *Climate change 2007: Impacts, adaptation and vulnerability. Contribution of Working Group II to the fourth assessment report of the Intergovernmental Panel on Climate Change*. Cambridge: Cambridge University Press.

IPCC. (2012). In C. B. Field, V. Barros, T. F. Stocker, D. Qin, D. J. Dokken, K. L. Ebi et al. (Eds.), *Managing the Risks of Extreme Events and Disasters to Advance Climate Change Adaptation. A Special Report of Working Groups I and II of the Intergovernmental Panel on Climate Change* (p. 582). Cambridge: Cambridge University Press.

IPCC. (2014). In C. B. Field, V. R. Barros, D. J. Dokken, K. J. Mach, M. D. Mastrandrea, T. E. Bilir, et al. (Eds.), *Climate change 2014: Impacts, adaptation, and vulnerability. Part A: Global and sectoral aspects. Contribution of Working Group II to the fifth assessment report of the Intergovernmental Panel on Climate Change* (p. 1132). Cambridge, UK and New York, NY, USA: Cambridge University Press.

Islam, M. M., Sallu, S., Hubacek, K., & Paavol, J. (2013). Limits and barriers to adaptation to climate variability and change in Bangladeshi coastal fishing communities. *Marine Policy, 43*, 208–216.

Klein, R. J. T., Midgley, G. F., Preston, B. L., Alam, M., Berkhout, F. G. H., Dow, K., et al. (2014). Adaptation opportunities, constraints, and limits. In C. B. Field, V. R. Barros, D. J. Dokken, K. J. Mach, M. D. Mastrandrea, T. E. Bilir, M. Chatterjee, K. L. Ebi, Y. O. Estrada, R. C. Genova, B. Girma, E. S. Kissel, A. N. Levy, S. MacCracken, P. R. Mastrandrea, & L. L. White (Eds.), *Climate change 2014: Impacts, adaptation, and vulnerability. Part A: Global and sectoral aspects. Contribution of Working Group II to the fifth assessment report of the Intergovernmental Panel on Climate Change* (pp. 899–943). Cambridge: Cambridge University Press.

Kusangaya, S., Warburton, M. L., van Garderen, E. M., & Jewitt, G. P. W. (2013) Impacts of climate change on water resources in Southern Africa: A review. *Physics and Chemistry of the Earth*, Parts A/B/C.

Mawere, M., Madziwa, B. F., & Mabeza, C. M. (2013). Climate change and adaptation in third world Africa: A quest for increased food security in semi-arid Zimbabwe. *The International Journal of Humanities & Social Studies, 1*(2), 14–22.

McGuigan, C., Reynolds, R., & Wiedmer, D. (2002). *Poverty and climate change: Assessing impacts in developing countries and the initiatives of the international community*, online, Available: www.odi.org/sites/odi.org.uk/files/odi-assets/publications-opinion-files/3449.pdf. Accessed February 4, 2016.

Mertz, O., Halsnæs, K., Olesen, J. E., & Rasmussen, K. (2009). Adaptation to climate change in developing countries. *Environmental Management, 43,* 743–752. doi:10.1007/s00267-008-9259-3.

Moser, S. C., & Ekstrom, J. A. (2010). *A framework to diagnose barriers to climate change adaptation*, online, Available: www.pnas.org/cgi/doi/10.1073/pnas.1007887107.

Moyo, B., Madamombe, E., & Love, D. (2005). A model for reservoir yield under climate change scenarios for the water-stressed City of Bulawayo, Zimbabwe. *Paper presented at the 6th WaterNet/WARFSA/GWP-SA Annual Symposium*, Swaziland, November 1–3, 2005. Amsterdam, Netherlands: WaterNet. http://hdl.handle.net/10568/21261.

Mtisi, S., & Prowse, M. (2012). *Baseline report on climate change and development in Zimbabwe*. Harare: Government of Zimbabwe.

Mubaya, C. P. (2010). *Farmer strategies towards climate variability and change in Zimbabwe and Zambia*, Ph.D. Thesis, University of the Free State: Bloemfontein.

Mugandani, R., Wuta, M., Makarau, A., & Chipindu, B. (2012). Re-classification of agro-ecological regions of Zimbabwe in conformity with climate variability and change. *African Crop Science Journal, 20*(s2), 361–369.

Nakashima, D. J., Galloway, M. K., Thulstrup, H. D., Ramos, C. A., & Rubis, J. T. (2012). *Weathering uncertainty: Traditional knowledge for climate change assessment and adaptation*. Paris: UNESCO.

New Zimbabwe, 25th January 2016, Dams about half full signalling water crisis, Minister Muchinguri–Kashiri says, New Zimbabwe Online.

Nicol, A., & Kaur, N. (2009). Adapting to climate change in the water sector. Overseas Development Institute (ODI), online, Available: http://www.odi.org/sites/odi.org.uk/files/odi-assets/publications-opinion-files/4118.pdf. Accessed February 12, 2016.

Nyong, A., Adesina, F., & Elasha, B. O. (2007). The value of indigenous knowledge in climate change mitigation and adaptation strategies in the African Sahel. *Mitigation and Adaptation Strategies for Global Change, 12,* 787–797.

Orlove, B., Roncoli, C., Kabugo, M., & Majugu, A. (2010). Indigenous climate knowledge in southern Uganda: The multiple components of a dynamic regional system. *Climatic Change, 100,* 243–265.

Patt, A., & Gwata, C. (2002). Effective seasonal climate forecast applications: Examining constraints for subsistence farmers in Zimbabwe. *Global Environmental Change, 12,* 185–195.

Roos, V., Chigeza, S., & van Niekerk, D. (2010). Coping with drought: indigenous knowledge application in rural South Africa. *Indilinga—African Journal of Indigenous Knowledge Systems, 9*(1), 1–11.

Speranza, C. I., Kiteme, B., Ambenje, P., Wiesmann, U., & Makali, S. (2010). Indigenous knowledge related to climate variability and change: Insights from droughts in semi-arid areas of former Makueni District, Kenya. *Climatic Change, 100,* 295–315.

Thornton, P. K., Jones, P. G., Owiyo, T., Kruska, R. L., Herrero, M., Kristjanson, P., et al. (2006). *Mapping climate vulnerability and poverty in Africa*. Report to the Department for International Development, ILRI, Nairobi.

Turner, N. J., & Clifton, H. (2009). "It's so different today": Climate change and indigenous lifeways in British Columbia, Canada. *Global Environmental Change, 19,* 180–190.

Unganai, L. S. (1996). Historic and future climatic change in Zimbabwe. *Climate Research, 6*, 137–145.

Unganai, L. S., & Murwira, A. (2010). Challenges and opportunities for climate change adaptation among smallholder farmers in southeast Zimbabwe. In: 2nd International Conference: Climate, Sustainability and Development in Semi-arid Regions. August 16–20, 2010, Fortaleza - Ceará, Brazil.

United Nations (UN). (2015). *Sendai framework for disaster risk reduction 2015–2030*. Geneva: UN.

United Nations Environment Programme (UNEP). (2013). *The emissions gap report 2013*. Nairobi: UNEP.

Wilbanks, T. J. (2005). Issues in developing a capacity for integrated analysis of mitigation and adaptation. *Environmental Science & Policy, 8*, 541–547.

WMO. (2014). *Record greenhouse gas levels impact atmosphere and oceans*. Press Release No. 1002. World Meteorological Organisation (WMO), online, Available: www.wmo.int/pages/mediacentre/press_releases/pr_1002_en.html. Accessed March 10, 2016.

World Bank. (2013). Turn Down the Heat: Why a 4°C Warmer World Must be Avoided, Climate Extremes, Regional Impacts, and the Case for Resilience, International Bank for Reconstruction and Development/The World Bank, Washington, DC.

ZIMSTAT. (2013). *Census 2012*: *National Report*. Zimbabwe National Statistical Agency (ZIMSTAT), Harare.

Chapter 7
Pastoralists Shifting Strategies and Perceptions of Risk: Post-crisis Recovery in Damergou, Niger

Karen Marie Greenough

Introduction

> It wasn't a little storm! It rained from three p.m. to five a.m. Cold! The cold catches a starving cow, she falls, her neck bends backward on her shoulder. Then you see water pours from her nose and she's dead. That night, between cows and calves, sixty died by morning. Just a few were left.[1]

Pastoralism in the Sahel depends upon variable, unpredictable, and patchy rain in one short season for vegetation and water throughout the year. Recurring ecological hazards—droughts, dry spells, flash floods and pests—are the norm, but climate change in West Africa is predicted to deliver even more variable rains with increased frequency and intensity of extreme events (Carabine and Lemma 2014; Sylla et al. 2016). Pastoralists' livelihood strategies, designed to manage variability and unpredictability, should equip them to manage climate change, however their adaptation to climate change is constrained by environmental, political and socioeconomic factors (Batterbury and Warren 2001). Much ongoing government policy, and ecological and socioeconomic changes constrict pastoral risk management strategies, constraining adaptive capacities and pushing households toward intolerable risk and adaptation limits. Opportunities to increase adaptive capacity are few and contain risky trade-offs.

This ethnographic study uses the Risk-based Framework of Klein et al. (2014: 905) to compare a pastoralist community's recovery from the 1984 drought-famine to their recovery from the double crisis of 2009–2010. The analysis gives insights into how pastoralist households' tolerance for certain risks has shifted over the years, allowing them to adapt management and coping strategies and embrace

[1]Boukari, a key informant interviewed in December 2014, male, about 55 years old (m, ~55) in 2014. All names have been changed.

K. M. Greenough (✉)
WASCAL, 06 BP 9507 Ouaga 06, Ouagadougou, Burkina Faso
e-mail: greenoughkm@gmail.com

W. Leal Filho and J. Nalau (eds.), *Limits to Climate Change Adaptation*,
Climate Change Management, https://doi.org/10.1007/978-3-319-64599-5_7

opportunities presented by socio-economic changes. Pastoralists in Niger accept more and more the sale of livestock, once an undesirable if not intolerable risk, and in this way expand the adaptation capacities of those who can afford this risk. Men also increasingly accept work outside of pastoralism, again with trade-off risks. Women, however have had much less opportunity to develop adaptation capacity. I focus on four livelihood strategies—livestock loans, cultivation, wage labor, including women's work, and buying fodder—to show how some strategies shift from coping toward management, entailing changes in perceptions of risk, as pastoralists respond to opportunities to expand adaptive capacities for recovery from extreme events. After detailing my methods, I discuss the Risk-based Framework in relation to pastoralist livelihood strategies. Then I detail changes that community individuals have perceived and that I observed between post-1984 and today, discussing how changes in strategies incur their own risks, managed by some households more easily than others.

Methodology

This chapter's information comes from on-going research with a community of Woɗaaɓe Fulɓe,[2] based in Tanout département (Damergou), Niger, with whom I have lived and worked since the mid-1990s. I began living in Tanout in 1985, when consequences of the 1984–1985 drought-famine were still evident. Then and early in the 1990s, I became acquainted with Woɗaaɓe families living in Tanout and Zinder towns. Throughout the latter half of the 1990s, I spent months at a time with the research community on the rangeland, then returned several times in the last fifteen years. They participated in both my masters (2002) and doctoral (2006–2007) thesis research. Over these decades, I've experienced with them both good and bad years, and heard many accounts of families' recoveries post-1984. In August 2012, when on vacation I visited for about a week with several households, and asked women and men about the 2009–2010 crises while both were still fresh in their minds. I recorded these conversations in my field notes (Emerson et al. 1995).

In December 2014, I met many more community families out in the northern rangeland, waiting for fields to open post-harvest. I lived with them for several weeks, following their annual winter migration into the cultivation zone. In addition to participation-observation and many conversations, I conducted sixteen interviews (Bernard 2002; Atkinson et al. 2003), mostly with people I knew well, from households with which I had experienced much.[3] Among other questions, I asked men and women, young and old, how many livestock they had lost during drought and storm, and whether and how many they had been able to add to their personal

[2]Fulɓe: Fulani or Peul; an ethnicity following varied livelihoods present throughout West Africa and beyond. The Woɗaaɓe are primarily nomadic, exclusive pastoralists.

[3]The few whom I do not know well were less comfortable and more reticent.

herds in the four and a half years after the storm. I recorded elder men's migration accounts, from early 2009 until late 2014, which included times and reasons why they sold livestock and what they purchased. Men and women discussed my questions with their older children, who helped them recall and estimate the numbers of cattle sold, abandoned, or killed during the drought and storm; everyone remembered clearly the few that survived the storm. I interviewed elder men and women about their experiences of 1984, and the differences they saw between 1984 and 2009–2010. Interview transcripts and fieldnotes were coded with MaxQDA for analysis. Though dealing with a small sample, the analysis in the research rests on long experience with this community, and in the Damergou rangeland, villages, and markets.

Pastoralism Within the Risk-Based Framework

In Chap. 16 of *Contribution of Working Group II to the Fifth Assessment Report of the IPCC,* Klein et al. (2014) offer a framework for analyzing a society's opportunities, constraints and limits to climate change adaptation. They define opportunities and constraints as factors that make it easier or more difficult, respectively, to plan and implement actions for adaptation to climate change. Actors reach limits when they cannot achieve their adaptation objectives in the face of intolerable risks. Pastoralists continually adapt to the risks and hazards of variable climate of arid and semi-arid lands, plus often unpredictable economic and policy environments. Although they do not consciously plan adaptation to climate change, per se, they know they must augment their adaptive capabilities to manage increased frequency and intensity of climate hazards in the face of constraints to their objective of maintaining livestock and livelihoods for the well-being of their households and communities.

Besides extreme climatic events, adaptation constraints include ecological degradation, both natural- and human-caused; demographic increases from births in both pastoralist and cultivator populations, and an influx of pastoralist refugees from Nigeria; intensified competition over natural resources; and rising grain prices. Demographic growth in cultivator populations drives field expansion, blocking seasonal migrations and access to resources, and rising grain prices and a perception of falling terms of exchange between pastoral products and grain calls food security into question. Pastoralists have few opportunities afforded them from outside their own socio-ecological systems and livelihood strategies. They and governments have long viewed each other with suspicion,[4] and both aid and development projects come to them rarely, the latter with very low success rates (Davies 2008; Nori et al. 2008). Damergou pastoralists have been able to engage two opportunities, however: wage labor, especially in Nigeria, and a nascent fodder market.

[4]Though some success has come through passing the Pastoral Code in 2010.

Household, Intolerable Risks, and Limits

Pastoral livelihoods are undertaken as household enterprises (de Bruijn 1995; Turner 2000; Greenough 2012), differing in resilience depending both on wealth (livestock) and the strategies that households are willing and able to undertake. Households are comprised of various individuals, whose agencies and activities are gendered. Wodaaɓe women have long had the responsibility of providing not only milk, but grain through milk sales, to households (Dupire 1962). Men are responsible for the livestock on which the households depend. Gender segregation is not absolute, but husbands and wives depend on each other to complete the work necessary for maintaining household viability (Greenough 2012). Within households, individuals practice strategies that deal with continual challenges, attempting to avoid the intolerable risk of losing too many livestock. Livestock loss may transpire through sales, with more or less recompense depending on market prices, through straying or theft, or through death, with little recompense. Crises push pastoralists toward the adaptation limit of losing so many livestock that the household must leave their pastoralist livelihood temporarily or definitively.

An adaptation limit may be "soft" (Klein et al. 2014: 919): a pastoralist household may give up their livelihood temporarily, for a season or a few years, while they settle to cultivate or move to town for wage labor. The longer they remain away, the adaptation limit progressively hardens. The increased frequency and intensity of extreme climatic events has increased the probability of intolerable risks that will lead to adaptation limits. With the idea that "[o]pportunities, constraints, and limits to adaptation develop along a dynamic continuum," (Klein et al. 2014: 907), the authors suggest that perceptions shift; once intolerable risks might seem more manageable over time, thus increasing actors' adaptation capacities.

Livelihood Strategies

Sahelian pastoralists live constantly with tolerable or manageable risks, as well as opportunities, requiring continual strategic management. Livelihood strategies are divided into "managing" strategies and "coping" strategies; the latter are more drastic measures used to deal with crises and disasters. The two are not dichotomous, for example some coping strategies are extensions of management strategies (Bollig 2006), and in recent years strategies once reserved for coping now approach management status. This adaptation reveals an important quality of all strategies: opportunistic flexibility (Niamir-Fuller 1999a), if a strategy does not work in one way, perhaps it will in another, or another will in its place. Both management and coping strategies comprise pastoralists' adaptive capacity in the face of variable climate. During and following extreme events, that capacity is challenged and often reduced, as coping strategies engender their own risks.

Besides much discussed and extremely important mobility (Niamir-Fuller 1999b), management strategies include maintaining peak herd size, livestock loans, and marketing livestock and their products. For resilient livelihoods, pastoralists try to build and keep their herds at peak levels (Roe et al. 1998; Krätli and Schareika 2010) The larger a herd a household maintains, the more livestock they will save through a crisis, and the sooner they will rebuild their herd, evident when examining the herd reconstitution of households of different wealth-statuses. Peak herd size is a critical objective of other livelihood strategies. In the most prevalent livestock loans, *haɓɓanaayi* (Dupire 1962; Bonfiglioli 1988), recipients borrow a young female, keep the young from one to three births, and benefit from the milk if a ruminant. Besides helping kin and neighbors, and solidifying social networks, the loaned animals are kept out of the market. Livestock are sold to purchase grain and other foodstuffs, clothing, household furniture and gear, other livestock, and to finance travel for wage labor migration. Women process milk from household and loaned ruminants, and sell or exchange surplus cultured milk, butter, and sometimes cheese. Though not a dairy product, repairing calabashes and gourds, important for milk collection and storage, has become a specialty of Woɗaaɓe women, and they repair calabashes for village women in exchange for grain, other foodstuffs, or cash.

Coping strategies often entail engaging tolerable risk to prevent intolerable risk. These include trekking with herds to new areas, diversifying livelihoods to cultivate or engage in wage labor, often migrating to cities, and selling livestock to buy fodder to feed the remaining livestock. When women do not have enough milk to sell, besides repairing calabashes, they pound grain to flour in villages and towns. As coping strategies become more long term, an individual is obliged to decide if the trade-off risks are tolerable or manageable. Can a household risk selling more livestock to buy fodder? Or is the risk of not buying fodder greater? When men leave their household livestock, they need to decide whether they can risk relying on another's herding skills. Men risk their own and their families' lives in Nigeria, and when young men leave the rangeland their continual accumulation of ecosystem and livestock knowledge is disrupted. Men who hire out for herding or field labor jeopardize their own self-esteem and social capital.

A Disaster and Two Crises

Disaster is extreme, defined by human mortality (Nelson and Finan 2009) or livelihood loss (Birkmann 2006). In the Sahel-wide 1984 drought, a rainy season with nearly no rain, livestock and people died from starvation and disease. Households fled to refugee camps, and lost their livelihoods temporarily or definitively. In subsequent years, Nigerien pastoralists managed drought years with migration and, especially since 2004, sales of livestock, including cattle, to buy fodder for their herds. In 2004–2005, the community trekked far south and bought anything made of stalks and straw, including empty granaries and old huts, to feed their herds through the dry season. In 2009–2010, no community members lost their lives from drought or inundating storm, and no livelihoods have yet been irrevocably lost, yet they experienced a double ecological crisis: two extreme events that have seriously challenged livelihood viability, and increased households' vulnerability to persistent ecological and socioeconomic hazards that degrade resilience and threaten full recovery to sustainable well-being.

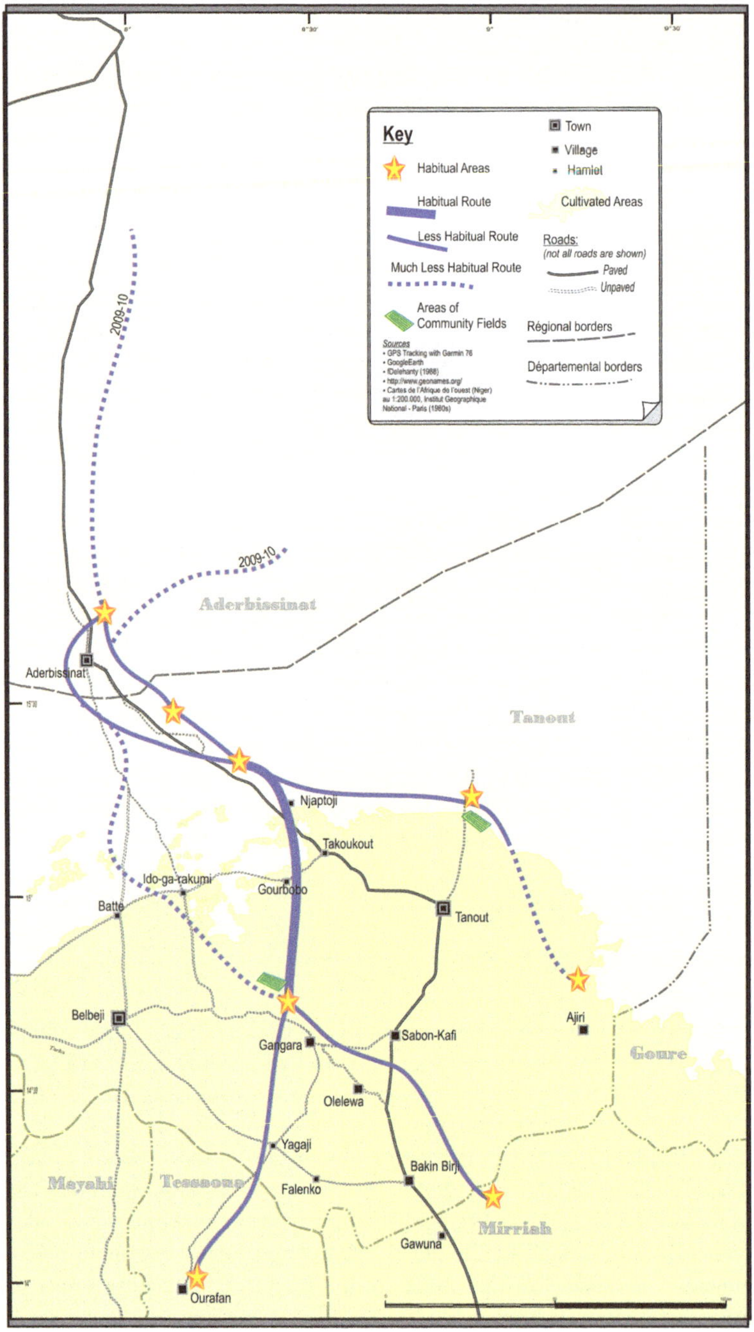
Key
Habitual Areas
Habitual Route
Less Habitual Route
Much Less Habitual Route
Areas of
Community Fields
Sources
• GPS Tracking with Garmin 76
• GoogleEarth
• fDelehanty (1988)
• http://www.geonames.org/
• Cartes de l'Afrique de l'ouest (Niger)
au 1:200.000, Institut Geographique
National - Paris (1980s)
Town
Village
Hamlet
Cultivated Areas
Roads:
(not all roads are shown)
Paved
Unpaved
Régional borders
Départemental borders
2009-10
2009-10
Aderbissinat
Aderbissinat
Tanout
Njaptoji
Takoukout
Ido-ga-rakumi
Gourbobo
Batte
Tanout
Ajiri
Goure
Belbeji
Gangara
Sabon-Kafi
Olelewa
Yagaji
Bakin Birji
Mayahi
Tessaoua
Falenko
Mirriah
Gawuna
Ourafan

In 2009, little rain fell on Niger, and pastoralists went looking for pasture to save their livestock. Some trekked south into conflict-ridden Nigeria; most, including the research community, trekked far north where they heard some grass had grown. Many communities from the east, south, and even Nigeria converged on the only grass available: tough, perennial *afasofi* (an *Andropogon* species), normally ignored by all livestock, except when very young and green. Contributing to the nascent fodder market, private merchants trucked dried grass, grain bran, cotton seed, and wheat chaff from Nigeria to Aderbissinat's market. Every 4 or 5 days, community men drove livestock, 3–4 days on foot, to sell in Aderbissinat, then paid trucks to haul purchased fodder to waiting donkeys that the men loaded for transport back to camp. "At least the market for livestock wasn't bad," Bukari told me; in other crises, livestock has sold for as little as 5–10% of normal prices. The men still sold much of their household herds, but could not provide the rest with enough fodder; they grew very weak from malnutrition. Men and women told me that the sharp afasoji stems pierced the cattle's stomachs from the inside, further degrading their health. Besides hunger, people and livestock suffered from intense heat and wind.

After a few scattered rainstorms that signaled the coming end of the dry season, the men heard at market that grass was growing not far to the south. They began to slowly move their households and livestock, abandoning the weakest animals. One afternoon, not far from their goal, a windstorm forced them to halt. Heavy rain began in the late afternoon and continued throughout the night. The weakened animals dropped into muddy water and most died from cold. Out of over 150 head in his household herd before the drought, Gojé (m, ~54) saved eight heifers. Matka (f, ~48) told me how she and her children, huddling under plastic, listened to their cattle splashing down into the water. Nearly all died. Her husband, returning from market in the morning, was stunned at their loss. People searched for the few living animals among the corpses, and hand-fed them grain. Some lived; others they sold cheaply before they died to butchers who came from Aderbissinat. The 2009–2010 drought took somewhat more toll in livestock sales and deaths than 2004–2005, but with the storm in 2010, the toll of the two crises combined caused almost as much livestock loss as 1984. "We still had livestock left," Gojé told me in answer to my question about the difference between the two crises. Several men told me they could have saved most of the livestock, if they had reached green grass before the storm.

Strategies Post-1984

When the rainy season failed in 1984, community households fled to two refugee camps. Maré (f, ~58) told me that those with no or very few remaining livestock, fled west to Tabalakh, near Tahoua. There they stayed for four years, receiving aid for two years. They cultivated dry season gardens and rainy season grain fields. Many also received a goat or sheep. She and her sister sold their nephew's surviving ram for traveling money, and with other women, made three trips west to Mali and Burkina Faso, where they sold herbal medicines. With the money they earned, they bought grain for their households and smallstock to restart their herds, including their nephew's. During the fourth rainy season, they headed back east with their small flocks. Maré's husband had left a loaned cow with his cousin; she had birthed two calves with which they restarted their herd. In Damergou, the men borrowed fields to cultivate. After about six years, their livestock multiplied enough that they abandoned cultivation.

Households with one or a few surviving cattle fled south to a refugee camp near Tessaoua, where they also received food aid and cultivated gardens and fields. They stayed for about a year, and then began making their way back north, cultivating in successively northern fields until they felt secure enough to return to exclusive pastoralism. With livestock born to their own surviving females and purchased livestock, some households could abandon cultivation after four years; others cultivated near Surutu into the mid-1990s, until they too returned to exclusive pastoralism. Both men and women emphasized the crucial role that wives played in post-1984 recoveries. Men also took the herds of urban and rural clients, in order to have enough milk for their families. Until herds were reconstituted enough that the women could sell milk, they went daily to villages to repair calabashes and pound grain in exchange for grain, cash, other foodstuffs, and clothes. The grain they brought home supplemented the men's small harvests. They also collected wild greens and other vegetables to cook for themselves and their children.

During the years immediately following 1984–1985, Bukari, his brother, and their cousins—the oldest sons in their households—traveled twice with their wives to Nigeria, Cameroon and Central African Republic, where they worked as guards and, with their wives, sold herbal medicines. They sent and then brought money home to their fathers to buy grain, smallstock and donkeys. Later, like other young households with too few livestock, their younger brothers moved to Zinder, where they lived into the 1990s, working as security guards and selling tea. They finally returned to the rangeland, but not permanently. During a bad year, the younger brothers may return, now to Kano, Nigeria, for a season or more.

Strategies Post-2010

The 2010 rainy season was very good; Boukari told me that the same storm that killed livestock also saved the surviving animals as good pasture grew from the rain. In subsequent years, especially in 2014, men bought fodder when pasture grass became scarce or lacked nutrition. Gojé credits the loans he gave before the crises with rebuilding his herd; people returned mother animals to him, often with young females as thank-you loans. These are now giving birth and re-establishing his herds. Some people told me loaned cows had been left with them past the births of their agreement, so that they could further benefit from the milk. A few households resorted off and on to cultivation after 2004, then several began cultivating in 2011, and all these households continued through 2014, harvesting at least a few months-worth of grain every year. Most cultivating households still migrate north with their livestock after planting; others have nearly settled in rangeland north of their fields on the cultivation zone border, leaving most of their livestock with brothers or other kin.

Instead of cultivation, some men resort to wage labor, either migrating to Nigerian cities (mostly Kano), working local harvests, or herding for villagers. Young men who had not already been traveling to work in Nigeria began in 2010 or subsequent years. Agola (m, 34), a generation younger than Gojé, told me that his three trips to work in Kano saved his herd, which he left with his uncle who managed them carefully. Though he sold livestock to finance his journeys, he was able to find good work and return with money, including cash to buy fodder in 2014. The few cows he had left after the storm have born calves. "If it wasn't for Nigeria," he told me, "some people wouldn't be here now. Just buying food would have finished their livestock. Now, if I'm careful, and I return to Nigeria a bit, in two years, I'll pull out of this." His two wives do not work in Kano, however, due to the insecurity. Matka's eldest sons made their first trip to Kano in 2014, but like other young men brought nothing back but a few gifts.

As after 1984, a few men took herding jobs in villages, while others took on field labor—a practice I had not heard of before among Woɗaaɓe. Women went to villages and towns to look for work repairing calabashes or pounding grain. Maama (f, 58) traveled twice to Tanout town in 2010, once with her eldest daughter, to work for a total of two and a half months. She was given grain and other foodstuffs for repairing calabashes. Later women were able to sell milk as well, but young women with no milk in 2012 looked for work pounding grain. As soon as the households entered the cultivation zone in 2014–2015, women began selling milk and looking for calabash repair work, though they told me how difficult this has become.

Perceptions and Observations of Change

In the last thirty to fifty years, households continue to rely on some management and coping strategies, like livestock loans, that have changed little. Except for post-1984 when, elders told me, people had no animals to give, loans have always acted as both manangement and coping strategies. Cultivation, a coping strategy when too many livestock are lost, has also changed little, except to become a more regular practice for many households. Most pastoralists in less arid climates, with less open rangeland, have adopted cultivation permanently (Moritz et al. 2009); however in the Sahel many pastoralists have been able work their way out of cultivation and other non-pastoral income-generating activities to return to exclusive pastoralism (Bonfiglioli 1988, 1990). Now, as many households become less able to return completely to exclusive pastoralism, the coping strategy shifts toward management, while assuming other risks. Men who still migrate with their herds hire villagers to weed their fields, risking careless weeding and a bad harvest. They also curtail their migration routes to stay closer to their fields, compromising the health and fertility of their livestock.

Ecological and socio-economic changes have both permitted and forced adaptations to other livelihood strategies. Bukari explained how the nutritive value of northern pastures has much diminished. On a grass-covered hill, which Bukari and others describe as once full of trees, he broke some grass stems to show they were empty of pith, which tells him they have little nutritive value. The lack of nutrition in northern pastures, even with a good rainy season, reduces cows' fertility and milk production. Degradation in the northern rangeland, due to lower rainfall and removal of trees and wood, increases the importance of southern pastures and harvest remains for livestock, and markets for grain purchases.

Selling cattle to buy grain and other foodstuffs, a strategy reserved for the late dry season during the 1950–1960s (Dupire 1962), when rainfall was generally better than the century norm and pastoralists had more milk, has increased ever since the 1970s and 80s droughts (Swift 1977), as rainfall has diminished (Hulme 2001; Herceg et al. 2007). Such sales had become common throughout the year by the 1990s; even selling "capital"—a cow with her calf (Swift 1977: 472)—is not uncommon when a household head without other high-priced livestock needs a large amount of cash. Now, pastoralist men also face a nearly annual decision whether to sell livestock, including cattle, to buy fodder, moving this relatively new coping strategy toward the management side.

In southern pastures and fields, cultivators have always raked up some grass and harvest remains to store for their own livestock. Over the last decade, however, they increasingly store more to sell to pastoralists once pastures are grazed over. There are two sides to this fodder situation. On the one hand, by buying fodder in 2004–2005 and 2009–2010, men could keep livestock alive in a way impossible in 1984. On the other hand, cultivators now expect pastoralists to buy fodder from them, and pastoralists worry that fields and pastures will be raked over before they arrive. Each winter, households race south with their herds—sometimes nearly forcing authorities to open fields, so their livestock can graze the remains before cultivators can rake them clean. Cultivators, meanwhile, with a new sense of tenure for anything in and surrounding their fields, complain that pastoralists eat their harvests, a competition that degrades relations that have never been good.

Selling livestock for grain plus fodder challenges the management strategy of maintaining peak livestock levels. If households cannot recover to optimum levels, the next crisis may hit them harder. They also will not have animals to loan and risk damaging social capital. In 2014–2015, Matka's household herd still consisted almost entirely of loaned animals, and Matka told me (confirmed in conversations with others) that her husband would sell young animals soon after their birth. The household migrates with other households far south in the dry seasons where grain is cheaper and there is no need to buy fodder—a move that pastoralists with more cattle view as risky—but their household is still very vulnerable. With no animals of their own, Matka's sons may end up dropping out of pastoralism as her brothers did after 1984.

As other strategies have shifted, women's work has simply become less remunerative. All women remarked on the decline in demand for their work, even their dairy products. Some blamed the increase of mechanical mills for the lack of pounding work, and the proliferation of plastic buckets and tubs for the decline in calabashes, and thus need for their repair. Degraded relationships between cultivators and pastoralists also affect women; villagers refused both their work and milk, simply because they are pastoralists. Lime, a widow (~55), told me that after 1984 she would fill a large calabash bowl with grain in exchange for work or milk, and bring it home to feed her household. Now villagers give only a little grain, if anything. Women most protested that instead of receiving grain for work or milk, they now receive only a few small coins. They perceive an increase in grain prices over the years that has outpaced the price of milk, and instead of taking home grain that they can use right away, must collect and save small amounts of money to purchase grain later.

The change that men most cited was the unfavorable exchange rate between livestock and grain. Both grain and livestock prices have risen over the years, but grain seems to have become more expensive. The devaluation of the franc CFA in 1992 and the construction of new marketplaces in Nigeria have increased livestock exports and raised their prices (Greenough 2010). The price of grain was held down for some years, and then allowed to rise with the regional market. Gojé remarked, "These years are completely upside down! They're frightening, because money isn't worth anything." Men are selling more livestock for more reasons, however, when fodder purchases are considered.

Though always an option for those who would or could not lead a pastoralist life, wage labor migration to cities has become more and more common since the 1974 drought (Bonfiglioli 1985, 1973 in western Niger), and has now become almost a rite of passage for most young men. They follow more experienced relatives to cities for a season, a year, or more, taking their wives and children once married. A few men are able to send remittances or buy livestock; most simply maintain themselves and their small households. Their absence, however, can allow their herds to multiply if left in good care.

Middle-aged heads of households worked harvests and herded village livestock near the villages, the latter strategy not practiced since post-1984. In 2014–2015, I observed a trend I had never seen before,[5] a marked absence of elder men managing household herds, and many women, young and old, left as de facto heads of households, doubling their work burden (Greenough 2012). When both husbands or widows' eldest sons work in fields or villages and younger sons leave for Nigeria, elderly mothers and grandmothers, or young women are left to manage households and herds with young boys. The cash or in kind income the men receive keeps them from having to sell livestock for grain, but livestock care may be jeopardized. Both men and women in these households seemed more distressed. Limé (f, ~55), told me with frustration how her son and his wife, recently returned from Nigeria, had brought nothing back. She described the desperation that divided the households: "Have you ever seen an elder man leave his household if it wasn't life and death necessity? When a woman is separated from her husband, she has no means."

[5]It was difficult to determine whether this had happened post-1984.

With higher livestock prices, community members, especially women, have not been able to purchase smallstock as they did post-1984. Only two people in the community were reported to have purchased livestock without selling an animal they already owned. Maama's son-in-law bought a donkey with money he earned in Nigeria, and his brother sent money to buy two sheep. This is very little when compared to reports of recoveries post-1984.

Conclusion

Researchers and especially policy makers have long predicted the collapse of mobile, exclusive pastoralism, and yet it has persisted with a resiliency surprising to some. Pastoralists adapt by relying on coping strategies more as management strategies, sometimes in response to new opportunities. In the cases described here, they have taken advantage of a demand for wage labor in Nigerian cities and, more recently, a growing market in fodder. In turning coping strategies into longer term management strategies, households take on risks perceived much less tolerable in times with fewer hazards, including following riskier migration routes, selling livestock year-round, and separation from herds, households, communities, and the rangeland. Households such as Matka's have persisted through hard times before, and may yet again. Like others, however, especially as they take on new risks, they are more vulnerable to hazards and constraints that reduce their adaptation capacity, especially to extreme events. Gender plays an important role, and with increasingly limited resource entitlements and adaptation choices (Klein 2014: 902), women have a harder time contributing to households' recoveries. This leaves many of them feeling powerless and more desperate, even as some of them take on double burdens of managing both households and herds.

Poorest households have always dropped out of pastoralism (Anderson and Broch-Due 1999), while those with more livestock, better able to manage risk, persist. Crucial variables remain, however. If the northern rangeland continues to degrade into desertification, pushing pastoralists to greater reliance on the cultivation zone and purchased fodder, the economic pressure of purchasing both grain and fodder and conflict over natural resources may drive more households past the threshold of viable pastoralism. If women are not able to build adaptation capacity and modify their income-generating activities, the household imbalance also threatens livelihood viability. Both Klein et al. (2014) and Batterbury and Warren emphasize the role of governance to provide "an 'enabling environment' to support … local responses" (2001: 3). Though some national governments like Niger and regional organisms such as ECOWAS have recognized the importance of pastoral mobility and pastoralism's significance to national economies, they have done little to address either the pastoral system as a whole, or the social-ecological system within which it operates. Without a systemic, holistic response, the escalation of hazards and crises, induced by climate and other changes, increases vulnerability and threatens even the strongest pastoral households' recoveries.

Acknowledgements The 2014–2015 research was supported by the West African Science Service Center for Climate Change and Adapted Land Use (WASCAL), and the German Ministry of Education.

References

Anderson, D. M., & Broch-Due, V. (Eds.). (1999). *The poor are not us: Poverty & pastoralism in Eastern Africa*. Oxford: J. Curry.

Atkinson, P., Coffey, A., & Delamont, S. (2003). *Key themes in qualitative research: Continuities and changes*. Walnut Creek, CA: AltaMira Press.

Batterbury, S., & Warren, A. (2001). The African Sahel 25 years after the great drought: Assessing progress and moving towards new agendas and approaches. *Global Environmental Change, 11* (1), 1–8.

Bernard, H. R. (2002). *Research methods in anthropology: Qualitative and quantitative approaches*. Walnut Creek, CA: AltaMira Press.

Birkmann, J. (2006). Measuring vulnerability to promote disaster-resilient societies: Conceptual frameworks and definitions. In J. Birkmann (Ed.), *Measuring vulnerability to natural hazards: Towards disaster resilient societies* (pp. 9–54). Tokyo: United Nations University Press.

Bollig, M. (2006). *Risk management in a hazardous environment: A comparative study of two pastoral societies*. New York: Springer.

Bonfiglioli, A. M. (1985). Évolution De La Propriété Animale Chez Les Wodaabe Du Niger. *Journal des africanistes, 55*(1–2), 29–37.

Bonfiglioli, A. M. (1988). *Dudal: Histoire De Famille Et Histoire De Troupeau Chez Un Groupe De Wodaabe Du Niger*. Cambridge: Cambridge University Press.

Bonfiglioli, A. M. (1990). Pastoralisme, Agro-Pastoralisme Et Retour: Itinéraires Sahéliens. *Cahiers des Sciences Humaines, 26*(1–2), 255–266.

Carabine, E., & Lemma, A. (2014). *The IPCC's fifth assessment report: What's in it for Africa?*. London: Overseas Development Institute.

Davies, J. (2008). Turning the tide: Enabling sustainable development for Africa's mobile pastoralists. *Natural Resources Forum, 32*(3), 175–184.

de Bruijn, M. (1995). A pastoral women's economy in crisis: The Fulbe in central Mali. *Nomadic Peoples, 36*(37), 85–104.

Dupire, M. (1962). *Peuls Nomades: Étude Descriptive Des Wodaabe Du Sahel Nigérien*. Paris: L'Institut d'Ethnologie, Musée de l'Homme.

Emerson, R. M., Fretz, R. I., & Shaw, L. L. (1995). *Writing Ethnographic Fieldnotes*. Chicago: The University of Chicago Press.

Greenough, K. M. (2010). Smallstock as cash crop, smallstock as Habbanayi: Fulbe exchanges in the 21st century. *Ethnology, 49*(2), 129–148.

Greenough, K. M. (2012). Women, men children & livestock: Partnerships and gendered negotiations in the Ful'be household livestock enterprise. *Gendered Perspectives on International Development*, v. 301.

Herceg, D., Sobel, A. H., & Sun, L. (2007). Regional modeling of decadal rainfall variability over the Sahel. *Climate Dynamics, 29*(1), 89–99.

Hulme, M. (2001). Climatic perspectives on Sahelian desiccation: 1973–1998. *Global Environmental Change, 11*(1), 19–29.

Klein, R. J. T., Midgley, G. F., Preston, B. L. et al. (2014). Adaptation opportunities, constraints, and limits. In C. B. Field, V. R. Barros, D. J. Dokken, et al. (Eds.), *Climate change 2014: Impacts, adaptation, and vulnerability. Part A: Global and sectoral aspects. Contribution of Working Group II to the Fifth Assessment Report of the Intergovernmental Panel on Climate Change* (pp. 899–943). Cambridge: Cambridge University Press.

Krätli, S., & Schareika, N. (2010). Living off uncertainty: The intelligent animal production of dryland pastoralists. *European Journal of Development Research, 22*(5), 605–622.

Moritz, M., Kyle, B. R., Nolan, K. C., et al. (2009). Too many people and too few livestock in West Africa? An evaluation of Sandford's thesis. *Journal of Development Studies, 45*(7), 1113–1133.

Nelson, D. R., & Finan, T. J. (2009). Weak winters: Dynamic decision-making in the face of extended drought in Ceará, Northeast Brazil. In E. C. Jones & A. D. Murphy (Eds.), *The political economy of hazards and disasters* (pp. 107–132). Lanham, MD: AltaMira Press.

Niamir-Fuller, M. (1999a). Introduction. In M. Niamir-Fuller (Ed.), *Managing mobility in African rangelands: The legitimization of transhumance* (pp. 1–9). London: Intermediate Technology Publications.

Niamir-Fuller, M. (Ed.). (1999b). *Managing mobility in African rangelands: The legitimization of transhumance*. London: Intermediate Technology Publications.

Nori, M., Taylor, M., & Sensi, A. (2008). *Browsing on fences: Pastoral land rights, livelihoods and adaptation to climate change. Drylands Issue Paper*. London: International Institute for Environment and Development.

Roe, E., Huntsinger, L., & Labnow, K. (1998). High-reliability pastoralism versus risk-averse pastoralism. *Journal of Environment & Development, 7*(4), 387–421.

Swift, J. (1977). Sahelian pastoralists: Underdevelopment, desertification, and famine. *Annual Review of Anthropology, 6,* 457–478.

Sylla, M. B., Nikiema, P. M., Gibba, P., et al. (2016). Climate change over West Africa: Recent trends and future projections. In J. A. Yaro & J. Hesselberg (Eds.), *Adaptation to climate change and variability in rural West Africa* (pp. 25–40). Switzerland: Springer International Publishing.

Turner, M. D. (2000). Drought, domestic budgeting, and changing wealth distribution within sahelian households. *Development and Change, 31*(5), 1009–1035.

Chapter 8
Political Limits to Climate Change Adaptation Practices: Insights from the Johannesburg Case

Karen Hetz

Introduction

In many cities in the global South, there is a dangerous convergence of urban development patterns and the effects of climate change, resulting in considerable exposure of urban population and businesses to already alarming and yet increasing environmental risks levels (UN-Habitat 2011). At the same time, substantial development challenges pre-exist in metropolitan cities in the global South. Political and societal support is required to effect sustained response to climate risks. In the global South, where developmental challenges remain considerable, this support is unlikely to be forthcoming, if developmental needs are not addressed together with climate change response (cp. Ziervogel et al. 2010). Also, we have insufficient insights into the extent to which adaptive practices are politically feasible or attractive, even though there may be possibilities to reconcile them with prioritised strategies towards major development constraints. During urban planning processes different development demands are negotiated. Planning is a political process that is required to correspond to societal conditions and needs to gain political legitimisation for planning policy and practices—including those in the field of climate change adaptation—in reference to the societal context and related development demands. Besides, planning is a potentially important means to address risks of climate change in cities as it has a catalyst function for the shaping

K. Hetz (✉)
Geography Institute, Humboldt Universität zu Berlin, Berlin, Germany
e-mail: karen.hetz@gmx.de

K. Hetz
School of Architecture and Planning, University of the Witwatersrand, Johannesburg, South Africa

K. Hetz
Environmental and Regional Sciences Department, University Trier, Trier, Germany

W. Leal Filho and J. Nalau (eds.), *Limits to Climate Change Adaptation*, Climate Change Management, https://doi.org/10.1007/978-3-319-64599-5_8

of related environmental risks (Wilson and Piper 2010). It also offers a holistic, integrative, and potentially long-term approach to urban development (Hartmann and Driessen 2013; Hutter 2007), which is required, if cities are to adapt pro-actively to the anticipated effects of climate change (Bicknell et al. 2009; Birkmann et al. 2012; Satterthwaite et al. 2009). Hence, local adaptation can be constrained perhaps considerably by political limits to realise adaptive options in urban planning.[1]

Despite a growing body of work that deals with adaptation barriers, also in the field of urban planning (e.g. Uittenbroek et al. 2013), it remains yet unclear to which extent adaptive planning options can be realised, considering the political nature of planning and the implied need to receive societal support for adaptive practices. Using Johannesburg, South Africa, as an illustrative case, the study investigates how adaptive practices are advanced or constrained by political realities of considerable development challenges in Johannesburg and it explores which implications these realities have for the possibility to realise urban adaptation practices. Providing empirical knowledge on the political limits of adaptive practices, the study aims to contribute to an emergent body of work that will progressively offer clearer insights into the extent to which adaptation practices are politically feasible, especially in cities with democratically constituted governments where negotiated prioritisation is required (Hetz 2015, 2016).

Framing Adaptation Limits

Despite—or perhaps rather due to—a growing body of literature on climate change adaptation challenges, no consistent definition of adaptation limits, barriers, or constraints is available. Often studies apply the different terms interchangeably, but there are also cases where distinctions are proposed for instance concerning the quality or durability of particular adaptation obstacles (e.g. Adger et al. 2009; Moser and Ekstrom 2010; Biesbroek et al. 2013; Eisenack et al. 2014; Fankhauser and McDermott 2014).

According to Klein et al. (2014: 907) adaptation barriers “make it harder to plan and implement adaptation actions”, but they do not completely prevent adaptive practices. Adaptation limits, contrarily, occur when unacceptable adaptive measures are required or no adaptation options are available at all (ibid). However, the boundary between adaptation barriers and limits, as suggested above, is rather fluid and we lack clear indicators for the empirical verification of either of them. Hence both options are being referred to as adaptation limit further on. Klein et al. (2014) further distinguish between hard and soft adaptation limits. The former occur when no acceptable adaptive practices are currently accessible and foreseeable, even

[1]In this study, adaptive planning options are policies or practices which have presumed adaptation benefits for the built urban environment.

when looking beyond the current planning horizon, while the latter occur temporarily and may be overcome (e.g. due to improved adaptation knowledge, changing risk perceptions, technological innovations, etc.) in the future (cf. Adger et al. 2009; Moser and Ekstrom 2010).

Origins of Adaptation Limits: A Focus on Societal Contexts

A growing number of studies on adaptation limits emphasise governance issues as the factor that hinders or complicates the adoption of adaptive options. These include, for instance, issues such as leadership and power, the lack of sufficient resources and capacities, and insufficient knowledge or risk awareness (Adger et al. 2009; Allen and Gunderson 2011; Tompkins and Adger 2003; Moser and Ekstrom 2010; Pasquini et al. 2013). Despite the availability of such studies, we do not have sufficient knowledge about the factors which make adaptive practices politically possible—or which constrain them. This may be due to the fact that in order to understand political barriers better, a close look at the societal context in which adaptation takes place is required. Despite a few exceptions (Dupuis and Knoepfel 2011), empirical research has marginalised this aspect.

What constitutes this societal context though? Two key components should be considered at least: First, collectively shared or strong value systems of a society, such as religious beliefs, historically derived norms or established cultural practices, can influence on the acceptance of new adaptive practices (Jones and Boyd 2011; Nielsen and Reenberg 2010; Garschagen 2011; Moser and Ekstrom 2010; Adger et al. 2009). Klein et al. 2014 remind us though that such cultural barriers are not well researched yet. Second, the developmental context is potentially important. In particular in the global south, not exclusively though, considerable pre-existing development challenges occur. Addressing them may be perceived to be more urgent by those affected, for instance poor living conditions, than responding to comparably distant issues of climate change, particularly if the related adaptive responses concern risk levels that are projected to be reached only in the future. In such situations, societal support for the adoption of adaptive options may be limited. This may particularly be true, if adaptive practices conflict with responses that address urgent development needs or imply inconvenient trade-offs, hence add yet another burden on society. In this paper, I focus on the developmental context of a society.

Development constraints have been taken into consideration as origin factors of adaptation barriers in a number of studies before (Pelling 2011; Moser and Ekstrom 2010). However, most of these studies have only considered the developmental context in terms of its implication for adaptive capacities, which are argued to be required for the design, mainstreaming, and implementation of adaptation strategies. Hence, the implication of the developmental context for the political possibility (or attractiveness) of adaptation practices requires more attention in current research.

Revi et al. (2014) suggest that theoretically a number of opportunities exist to realise synergistic options that address development constrains and adaptation needs simultaneously (see also Klein et al. 2005; Tompkins and Adger 2003; Schipper 2007; Huq and Reid 2004; Picketts et al. 2014; Sharma and Tomar 2010; Roberts 2010). Applying synergistic responses can be politically attractive, because they open up opportunities to implement adaptive practices under the umbrella of particular developmental strategies. No additional argumentation for the support of adaptive practices may be required in such cases. Although the conceptual argumentation to support this proposition is convincing, the empirical evidence for the existence of sufficient synergy possibilities in practice is limited. Besides, we have insufficient insights into the extent to which adaptive practices are politically feasible or attractive. In particular, we lack case studies demonstrating how to reconcile climate adaptation with prioritised strategies towards major development constraints.

The Case Study: Urban Planning in Johannesburg

Johannesburg is an interesting case for studying political barriers to climate change adaptation for a number of reasons. It is prone to environmental risks of climate change, such as urban flooding caused by heavy precipitation events and there is a strong political commitment to fundamentally revise current planning policy and practices in order to promote a more resilient, sustainable, and liveable city in Johannesburg (CoJ 2011b; Tau 2013; Fowler 2013). Climate change resilience objectives have been mainstreamed into the strategic planning frameworks that guide day-to-day urban development decisions (CoJ 2011b, 2013). Already today, floods occur regularly during rainfall events and they are likely to further increase in frequency and magnitude due to climate change in the future. The continued loss of urban wetlands due to urban sprawl, as well as continued development in flood risk prone areas further contribute to raising flood risk levels in Johannesburg (Storie et al. 2012; JRA 2007).

However, in terms of climate change related environmental risks, Johannesburg does not represent an extreme case and the current and projected environmental risks do not over-dominate other development issues. Instead, Johannesburg arguably is an extreme case of a metropolitan city with high level of urban divide: it is characterised by a high degree of social polarisation and considerable socio-spatial disparities. Although the juxtaposition of multiple co-existing urban dynamics and the related development constraints is complex and to some extent location specific, there are similarities between Johannesburg and other metropolitan cities in the global South. As the UN-Habitat (2010) puts it,

> Cities as diverse as Nairobi, Buenos Aires, Johannesburg, Mexico City, and Rio de Janeiro are similar in that pockets of wealth and poverty co-exist in close proximity […]. Examples such as these highlight the large disparities between better-off minorities and the many poor, which are also reflected in different degrees of access to […] facilities, public goods, transportation and open space in most cities in the developing world (pp. 52–53).

Like in other cities in the global south, planners in Johannesburg face a situation in which pro-poor strategies and economic-growth approaches need to be mediated in order to meet the different development demands in the city. Climate change adaptation is yet another issued to accommodate. Insights gained in this study on the possibility to link climate change adaptation with responses to development issues of urban divide in Johannesburg can hence be inspiring for researchers and practitioners that work in similar contexts.

Methodology

This paper presents key results of a comprehensive study on climate change adaptation practices that was undertaken between 2012 and 2015. The study has an inductive-qualitative design that supports explorative research. A mixed method approach included expert interviews, a qualitative analysis of planning documents, and visits on development sites and risk areas in Johannesburg. Such a mixed method approach promotes complementary data collection and helps to ensure construct validity (Yin 2014). Expert interviews were conducted mainly to capture planners' subjective perception about limiting factors for adaptive practices. They also helped to complete information gained from the analysis of planning documents that reveal political adaptation objectives—both explicit and implicit—, their (rhetorical) linkage with sustainable urban development issues in Johannesburg, and their translation into programmes and projects (Hetz 2015).

In total, 67 planning actors were interviewed. The group of experts comprises governmental planning officials in the political and executive field that work either at the municipal, the provincial, or the national governmental tier—although the majority of interviewees are local government officials—as well as representatives from the private sector, which were involved in private urban development projects in Johannesburg or worked for local government as consultants. A number of strategies and assessments (including the vulnerability assessment of Johannesburg) were conducted by private sector actors on behalf of the government, for instance. Planning documents considered in the analysis included the development planning frameworks as well as budgeting documents, political speeches, and press statements. In addition, access to the city's GIS-based planning database was granted by the city. It provides information on spatial development areas, spatial programme framings, and approved land use change and development applications (Hetz 2015).

Flood risks, which are shaped both by urbanisation patterns and climate effects, were used as an illustrative example to study political barriers to the adoption of adaptive planning options in Johannesburg.

As I highlighted in previous work (Hetz 2015), the study cannot provide a generalised answer as to what the political response space for adaptive planning is likely to be. This is because the chosen case study approach considerably limits the possibility of making generalised statements based on the research findings (Flyvbjerg 2006): Different planning responses may be politically possible in different local context situations, as factors such as societal norms and culture—or development realities, in terms of which planning practices need to gain legitimisation—vary. Acknowledging that there are clearly limits to the case study approach in terms of the generalisability of findings, the study is intended to contribute to a emergent corpus of work which will offer progressively clearer insight into the extent to which adaptive planning practices are politically possible in cities with democratically constituted governments where negotiated prioritisation is required (Hetz 2015: 106).

Tracing Political Adaptation Barriers

The first subchapter presents examples of land use planning, to investigate whether and how synergies between development priorities and adaptive options are being realised and whether additional adaptive options can perhaps be politically legitimised and implemented. In this study, options are understood to be politically legitimate if planning actors perceive them to be feasible or attractive in view of expected societal reactions or demands. In the second subchapter, I take an example of a prioritised development programme—the provision of give-away houses for urban poor. Planners recognise that the current practice needs to change substantially in order to be more environmentally sustainable. Against this background, I examine the extent to which it is possible to adopt alternative adaptive options and dismiss established practices.

Land Use Planning

In many fields and sectors, technological innovations are required to advance the implementation of climate change adaptation objectives. An innovative private sector can provide such solutions (Pow and Neo 2013). I was curious whether the private sector offers housing solutions with adaptive benefits in the framework of urban development projects, which government would strategically use to realise adaptive land uses in Johannesburg. Two large-scale private urban development projects in Johannesburg, which are marketed with ecological reasoning, were looked at. It was presumed that due to the use of environmental arguments to advertise the projects, environmental issues were perhaps taken more into consideration than in conventional projects. Both projects comprise extensive housing developments for upmarket segments. In the past, residential developments, which addressed the same income-bracket as the illustrative projects, have been characterised by low-density housing designs, large land consumption and little consideration of ecological issues, such as wetland preservation.

Interestingly, the urban design of the two cases differs considerably. In the first case, Steyn City, an extensive naturally kept park landscape is integrated into the land use plan, covering wetland areas on site and providing comparably large buffers between them and the developed areas. In addition, medium-density housing solutions were introduced, which make up about 45% of the planned housing units in the examined project area (Hetz 2016: 10). Linking medium-density designs with the provision of extensive green open spaces is perhaps indeed an innovation in South Africa's housing market.[2] In the second example, Waterfall City, the conventional low-density design was reproduced; instead of providing new housing concepts, the developers serve an established market demand for free standing town houses. Also a park landscape is being advertised and has been provided. However, the green open space areas are substantially smaller than in the Steyn City case and are—despite a few exceptions—limited to riparian zones along the rivers, many of which fall into the 1:100 year flood line (Hetz 2016). Legally, no development should take place in these areas anyways. Most importantly though, there are indications that development areas overlap with areas that the city presumes to be wetlands (ibid.).

Government-internal documents that are contained in the projects' planning file revealed that in both projects no binding development conditions in terms of minimum density requirements had been imposed onto any of the two developers. Instead, the different outcomes in terms of medium-density housing designs as well as the provision of green open spaces, which can perform adaptive functions, is a result of different development motivations driving the two projects. In the case of Steyn City, the described outcome was reported to have been pushed by the investor for a number of investment related reasons. These include his believe in the demand for urban residences that include safe and secure, and well-kept garden landscapes on the fenced development premises in which residents can enjoy a number of outdoor activities, such as equestrian and golf. In a crime burdened city such as Johannesburg, this certainly appears to be an attractive product for high-income clients. In the case of Waterfall City, developers reported to have been confronted by government with a couple of demands to change their land use plans during the development application process in order to better accommodate green open spaces in the overall urban design. Various planners outlined that the developer used existing loopholes in the environmental legislation strategically though—with consent from provincial government—to get development permission for the project.

It also became apparent that both project sites overlap with a number of intact wetlands. In Waterfall City, a site is affected by development which overlaps with one of the last areas of pristine grassland that is endemic to the Gauteng province (CoJ 2007, 2011a). From an ecological point of view, development permission should not have been granted in both cases. The positive contribution of the undeveloped green open space areas to manage flood risks was recognised prior to the start of the projects and there is environmental legislation available that could have prevented the development of the sites (CoJ 2009, 2011a). Various municipal

[2]Detailed examples of the housing designs are published online (Steyn City 2016).

planners that work at the executive level criticised the outcome of the projects due to its environmental implications but mentioned that the projects had been pushed by higher government tiers and that there were no possibilities to further impose constricting development conditions.

The two cases indicate that at the city scale, developments continue to take place at the expense of the preservation of the few remaining wetland areas in Johannesburg. Governmental decision-makers continue to tolerate land use practices which are known to continue to further increase flood risks of climate change (e.g. by not establishing and enforcing minimum density standards). Granting permission to develop ecologically sensitive areas that perform valuable adaptive functions, they even actively support elements of these practices. An explanation can indeed be found in the political development realities. One has to widen the focus to infrastructure challenges in the city, for instance, in this regard.

Infrastructural systems are in concerning states in Johannesburg, in almost all fields (roads, sewer, water, stormwater), due to aging components, defects, or insufficient capacities. Already today, these shortcomings have started to jeopardise the achievement of economic growth and social distribution objectives in Johannesburg. Hence, "there has been a [political] recognition that we need to invest in our existing infrastructure systems, to replace them and not allow them to collapse and decay anymore" (planner A—interview 2013). The available municipal budget falls short to adequately respond to the related investment needs though. Against this background, planners interviewed illustrated how government is using the large scale projects strategically to fill investments gaps in their infrastructure budget.

Developers are obliged to pay an infrastructure contribution which is intended for the adjustment of infrastructure systems to additional capacity requirements that arise in the course of a new development, say a residential area. In the case of the two large-scale projects, these contributions imply considerable investments, which profit the surrounding neighbourhoods, including marginalised areas, too. In the first project, Steyn City, the provision of an additional water pump station—amongst other infrastructure investments—benefits a marginalised township located north of the project site, named Diepsloot. The township had been underserviced for many years. In the second project, Waterfall City, a number of infrastructural projects took place, including the upgrade of a strategic highway interchange, which connects a major development priority corridor with the national highway. Investments such as these contribute to further unlock development potentials in the wider sub-region of the projects. In both projects, government entered into an agreement with the developers according to which they would not only fund, but also plan and implement infrastructural developments. Hence government would not have to apply own capacities to realise a number of developmental benefits.

One may hence conclude, that the Steyn City project demonstrated how synergistic planning options which link development demands and adaptive land uses can be realised at the project scale. However, both projects also showed that the realisation of synergistic adaptation responses is not strategically promoted by government. Instead, synergies between various development interests of actors in the private sector and governments' need for investment to push its development agenda are preferably realised, at the expense of adaptive land uses.

Housing

South Africa provides give-away houses for households with an income of less than 3.500 ZAR. The beneficiaries must be South African and not have benefited from a give-away house before. This practice is the key response to today's informal growth in South Africa. Initially, it was motivated by restitution efforts after Apartheid in view of forced evictions and limited housing provision for coloured and black population in the townships. The give-away housing practice has various welcomed social development effects, but at the same time, it is problematic for a number of reasons. Despite an impressive number of housing units which have been handed over to beneficiaries since the introduction of the programme in the 1990s,[3] for instance, government's housing provision cannot keep pace with informal growth in Johannesburg. In addition, poor households lack alternative formal shelter options. Consequently, many of them continue to settle informally and often do so in areas prone to flooding. In fact, most areas at highest risk of flooding in Johannesburg are in informal settlements (EMS 2011). Another problematic aspect is the established low-density design of give-away houses and the fact that consequently, give-away housing settlements have a large spatial footprint. At the same time developable land is increasingly becoming scarce or too expensive for social housing in Johannesburg. As a consequence, planners indicated during the interviews that there are already cases in which eco-system areas (such as wetlands) or risk prone areas have been made available for give-away houses developments (planner B—interview 2014; planner C—interview 2013; cf. Phalatse 2011).

There is broad recognition that the current give-away housing practice needs to be changed. As one government official at the provincial tier put it, "there are many voices going up, even in government, to say that we need to revise our housing policy because it is not sustainable" (planner D—interview 2013). In Johannesburg, a couple of alternatives to the give-away housing practice are currently being tested or discussed. Some include the integration of self-help mechanisms, some of which are promising adaptive planning options (for further details see Hetz and Bruns 2014). Here, I'd like to focus on another option, which has been discussed within government in South Africa for a number of years already, namely the attempt to shift from free standing houses to a medium-density design. Such an approach is advisable, due to the smaller spatial footprint it implies. If implemented in such a way that developable areas are strategically densified while ecological areas are preserved instead—and integrated intelligently into the overall spatial fabric, as

[3]More than 3 million give-away houses were provided since the start of the RDP programme (NPC 2012: 242).

envisaged by Register (2006) for instance—this approach, too, could lead to adaptive benefits which the current practice does not provide.

Besides financial complications to fund medium-density designs, for instance in cooperation with the private sector, planners find this option to be politically very difficult to implement. The group of qualifiers for a give-away house, which the Gauteng Department of Housing estimates to include about 90% of the informal settlers in Johannesburg (GDH 2005), expects to receive a house eventually and to profit from a number of benefits associated with the life in a formal single-family house and land ownership. One of such benefits is the possibility to rent out space for informal shelter on the property or to use the premises to run small businesses (such as spaza shops and barber services) in order to sustain their livelihood. Shifting planning practices from these housing options towards medium-density houses would eliminate—or at least complicate—such options of income generation. Besides, the "give-away houses"-programme that was started in the 1990s in the framework of the post-apartheid Reconstruction and Distribution Programme (RDP), until today, remains strongly tied to the rhetoric of apartheid restitution in the public discourse (Hetz and Bruns 2014). Due to its established link to the collective commitment to overcome the ills of the apartheid past, the potential beneficiaries—often black African population that has suffered tremendously under apartheid politics—are reported to claim to hence have a right to get these give-away houses, although restitution could be achieved through other means, too. The situation resembles what Konrad and Thum (2014) portrait as a Samaritan's dilemma of the state that has started to provide social benefits in a particular way at one point in time and finds itself trapped in this very practice a few years later. Although there have been slight modifications in the design of give-away houses (such as the shift from detached to semi-detached or row houses), the concept of providing land ownership has not changed and the spatial footprint of the give-away housing settlements has not decreased significantly. Hence, the problematic aspects these settlements have in terms of climate change adaptation needs have not been eliminated. Despite possible alternatives, a planning practice is therefore being continued that contributes to further shape flood risks of climate change in Johannesburg as indicated above.

Discussion

The development dynamics in Johannesburg are complex. Various governmental programmes have resulted in improvements of living conditions since the end of apartheid. Nonetheless, a decade after the political liberation, Johannesburg is still a highly divided city. Its division mirrors in the mushrooming of informal settlements and a high level of urban divide. As for many marginalised households, living conditions have not improved substantially since the end of Apartheid, and political decisions-makers are increasingly pressured to further present considerable development results in Johannesburg shortly. As one political planning actor put it,

> there is a serious level of discontent on the ground and we do face protests all the time. It is not the kind of mass protest of the Arab spring, but there are certainly sporadic outbreaks of protests throughout the country, virtually on a daily basis (planner E—interview 2013).

In this situation, the prioritised support of economic-growth by infrastructure and land use related planning means, is a logical consequence, linked to both attempts to shape growth as inclusive as possible and to the continuation and simultaneous prioritisation of social programmes and investments in the marginalised areas of the city. Against this background, and in view of scarce capacities, it may be understandable that the observed planning practices had been motivated primarily by considerations of pre-existing development demands and in view of the related development priorities of urban economic growth generation and pro-poor programmes that aim to boost wealth creation and the improvement of living conditions in marginalised areas in Johannesburg.

Besides, we also saw that planners are confronted by entrenched expectations of urban poor to benefit from a specific planning programme—the give-away housing practice. As claims raised during social protest often concern shelter and land demands, the abolishment of the give-away housing practice may further jeopardise the social peace in the city. Households of working-poor, those being dependent on income generated in the informal sector or the unemployed, for instance, are likely to not achieve the same benefits which the give-away housing programmes offers by own means. Although it is increasingly becoming apparent that not all qualifying households will receive a give-away house eventually (because government does not have the capacities to deliver the required number of houses), the allocation process has become "a lottery for houses" (planner F—interview 2013), with a positive political effect: It gives poor households the hope to ultimately be the owner of land and formal shelter at last. Occasional ceremonies in the townships during which politicians hand over keys of give-away houses to selected beneficiaries appear to have become a political means to try and maintain peace in the township. As such, the give-away housing programme can be interpreted as an example where a lack of possibilities to reconcile alternative practices with entrenched development expectations of a politically important[4] social group constitute an political adaptation limit that is perceived to be impossible to overcome in the current development situation.

Altogether, the stated observations suggest that there are limited possibilities in Johannesburg to realise adaptive options. In the case of the presented examples, adaptive options are either perceived to be politically unattractive due to unsolved conflicts between adaptive measures and development expectations or adaptive options imply opportunity costs for other prioritised development needs. Besides,

---

[4]The group of beneficiaries can be interpreted as being rather politically important, because their dissatisfaction about living conditions in post-apartheid South Africa has begun to become a considerable threat to social peace. In view of their social protests which originate in the marginalised townships and which have already turned violent in many cases in the past, it can be argued to be a collective interest to satisfy the demand for give-away houses or at least give the impression that the demand is being satisfied.

synergistic options between development priorities, which may help justifying adaptive practices in view of pre-existing development constraints, are not realised in the presented cases, either due to a lack of such opportunities or lacking political will to inconveniently push them.

Bai (2007) put forward the readiness argument, stating that climate change *mitigation* requirements would play out at the global scale and fail to receive the required attention of an urgent local issue, particularly if considerable development constraints occur locally. Hence, mitigation options are difficult to motivate at the local level; or, in other words, local conditions are not 'ready' for their implementation. Based on insights gained in Johannesburg one may propose that in terms of climate change *adaptation*, too, a readiness argument can be made: in comparison to a number of other pre-existing development constraints, climate change adaptation needs are perceived to be less urgent *today* (Parnell et al. 2007; Roberts 2010; Ziervogel and Parnell 2014) and are not perceived to be the biggest threat to the achievement of development objectives (yet). Hence, the full implication climate change effects may have in the future are a too distant concern in comparison with today's development constraints, which are hence given priority in today's planning practices, as observed in the Johannesburg case.

Conclusion

Using urban planning in Johannesburg as an illustrative case and planning practices in the field of land use planning and housing as examples, the paper explores which adaptive planning options are politically feasible or attractive in the context of development constraints of a high level of urban divide, and why. Empirical observations revealed that (1) it is difficult to find opportunities to realise synergistic options between prioritised development programmes of social distribution and economic growth promotion and climate change adaptation and that (2) possible synergistic options are not strategically used by government. Hence, the Johannesburg case provides an initial indication that perhaps a rethink of synergy argumentation (Revi et al. 2014; Schipper 2007) is needed: there may either be far fewer synergy opportunities in practice than suggested by conceptual argumentation or the political attractiveness of realising synergistic adaptation options is smaller than assumed due to alternative possibilities to realise synergies between other, prioritised development objectives instead. If further studies verify the paper's findings, a rethink of the international policy advice to promote particularly synergistic adaptation options is required.

In view of entrenched development expectations and the hope about the speedy improvement of living conditions after the political liberation, governmental actors feel urged to demonstrate action and to be able to present positive development results that meet post-apartheid South African society's development expectations. In the Johannesburg case, these societal realities, in reference to which planning practices need to be politically justified, represent a considerable political adaptation limit. The observed limits are examples of soft adaptation limits (Adger et al. 2009; Klein et al. 2014): Changes in development conditions and/or societal expectations towards planning actions can influence on the political attractiveness or feasibility of available adaptive options in urban planning. Although the limits are 'soft', they are not less significant in their constraining effects than irremovable 'hard' adaptation limits: As results of the Johannesburg case show, soft political adaptation limits can reduce adaptive practices considerably to re-active practices and short-term approaches. In Johannesburg, soft adaptation limits even led to a lock-in situation in which the adoption of adaptive practices is perceived to be politically impossible, at least in the current societal situation.

The insights gained in this study suggest that empirical studies should pay more attention to the societal context in order to better understand why political adaptation limits occur. Such an understanding can help to advance the effectiveness of adaptation programmes. Above all, a main observation made in Johannesburg is that it is extremely difficult to realise pro-active and long-term responses to anticipative effects of climate change. Such responses are required though if cities are to be successfully adapted. Currently, responses to flood risks are reduced to coping mechanism, while at the same time planning practices are continued in Johannesburg which further increase flood risks. The study raises the questions whether and how anticipated risks of climate change can be addressed pro-actively today, even in a highly polarised city where considerable pre-existing development challenges are prioritised. The Johannesburg case suggests that finding workable ways to improve political legitimisation possibilities for responses to *anticipated* risks of climate change, perhaps through the adoption of new governance elements such as future councils (Oxford Martin Commission for Future Generations 2013; WBGU 2011), is an important key to successful urban adaptation.

Acknowledgements This research has been partly funded by the German Federal Ministry of Education and Research (BMBF) under the Funding Reference Number 01 LN 1316 A.

References

Adger, W., Dessai, S., Goulden, M., et al. (2009). Are there social limits to adaptation to climate change? *Climatic Change, 93,* 335–354.

Allen, C. R., & Gunderson, L. H. (2011). Pathology and failure in the design and implementation of adaptive management. *Journal of Environmental Management, 92*(5), 1379–1384.

Bai, X. (2007). Integrating global environmental concerns into urban management. The scale and readiness arguments. *Journal of Industrial Ecology, 11*(2), 15–29.

Bicknell, J., Dodman, D., & Satterthwaite, D. (Eds.). (2009). *Adapting cities to climate change. Understanding and addressing the development challenges*. London: Earthscan.

Biesbroek, G. R., Klostermann, J. E. M., Termeer, C. J. A. M., & Kabat, P. (2013). On the nature of barriers to climate change adaptation. *Regional Environmental Change, 13*(5), 1119–1129.

Birkmann, J., Schanze, J., Müller, P., & Stock, M. (Eds.). (2012). *Anpassung an den Klimawandel durch räumliche Planung - Grundlagen, Strategien, Instrumente*. Hannover: ARL.

City of Johannesburg (CoJ). (2007). *Draft Environmental Management Framework (EMF) report for the Kyalami/Modderfontein KM EMF area*. Johannesburg: CoJ.

City of Johannesburg (CoJ). (2009). *Wetland audit for the City of Johannesburg. Draft report.* Johannesburg: CoJ.
City of Johannesburg (CoJ). (2011a). *Bioregional plan for the City of Johannesburg. 2nd draft.* Johannesburg: CoJ.
City of Johannesburg (CoJ). (2011b). *Growth development strategy. Joburg 2040.* Johannesburg: CoJ.
City of Johannesburg (CoJ). (2013). *Integrated development plan 2012–16, 2013 Review.* Johannesburg: CoJ.
Dupuis, J., & Knoepfel, P. (2011). Les barrières à la mise en œuvre des politiques d'adaptation au changement climatique: le cas de la Suisse. Swiss Political Science Review. *Swiss Political Science Review, 17*(2), 188–219.
Eisenack, K., Moser, S. C., Hoffmann, E., et al. (2014). Explaining and overcoming barriers to climate change adaptation. *Nature Climate Change, 4*(10), 867–872.
Emergency Services Department (EMS). (2011). Implementation report on flood risk monitoring to prevent flooding in the city. Presented to the CoJ Mayoral Committee. Johannesburg: EMS.
Fankhauser, S., & McDermott, T. K. J. (2014). Understanding the adaptation deficit: Why are poor countries more vulnerable to climate events than rich countries? *Global Environmental Change, 27,* 9–18.
Flyvbjerg, B. (2006). Five misunderstandings about case-study research. *Qualitative Inquiry, 12* (2), 219–245.
Fowler, T. (2013). Presentation by the Johannesburg City Manager on resource resilience in Johannesburg. Metropolis Congress "Caring Cities" on the 18 July 2013 in the Hilton Sandton Hotel, Johannesburg.
Garschagen, M. (2011). Resilience and organisational institutionalism from a cross-cultural perspective. An exploration based on urban climate change adaptation in Vietnam. *Natural Hazards*. doi:10.1007/s11069-011-9753-4.
Gauteng Department of Housing (GDH). (2005). *Registration of the informal settlements in Gauteng*. Johannesburg: GDH.
Hartmann, T., & Driessen, P. P. (2013). The flood risk management plan. Towards spatial water governance. *Journal of Flood Risk Management.* doi:10.1111/jfr3.12077.
Hetz, K., & Bruns, A. (2014). Urban planning lock-in: Implications for the realization of adaptive options towards climate change risks. *Water International, 39*(6), 884–900.
Hetz, K. (2015). Political adaptation barriers to urban planning. Origins and implications for the adoption of adaptive planning options for climate change challenges in Johannesburg, South Africa. Doctoral dissertation submitted to the Faculty of Mathematics and Natural Sciences at Humboldt Universität zu Berlin on the 28th October 2015.
Hetz, K. (2016). Urban ecosystem-based adaptation in eco-city projects: Opportunities and political implementation barriers in Johannesburg, South Africa. Unpublished paper.
Hutter, G. (2007). Strategic planning for long-term flood risk management. Some solutions for learning how to make strategy at local and regional level. *International Planning Studies, 12* (3), 273–289.
Huq, S., & Reid, H. (2004). Mainstreaming adaptation in development. *Institute of Development Studies Bulletin, 35*(3), 15–21.
Johannesburg Roads Agency (JRA). (2007). *Integrated stormwater management plan.* Johannesburg: JRA.
Jones, L., & Boyd, E. (2011). Exploring social barriers to adaptation: Insights from Western Nepal. *Global Environmental Change, 21*(4), 1262–1274.
Klein, R. J. T., Midgley, G. F., Preston, B. L., et al. (2014). Adaptation opportunities, constraints, and limits. In *Climate change 2014: Impacts, adaptation, and vulnerability. Part A: Global and sectoral aspects. Contribution of working group II to the fifth assessment report of the intergovernmental panel on climate change* (pp. 899–943). Cambridge University Press, Cambridge.

Klein, R. J. T., Schipper, E. L. F., & Dessai, S. (2005). Integrating mitigation and adaptation into climate and development policy: Three research questions. *Environmental Science & Policy, 8* (6), 579–588.

Konrad, K. A., & Thum, M. (2014). The role of economic policy in climate change adaptation. *CESifo Economic Studies, 60*(1), 32–61.

Moser, S. C., & Ekstrom, J. A. (2010). A framework to diagnose barriers to climate change adaptation. *Proceedings of the National Academy of Sciences, 107*(51), 22026–22031.

Nielsen, J. Ø., & Reenberg, A. (2010). Cultural barriers to climate change adaptation: A case study from Northern Burkina Faso. *Global Environmental Change, 20*(1), 142–152.

National Planning Commission (NPC). (2012). *National Development Plan. Vision for 2013.* Pretoria: NPC.

Oxford Martin Commission for Future Generations. (2013). *Now for the long term.* Oxford: University of Oxford.

Parnell, S., Simon, D., & Vogel, C. (2007). Global environmental change: Conceptualising the growing challenge for cities in poor countries. *Area, 39*(3), 357–369.

Pasquini, L., Cowling, R. M., & Ziervogel, G. (2013). Facing the heat: Barriers to mainstreaming climate change adaptation in local government in the Western Cape Province, South Africa. *Habitat International, 40,* 225–232.

Pelling, M. (2011). *Adaptation to climate change: From resilience to transformation.* London: Routledge.

Phalatse, L. (2011). Flooding at the City of Johannesburg. Johannesburg, Presentation given to the Johannesburg Roads Agency Risk and Finance Board Committee on the 27th January 2011. The presentation was made available to the author.

Picketts, I. M., Déry, S. J., & Curry, J. A. (2014). Incorporating climate change adaptation into local plans. *Journal of Environmental Planning and Management, 57*(7), 984–1002.

Pow, C. P., & Neo, H. (2013). Seeing red over green: Contesting urban sustainabilities in China. *Urban Studies, 50*(11), 2256–2274.

Register, R. (2006). *Ecocities. Rebuilding cities in balance with nature.* Gabriola, BC: New Society Publishers.

Revi, A., Satterthwaite, D., Aragon-Durand, F., et al. (2014). Towards transformative adaptation in cities: The IPCC's fifth assessment. *Environment and Urbanisation, 26*(1), 11–28.

Roberts, D. (2010). Prioritizing climate change adaptation and local level resilience in Durban, South Africa. *Environment and Urbanisation, 22*(2), 397–413.

Satterthwaite, D., Huq, S., Reid, H., et al. (2009). Adapting to climate change in urban areas: The possibilities and constrains in low- and middle income nations. In J. Bicknell, D. Dodman, & D. Satterthwaite (Eds.), *Adapting cities to climate change* (pp. 3–51). London: Earthscan.

Schipper, L. (2007). Climate change adaptation and development: Exploring the linkages. Tyndall Centre for Climate Change Research. Working paper 107. www.tyndall.ac.uk/content/climate-change-adaptation-and-development-exploring-linkages. Accessed on the 10 March 2016.

Sharma, D., & Tomar, S. (2010). Mainstreaming climate change adaptation in Indian cities. *Environment and Urbanization, 22*(2), 451–465.

Steyn City. (2016). Project website. www.steyncity.co.za. Accessed on the 10 March 2016.

Storie, J. M., Ngie, A., & van der Linde, T. (2012). Potential for urban transformation through the utilisation of stormwater infrastructure: Consideration from the Gauteng City-Region. FloodRisk2012. Rotterdam, November 2012.

Tau, M. P. (2013): 2013 State of the city address by the Mayor of the City of Johannesburg. http://www.joburg.org.za/images/pdfs/coj%20state%20of%20the%20city%20address%20soca%202013.pdf. Accessed on the 10 March 2016.

Tompkins, E. L., Adger, N. W. (2003). Building resilience to climate change through adaptive management of natural resources. Tyndall Centre for Climate Change Research. Working Paper 27. www.tyndall.ac.uk/content/building-resilience-climate-change-through-adaptive-management-natural-resources. Accessed on the 10 March 2016.

Uittenbroek, C., Janssen-Jansen, L. B., & Runhaar, H. A. C. (2013). Mainstreaming climate adaptation into urban planning. Overcoming barriers, seizing opportunities and evaluating results in two Dutch case studies. *Regional Environmental Change, 13,* 399–411.

UN-Habitat. (2010). *State of the world's cities 2010/2011. Bridging the urban divide.* London: Earthscan.

UN-Habitat. (2011): Cities and climate change. Global report on human settlements 2011. Nairobi: United Nations.

Wilson, E., & Piper, J. (2010). *Spatial planning and climate change.* New York: Routledge.

Wissenschaftlicher Beirat der Bundesregierung für Globale Umweltveränderungen (WBGU). (2011). *Welt im Wandel: Gesellschaftsvertrag für eine Große Transformation.* Berlin: WBGU.

Yin, R. (2014). *Case study research. Design and methods.* London: Sage.

Ziervogel, G., Shale, M., & Du, M. (2010). Climate change adaptation in a developing country context: The case of urban water supply in Cape Town. *Climate and Development, 2*(2), 94–110.

Ziervogel, G., & Parnell, S. (2014). Tackling barriers to climate change adaptation in South African coastal cities. In G. P. Smith & B. C. Glavovic (Eds.), *Adapting to climate change: Lessons from natural hazard planning* (pp. 57–73). Dordrecht: Springer.

Chapter 9
Constraints and Limits to Climate Change Adaptation Efforts in Nigeria

Idowu O. Ologeh, Joshua B. Akarakiri and Francis A. Adesina

Introduction: Climate Change and Its Impacts in Nigeria

Climate change is impacting the society and ecosystems in a broad variety of ways. For example, climate change can increase or decrease rainfall, influence agricultural crop yields, affect human health, cause changes to forests and other ecosystems, or even impact our energy supply (United States Environmental Protection Agency—USEPA 2014). According to the Intergovernmental Panel on Climate Change—IPCC (2007), the extent of climate change effects on individuals and nations will vary over time and with the ability of different societal and environmental systems to mitigate or adapt to change. Nigeria is experiencing adverse climate conditions with negative impacts on the welfare of millions of people. Persistent droughts and flooding, off season rains and dry spells have sent growing seasons out of track, in a country dependent on a rain-fed agriculture. The impacts are alarming with lakes drying up and a reduction in river flow in the arid and semi-arid region. The result is reduced water supplies for use in agriculture, hydro power generation and other users. It is widely believed that climate change is responsible for all these as reported in the 4th IPCC Assessment Report, which suggests that Africa will be worst hit by the effects of climate change of which Nigeria is a part.

I. O. Ologeh (✉) · J. B. Akarakiri
African Institute of Science Policy and Innovation (AISPI),
Obafemi Awolowo University, Ile-Ife, Nigeria
e-mail: iologeh@gmail.com

J. B. Akarakiri
e-mail: jakarakiri@yahoo.co.uk

F. A. Adesina
Department of Geography, Obafemi Awolowo University, Ile-Ife, Nigeria
e-mail: faadesin@yahoo.com

W. Leal Filho and J. Nalau (eds.), *Limits to Climate Change Adaptation*,
Climate Change Management, https://doi.org/10.1007/978-3-319-64599-5_9

It is of serious concern in Nigeria with population growth rate that is one of the highest in the world. The farming seasons are increasingly unstable, even when the onset and retreat fall within the long-term averages, growing season may be truncated by periods of intra-seasonal droughts (Adesina and Odekunle 2011b). Nigeria like many other African countries is already beset by ecological problems such as flooding and infestations of pests which have been directly or otherwise linked to climate change (Adefolalu 2007; Ikhile 2007). The southern part of Nigeria, a large proportion of which have fairly long rainy season and high annual rainfall, has been experiencing distorted rainfall patterns particularly with respect to onset and retreat, while guinea savannah is facing increasing temperatures. The ecological zones from the Sudan Savanna northwards face the threat of desert encroachment at a very fast rate per year due to a reduction in the amount of surface water, flora and fauna resources (Federal Ministry of Environment—FME 2010; Obioha 2009). Farmers in most parts of the country are confronted with one challenge or the other, which reduces agricultural productivity, increases hunger, poverty, malnutrition and diseases (Obioha 2009). Thus, the less dependent an agriculture system is on climate the less exposed it is. In this respect a well irrigated faming system will be less exposed to drought than a rain-fed system since its immediate source of water is not necessarily rainfall (Adesina and Odekunle 2011a).

In Nigeria, the main direct effects are changes in temperature, precipitation, length of growing season and timing of extreme or critical threshold events relative to crop development. Availability of climate information to farmers, understanding farmers' vulnerability to the impact of climate change and the ways climate change and variability impacts on farming activities as well as farmers' adaptation to these impacts is becoming increasingly important for sustainable adaptation (Nigerian Environmental Study Team—NEST 2011). Flooding has been identified as one of the major factors that prevent Africa's growing population of farmers from escaping poverty and stands in the way of United Nations 2020 goal of achieving significant improvement in the lives of rural dwellers (Adebayo 2009). This is because many African farmers lack the infrastructure to withstand extreme weather conditions (Adelekan 2009)

Nigerian communities are experiencing extreme weather conditions as a result of increasing temperature and an associated changing climate (Olaniran 2002; Ayoade 2003; Odjugo 2010). Communities with different livelihoods experience different types of climate risks. For instance, the communities along the coastline are confronted with the risk of ocean surge and sea level rise while the wetland communities are faced with the risk of pests and flooding (NEST 2011). Ayoade (2003) recorded a slight drop in air temperature within the late 1940s and early 1950s in Nigeria. According to Mabo (2006) and Ikhile (2007), a sharp increase in temperature between 1971 and 2005 could be linked to the effect of climate change and its associated global warming. Temperature anomalies confirm the facts that global warming is unequivocal (IPCC 2007). According to Odjugo (2010), increasing temperature is already present in Nigeria. Odjugo and Ikhuoria (2003) and Adefolalu (2007) report that increasing temperature and decreasing rainfall in the semi-arid region of Sokoto, Katsina, Kano, Nguru, and Maiduguri may have resulted in the increasing evapotranspiration, drought and desertification in Nigeria.

Also, rainfall data analysis undertaken by Adejuwon (2002) indicates rainfall decline in several locations in Nigeria thereby authenticating this effect of climate change. Adejuwon (2004) also linked up reduced agricultural productivity and low income associated with Nigerian farmers with fluctuation in available moisture pattern. Climate change also influences the existing vegetation type, which favors cattle production in many southern parts of Nigeria (Ayoade 2004) and are plausible reasons for the ubiquitous face-off between crop farmers and pastoralists.

According to UFCCC (2014), about a third of greenhouse gas emissions which contribute to climate change are from land use, land use change and forestry. Agriculture is a dominant factor in this. In turn, climate change affects agriculture more than any other sector, increasing risks of crop failures and livestock losses and threatening food security. The decline in crop yields, especially in Africa, could leave hundreds of millions without the ability to produce or purchase sufficient food. Warming may also induce sudden shifts in regional weather patterns that would have severe consequences for water availability and flooding in tropical regions. Also, the impact of sea level rise could be catastrophic in many developing countries (Susmita et al. 2007).

Nigerian farmers have been experiencing reduction in their harvests particularly over the last five decades and this is traceable to the negative impacts of climate change. For example, maize farmers in South-West Nigeria experienced crop failure twice in 2013 due to erratic rainfall (Farmers Academy Ede—FAE 2013). Also, South-Western Nigeria has experienced change in rainfall patterns in the last two decades leading to unpredictable yields (Adelekan 2009). Also, in northern Nigeria, farmers are experiencing untold hardship and desert encroachment is threatening their livelihood; the Lake Chad for example, which used to be a vibrant lake is now mere swamps in many of its parts (UNEP 2008). In south-east Nigeria, flooding and erosion have been the norm for over twenty years (Nchuchuwe and Adejuwon 2012). These incidences have led to poor crop yields, which ultimately have led to poorer livelihoods. With perceived long-term changes in climatic parameters, particularly temperature and rainfall, farming is expected to become increasingly unprofitable. Vulnerability to climate change shocks, particularly extreme events such as droughts and floods, varies by location, indicating the need for climate-smart agriculture and location-specific policy responses.

The agricultural sector contributes a substantial percentage of the Nigerian Gross National Product (GDP) and majority of the rural populace are employed in this sector. The dominant role of agriculture makes it obvious that even minor negative shift in climate can cause devastating socioeconomic consequences; hence this study is a response to what the country needs to do to alleviate the effects of climate change on farmers and so improve food production in the face of a changing climate. Climate-related impacts are occurring across the six geo-political zones of the country and across many sectors of the economy. The federal government is already preparing for the impacts of climate change through "adaptation," which is planning for the changes that are expected to occur.

Climate Change and Adaptations in Nigeria

"Adaptation" can be defined as a means of anticipating the adverse effects of climate change and taking appropriate action to prevent or minimize the damage they can cause, or taking advantage of opportunities that may arise (European Commission—EUROPA 2015). It has been proved that well planned, early adaptation action saves money, properties and lives. Adaptation to climate change is not new; history has told how human societies have repeatedly demonstrated strong capacities for adapting to different climates and environmental changes whether by migration, improved crop varieties or building different types of shelter (Adger et al. 2007). However, the current rate of global climate change is unusually high compared to past changes that society has experienced. In an increasingly interdependent world, negative effects of climate change on one population or economic sector can have repercussions around the world (United States Global Change Research Program—USGCRP 2009).

The resilience and adaptive capacity of a country is dependent on its level of development with respect to economy and political stability. To develop adaptation strategies, there is need for a clear understanding of local vulnerabilities to climate change and critical thresholds. Adaptation strategies should be sufficiently flexible to take care of future possible changes in climatic parameters, which may call for quick review of plans; the plan should be reviewed from time to time. The United Nations Framework on Climate Change (UNFCCC) is an umbrella under which the global community has assembled to address issues associated with climate change. Nigeria is part of the global community and is involved in the various negotiations on actions to reduce the manifest as well as anticipated impacts of climate change. Under the UNFCCC, Nigeria as a nation along with other developing countries is to focus on adaptation to make them able to cope with the new extremes in climatic regimes. They are also to focus on environmentally as well as economically sustainable development (Adesina and Odekunle 2011a). The developing countries' development is being impaired by climate change effect although these countries produce only a small fraction of the greenhouse gases (GHG) that is causing climate change. They also have weak adaptive capacity because of their poor economies, making them highly vulnerable to climate change impacts.

Historically, agriculture has shown a considerable capacity to adapt to changing conditions. If climate change is gradual, the adjustment may go widely unnoticed, and the process is one largely of autonomous adjustment (Reilly 1996). Many adaptations occur autonomously without the need for conscious response by farmers and agricultural planners. However, it is likely at least in some or most parts of the area that the rate and magnitude of climate change may exceed that of normal change in agriculture that specific technologies and management styles may need to be adopted to avoid the most serious of effects (Parry et al. 2001). Thus areas likely to be most vulnerable to climate variability can be spared from its impacts through implementation of appropriate adaptation measures. Such methods include new traits, varieties and crops, water management and irrigation, increase in farm land, improved production management and practices, access to information, marketing and supply chains.

In Nigeria, many possible adaptation strategies for agriculture are available, these vary according to the local conditions and the particular farming system. The options range from routine management alteration by farmers themselves, such as changing timing of operations, adoption of conservation tillage practices and diversification in production systems, to investment of funds by public agencies in order to develop or improve irrigation schemes, modification of farm support programs and development of new plant varieties. Some impact studies have suggested 'adaptations' involving contractions and expansions of agricultural zones. For producers, this would mean that some farmers would relocate, or others would cease operations in some locations (or completely change the type of farming), whereas in other locations there would be new farmers or new types of farming by existing land owners (Adesina and Odekunle 2011b). While adaptations can be planned at the farm level, the term 'planned adaptation' is generally used to refer to actions taken by governments as a conscious policy response. Possible planned adaptations involving governments include encouragement of technological adaptations, such as crop development and early warning systems, promotion of land and water use options, assistance with changes in diversification or intensity of production, and changed financial support in established programs and ad hoc compensation.

Wisner et al. (2004) report that the vulnerability of agriculture is not determined by climate change per se but by the combination of mitigation (efforts to reduce or prevent) and adaptation (capacity to cope with) to environmental change. While the coping capacity and degree of exposure is related to environmental changes, they are both also related to changes in societal aspects such as land use and cultural practices. This study discusses the challenges of agricultural adaptation to climate change in Nigeria. Most studies on climate change and agriculture in Africa have tended to concentrate on actual and projected impacts as well as farmers' coping/adaptation strategies (Adejuwon 2006; FAO 2007). There has been little or no work in the area of constraints and limits to adaptation. This study is therefore focused on filling this gap. Consequently, the research was embarked on to acquire data through the use of questionnaire survey and review of relevant literature.

Objective

The major objective of the study is to assess the various constraints and limits to climate change adaption efforts in Nigeria. To achieve this, the study focused on the indigenous adaptation strategies employed by the farmers, the government assisted adaptation programs for farmers, reasons for non-adoption of some of these programs by farmers and policy formulation preferring solutions to the constraints.

Methodology

The study was designed to assess the Nigerian climate change adaptation measures and their constraints and limits. The scope of the study is restricted to small-scale farmers only because they produce majority of the food in the country and agriculture is their sole source of living; the challenges they face will affect national food security and their socio-economic status. However, though the sector is male dominated, the study is designed to be gender responsive as it considered both male and female stakeholders in the agricultural sector. The study area is the Guinea Savanna belt of Nigeria because it is the food basket of Nigeria with great potentials for agricultural development (Adegbola and Onayinka 1976). Four out of the six geo-political zones in Nigeria are within the Guinea Savana vegetation belt and their small-scale farm clusters were surveyed for this study. In each geo-political zone, the state with the highest food production record was sampled for the study. The following states are the major food producing states in their geo-political zone and their small holding farms are considered in this study: North-Central—Benue State, South-East—Imo State, North East—Kaduna State, and South-West—Oyo State.

The states with the highest yields were chosen because they are known for bumper harvests and reduction in yields as a result of climate change can be easily noticed and measured. Background information on indigenous climate change mitigation and adaptation strategies was acquired through structured interviews of farmers, extension workers and technical experts. Six hundred (600) questionnaires were purposively distributed to small holding farmers across the study area. The questionnaires were distributed majorly through the Agricultural Extension Workers in each state. The questionnaires were designed to collect information on farmers' biometric data, years of experience and skill, farm size, types of crops cultivated, educational level and number of laborers. It also sampled their opinions on climate change, its effects on their harvests, adaptations and challenges.

Also, sixty (60) questionnaires in all were administered to the technical experts in the Crop Production Department in the Federal Ministry of Agriculture, Abuja and to the State Ministry of Agriculture in Benue, Imo, Kaduna and Oyo States to assess the various adaptation strategies adopted and their level of relationship with the Nigerian Meteorological Agency (NIMET) for geo-data. Ten (10) technical experts from NIMET were also sampled to measure the contribution of their agency towards rural agriculture and mode of linkage with the users of their products. The results are analyzed using Microsoft Excel spread sheet 2013 and presented in frequencies and percentages.

There are some limitations associated with the research methodology. The major limitation of the study is that majority of the farmers are illiterate and can either read or write in English. The questionnaire was interpreted and administered verbally to their local languages and their responses are interpreted back to English and documented by the extension workers. The problem with this method is the error of misinterpretation; the study relied on the capacity of the extension workers to interpret the questions in its context to the farmers' language and back to English.

Another limitation is that not all questionnaires were retrieved and some of the retrieved questionnaires were not properly and completely filled. The study was constrained with limited number of educated extension workers. Since the questionnaire administration requires interpretation, only tertiary educated extension workers were employed for the study and they are under staffed especially in Imo state, thus the need to employ independent field workers to support them. The number of extension workers in the various states cannot go round the farmers and the available ones are not well motivated.

Demographic and Socio-Economic Characteristics of Farmers

A total of six hundred (600) questionnaires were distributed to small scale farmers in Benue, Imo, Kaduna and Oyo States but five hundred and fifty (550) questionnaires were retrieved. The respondents are from 55 villages and towns from 12 Local Government Areas (LGA) across the four States. Also, ten farmers from each state were interviewed in order to observe their expressions and detailed information. The average age of respondents is 37 years with majority falling between the brackets of 31–40 years (Table 9.1). This age distribution justified the reason why the farmers were most vulnerable; this is the stage at which they are trying to raise a family and build a career. If the farming profession continues to fail them economically as it has in recent years, in line with the words of Hassan and Nhemachena (2008), there will be constrained economic opportunities. As a result, most of the young men may resort to motor cycle riding to survive, others may experience delayed marriage or find it difficult to feed their family. The last option is most viable as the study shows that most of the respondents (84.5%) are married.

Twenty-eight percent (28.2%) of respondents are female, while seventy percent (70.2%) are male; the women farmers are more in Oyo and Enugu States. Field observation showed that women are majorly laborers rather than farm owners or tenants confirming the statement of Annon (2006) who said women are responsible for carrying out 70% of agricultural labor. The result also shows that nearly 69% of

Table 9.1 Age distribution of farmers. *Source*: Field survey

Age group	Frequency	Percentage
10–30	78	14.18
31–40	173	31.45
41–50	161	29.27
51–60	98	17.82
61–70	27	4.91
>70	2	0.36
DNI	11	2.00
Total	550	100

the women rented the lands; ten percent (10%) leased it while the remaining 21% owned/inherited the lands. This result concurs with the work of Mamadou (2013) who said women have limited access to land.

Majority of the farmers (71.8%) owned one to two farms, but some farmers in Kaduna and Benue states owned two to three farms in different locations with most farms (47.3%) ranging from two to four hectares. Almost half of the respondents (46.9%) owned their farmlands, while twenty-two percent (22%) rented their source of livelihood. The crops common to the farms surveyed are cassava, maize, yam, tomatoes, millet, rice, soybeans, vegetables and beans.

Demographic and Socio-Economic Characteristics of Technical Experts

Forty-nine (49) technical experts (out of 60) from 14 government agencies, ministries and Local Government Areas (L.G.A.s) responded to the administered questionnaires. These include experts from the Federal Ministry of Agriculture, NIMET and various state and local government agricultural bodies. Sixty-nine percent (69%) are men; 72.4% are within the age bracket of 31–50, which is actually the active age group, thus explaining the fact that 93.1% are married. The educational qualifications of the respondents are evenly distributed (see Table 9.2); 39% have bachelor degrees (B.Sc.), 30% have higher diplomas (HND) and 30% possess ordinary diplomas (OND). 47.4% out of the bachelor degree holders have master's degrees, while 10.5% are doctorate degree holders. In relation to specialization, 61.2% respondents have agricultural background, 16.3% have meteorological/Geography background, 8.2% are space specialist, while 4% have social science background. Others (10.2%) are support staff.

Table 9.2 Educational qualification of technical experts. *Source*: Field survey

Highest qualification	Frequency	Percentage
OND/NCE	15	30.61
HND	15	30.61
B.Sc., B.A., B.Agric.	19	38.77
Total	49	100
University qualifications	Frequency	Percentage
BSc	8	42.11
MSc	9	47.37
PhD	2	10.53
Total	19	100

Farmers' Perceived Changes as a Result of Climate Change

The farmers are asked whether or not they are affected by climate change; most of them do not understand what climate change is. Climate change was explained to them in terms of increase or decrease in temperature, rainfall and humidity. Almost all the respondents are affected as 97.8% respondents signified that they are experiencing erratic rainfall, while 97.5% indicated there is much increase in temperature. Also 78.9% respondents disclosed that they are experiencing changes to water content in the air (humidity). This is in agreement with the findings of NEST (2011) that in Nigeria, the main direct effects of climate change are changes in temperature, precipitation, length of growing season and timing of extreme or critical threshold events relative to crop development.

The farmers were given a table of ten parameters that are affecting their farms due to the effects of climate change, the three parameters that topped the list are soil infertility, increase in temperature and pests and diseases (Table 9.3). This is in line with the sayings of Lybbert and Sumner (2010), that temperature increase is an indication of climate change, and also in agreement with NEMA (2012) that stated climate change causes extreme temperature. Also the result is in relation with the words of Adegoke et al. (2014), who stated that weeds, pests and fungi thrive under warmer temperatures, wetter climates and increased carbon dioxide (CO_2) levels.

All of the technical experts concur to the fact that climate is changing and is evident in high temperature and flooding and that their agencies have incorporated various mitigation and adaptation methods like public campaign and ridge construction in their operations especially against flooding. Majority (86%) of the technical experts surveyed are agriculture specialist who are conversant with the use of and promotion of meteorological data. Others are crop specialists (4%), climatic data providers (4%), agricultural extension workers (4%) and planting advisors (13.8%).

Table 9.3 Farmers' perception of climate change effects. *Source*: Field survey

Perceived changes	Frequency	Percentage
Pest and disease (3)	439	79.8
Erosion	343	62.4
Soil fertility (1)	492	89.5
Access to water/availability	435	79.1
Flooding	259	47.1
Crops grown	345	62.7
Change in temperature (2)	466	84.7
Land degradation	294	53.5
Crop failure	349	63.5
Drought	297	54

Farmers' Indigenous Climate Change Adaptations

Parry et al. (2001) stated that the rate and magnitude of climate change may exceed that of normal change in agriculture and that specific technologies and management styles may need to be adopted to avoid the most serious effects. The farmers are asked to indicate what management styles they adopted to avoid serious effect and how they respond to their challenges (climate change adaptation) indigenously. Twenty one percent (21.54%) indicated they practice land fallowing to allow the land gain its richness. This practice is common in Kaduna but land fallowing is not widely practiced in Imo because of restricted access to land. IPCC (2007) has explained that adaptations to climate change will depend on the ability of different societal and environmental systems to mitigate or adapt to change as in the case of Kaduna practicing land fallowing as an adaptation strategy because of availability of land. Eighteen percent (18.88%) use pesticides to regulate the invasion of pests and diseases on their farms while 13.30% enrich the farmland using both organic and chemical fertilizers. Some farmers rely on government intervention (the farmers that depend on government intervention, though they come late, are mostly poor and cannot practice any adaptation strategy if not assisted by the government) while others diversify their cropping pattern to adapt to climate change effects. seventy-six percent (76%) practice mixed cropping (if some crops fail, others may survive), and 64% practice crop substitution (replacing water loving crops with cassava which is climate resistant) (Table 9.4).

Other captivating indigenous adaptation methods practiced by the rural farmers are swamp farming in Oyo State, irrigation and application of neem seed in Kaduna State. Farmers that have access to river banks practice swamp farming to have constant access to water. Also, those who do not have this opportunity, practice irrigation. They fetch from the rivers or wells and water their crops. In all communities surveyed in Kaduna State, the farmers use neem seed as pest control and these seeds according to them are more efficient and cheaper than pesticides. The

Table 9.4 Farmers' indigenous adaptations to climate change effects. *Source*: Field survey

Indigenous adaptation	Frequency	Percentage
Swamp farming	5	1.33
Irrigation	32	8.51
Mixed cropping	37	9.84
Crop substitution	20	5.32
Early maturing seed	16	4.26
Tolerant seed	3	0.80
Ridge making	24	6.38
Fertilizer application	50	13.30
Land fallowing	81	21.54
Use of pesticides	71	18.88
Government intervention	37	9.84
Neem seed	37	9.84

result is in line with the findings of Adger et al. (2007) who listed adaptation strategies as migration, improved crop varieties or building different types of shelter, while Anríquez and Stamoulis (2007) adaptation list includes changing timing of operations, adoption of conservation tillage practices and diversification in production systems, improvement of irrigation schemes, modification of farm support programs, and development of new plant varieties. Also since most of the surveyed farmers are rural farmers, only three percent (3%) of them adopted foreign technology on their farms in terms of renting tractors and other mechanized farm implements.

Government Assisted Adaptation Strategies and Limitations

The federal government is ever ready to alleviate the impacts of climate change through various adaptation strategies. The government through various Agricultural and related agencies has been planning for the changes that are occurring and expected to occur to the agricultural sector through climate change (Federal Ministry of Agriculture 2014). The response from technical experts from the Federal Ministry of Agriculture revealed the various adaptation programs designed by the federal government for farmers in order for them to cope with the effects of climate change. Some of their efforts are provision of improved crop varieties, fertilizers, irrigation schemes and geo-data. They also assist with soft loans, farming machinery and ready markets. There are a variety of these efforts in place, and these efforts are acknowledged by farmers and held in importance (Table 9.5). These government initiatives are disseminated among farmers through farmers' interactive groups (29.1%), village meetings (20%), age bracket groups (12.9%) and farmers' cooperatives (9.3%).

The most controversial government adaptation plan introduced to rural farmers is geo-data. According to technical experts from NIMET, the agency has made a lot

Table 9.5 Government adaptation acknowledged by farmers and importance. *Source*: Field survey

	Acknowledged		Importance	
	Frequency	Percentage (%)	Frequency	Percentage (%)
Improved crop varieties	541	98.36	312	56.72
Fertilizers	532	96.72	528	96.00
Irrigation	112	20.36	100	18.18
Geo-data	161	29.27	68	12.36
Soft loan	319	58.00	538	97.82
Farming machinery	40	7.27	120	21.82
Ready market	383	76.60	520	94.55

of climatic information available for farmers, through publications, radio and television programs, workshops and seminars, but they cannot ascertain whether or not it is being utilized by the farmers. They ascertain that agricultural technical experts and stakeholders are often invited to their workshops and seminars and this information is available to them. They also stated that climatic data are disseminated weekly through agricultural extension workers majorly and also through interaction with farmers and public campaigns. But no one can specify who is in charge of promoting such information, and whether or not they get to the end users. This finding is in agreement with the work of Ologeh (2016) that there exists a large information gap among critical stakeholders of climate change for agricultural adaptations. Also, the work of Oni (2008) confirms that inconsistent, uncoordinated and inappropriate policies, and fragmented and overlapping government institutions are impediments for agricultural development in Nigeria.

On the receiving end, some of the farmers (40%) claimed they have neither heard about, have access to, nor make use of any climatic data. The remaining 60% indicated that they have access to and also made use of climatic data such as rainfall data (58%), temperature (50%), sunlight (29%), wind speed (25%) and humidity (23%). The rainfall forecast they easily heard on radio but others (including tabulated rainfall data) they obtained from other sources including paid sources with access fee of 1000 naira or more. The climatic data is useful for farmers because it improved their yields (57.8%), increased their overall production output (21%) but increased their cost of farming (29.1%) (Table 9.6).

Although climatic data has advanced agricultural production and forecasting of planting and yield trends in developed countries (Kaspar et al. 2003), majority of farmers in Nigeria are losing out on the benefit of this information to their production. NEST (2011) indicated that availability of climatic data to farmers is becoming increasingly important for sustainable adaptation the efficacy of such alliance. NIMET though has been fulfilling its mandate as an agency in producing information that can help farmers but our study shows that the information rarely reaches the farmers. There is a big gap between climatic information providers and users that need to be urgently bridged. The bridge will allow free flow of information from providers to users, thereby making both ends effective; the expert will have his product being used and producing results while the farmers will have their crops monitored and protected to avoid loss.

However, the government adaptation initiatives are faced with challenges and these adaptation limits are majorly due to lack of finance especially as the country is

Table 9.6 Importance of climatic information to crop production. *Source*: Field survey

	Frequency	Percentage (%)
Improve yield	318	57.8
Increase production	120	21.8
Reduce production	23	4.2
Increase cost of farming	160	29.1
Make the work more tedious	36	6.5

Table 9.7 Government adaptation limits. *Source*: Field survey

Challenges	Frequency	Percentage
Technical ability of the farmers	1	3.4
Inadequate fund	5	17.2
Lack of access to timely and update information by the farmers	1	3.4
Poor communication between farmers and extension workers	4	13.8
Refusal of farmers to accept update and innovation	7	24.1
Lack of infrastructure	1	3.4
Lack of extension professional who are qualified, motivated, committed and responsive to ever changing climate	2	6.9
Poor road	4	13.8
Lack of motivation for extension agents, development	3	10.3
Farmers illiteracy	3	10.3
No vehicle for mobility	1	3.4
Failure of farmers to attend meetings	1	3.4

going through economic recession. Also, some extension workers are not faithful to their duties thereby sabotaging government efforts. The farmers also pose limitation to adaptation through their actions of cultural and religious sentiments, illiteracy, language barrier and unwillingness to change mode of farming (Table 9.7). The major challenges confronting technical experts in delivering adaptation to farmers are inadequate funds, refusal of farmers to accept update and innovation, poor communication between farmers and extension workers, dilapidated roads, lack of motivation for extension agents, poor development and farmers' illiteracy.

Conclusion

Climate change effects are evident in the Nigerian agricultural sector. The farmers however have indigenous ways of adapting to climate change which though may not be as effective as foreign technologies, are able to protect them from the adverse effects of climate change. The government through its various agricultural agencies are also supporting farmers through the provision of fertilizers, improved crop varieties and irrigation schemes. Some farmers nonetheless are still vulnerable to climate change effects because they are too poor to adopt any indigenous adaptation strategy or their indigenous adaptation methods are not effective. Others are vulnerable because they refuse to adopt government adaptation initiatives, which are free and effective. This calls for advanced government efforts in training and enlightening the farmers on the importance of adaptation through national radio programmes targeted at sharing information about climate change, its impacts, mitigation and adaptation on both local and national stations as well as focused group trainings. Also, the efficacy of indigenous adaptation methods must be enhanced through research, and having verified the importance of geo-data as a climate adaptation tool, the government should make it affordable and accessible to the farmers. There is need to motivate extension workers; one of the challenges faced by policy experts and researchers is how to convey their products and research questions to the farmers respectively. In order to alleviate poverty among the farmers, climate finance programs should be developed to focus on improving livelihoods and income so that the farmers will be motivated to accept government adaptations programs. Finally, government should formulate a policy enforcing the results of researches conducted by government agencies using government funds to be implemented within one year of result publication.

Further studies can be carried out on a larger scale by surveying every state in Nigeria, developing the technology behind the indigenous adaptations and determining the sustainability of climate adaptation during economic recession.

References

Adebayo, I. C. (2009). Dynamics of technology adoption in rural-based Cassava Processing Enterprises in South-West Nigeria. *International Journal of Agricultural Economics and Rural Development, 2*(1), 10–36.

Adefolalu, D. O. (2007). *Climate change and economic sustainability in Nigeria*. Paper presented at the international conference on climate change and economic sustainability held at Nnamdi Azikiwe University, Awka, Nigeria. 12–14 June 2007.

Adegbola, A. A., & Onayinka, E. A. O. (1976). A review of range management problems in the Southern Guinea and derived Savanna Zones of Nigeria. *Tropical Grasslands, 10*(1), 41–51.

Adegoke, J., Araba, A., & Ibe, C. (2014). *National agricultural resilience framework*. Abuja: Federal Ministry of Agriculture and Rural Development.

Adejuwon, J. O. (2002). *Global warming and climate change: The global efforts towards reversing the trend*. Paper presented at a climate change workshop in University of Ibadan conference centre, Ibadan. February 2002.

Adejuwon, J. O. (2004). *Impacts of climate variability and climate change on crop yield in Nigeria*. Paper presented at the Stakeholder's workshop on assessment of impacts and adaptation to climate change, conference centre, Obafemi Awolowo University, Ile-Ife 20–21 September 2004

Adejuwon, J. O. (2006). Food crop production in Nigeria. II: Potential effects of climate change. *Climate Research, 32*, 229–245.

Adelekan, O. A. (2009). Vulnerability of poor urban coastal communities to climate change in Lagos, Nigeria. In *Fifth urban research symposium, Lagos*.

Adesina F. A., & Odekunle, T. O. (2011a). Climate change and adaptation in Nigeria: Some background to Nigeria's response—Part I. In *2011 international conference on environmental and agriculture engineering IPCBEE* (Vol. 15). Singapore: IACSIT Press

Adesina, F. A., & Odekunle, T. O. (2011b). Climate change and adaptation in Nigeria: Some background to Nigeria's response—Part III. In *2011 international conference on environmental and agriculture engineering IPCBEE* (Vol. 15). Singapore: IACSIT Press

Adger, W. N., Agrawala, S., Mirza, M. M. Q., Conde, C., O'Brien, K., Pulhin, J., et al. (2007). Assessment of adaptation practices, options, constraints and capacity in climatechange 2007: Impacts, adaptation, and vulnerability. Contribution of working group II to the fourth assessment report of the intergovernmental panel on climate change.

Annon. (2006). *National gender policy; Federal Ministry of Women Affairs and Social Development* (pp. 7–14). Amana Printing Limited: Kaduna.

Anríquez, G., & Stamoulis, S. (2007). Rural development and poverty reduction: Is agriculture still the key? *Journal of Agricultural and Development Economics, 4*(1), 5–46.
Ayoade, J. O. (2003). *Climate change*. Ibadan: Vantage Publishers.
Ayoade, J. O. (2004). *Introduction to climatology for the tropics* (2nd ed.). Ibadan: Spectrum Books.
European Commission—EUROPA. (2015). *Climate action: Adaptation to climate change*. Retrieved April 15, 2015 on http://ec.europa.eu/clima/policies/adaptation/index_en.htm.
Farmers' Academy Ede—FAE. (2013). *Our crops are failing*. Osogbo: Osun Press.
Federal Ministry of Agriculture and Rural Development. (2014). National Agricultural Resilience Framework. A report by Advisory Committee on Agricultural Resilience in Nigeria. Edited by Jimmy Adegoke, Adebisi and Chibi Ibe
Federal Ministry of Environment Abuja, Nigeria (Special Climate Change Unit) National Environmental, Economic and Development Study (Needs) for Climate Change in Nigeria (Final Draft) September 2010. Retrieved on April 10, 2015 on https://unfccc.int/files/adaptation/application/pdf/nigerianeeds.pdf
Food and Agricultural Organization FAO. (2007). *Climate change and food security*. Rome: FAO.
Hassan, R., & Nhemachena, C. (2008). Determinants of African farmers' strategies for adapting to climate change: Multinomial choice analysis. *AfJARE, 2*(1), 83–104.
Ikhile, C. I. (2007). Impacts of climate variability and change on the hydrology and water resources of the Benin-Owena River Basin. Ph.D. thesis submitted to the Department of Geography and Regional Planning, University of Benin, Benin City, Nigeria.
Intergovernmental Panel on Climate Change-IPCC. (2007). Summary for policymakers. In *Climate change 2007: Impacts, adaptation and vulnerability. Contribution of working group II to the fourth assessment report of the intergovernmental panel on climate change* (p. 17). Cambridge: Cambridge University Press.
Kaspar, T. C., Colvin, T. S., Jaynes, B., Karlen, D. L., James, D. E., & Meek, D. W. (2003). Relationship between six years of corn yields and terrain attributes. *Precision Agriculture, 4*, 87–101.
Lybbert, T., & Sumner, D. (2010). *Agricultural technologies for climate change mitigation and adaptation in developing countries: Policy options for innovation and technology diffusion*. Geneva: International Centre for Trade and Sustainable Development (ICTSD).
Mabo, C. B. (2006). Temperature variation in Northern Nigeria between 1970 and 2000. *Journal of Energy Environment, 19*(1), 80–88.
Mamadou, A., & Salaou, A. (2013). Women's land rights in a changing climate: A case study from Maradi, Niger. *A New dialogue: Putting people at the heart of global development*. Dublin: Hunger-Nutrition-Climate Justice.
National Emergency Management Agency—NEMA. (2012). *South–South lost N2.51trn to floods in 2012*. Retrieved December 10, 2012, from http://nema.gov.ng/south-south-lost-n2-51trn-to-floods-in-2012-says-nema/.
Nchuchuwe, F., & Adejuwon, K. (2012). The challenges of agriculture and rural development in Africa: The case of Nigeria. *International Journal of Academic Research in Progressive Education and Development, 1*(3), 43–45.
Nigerian Environmental Study/Action Team—NEST. (2011). Reports of research projects on impacts and adaptation. By Building Nigeria's Response to Climate Change (BNRCC) project, Ibadan, Nigeria
Nworgu, B. G. (2006). *Educational research: Basic issues and methodology* (2nd ed.). Nsukka: University Trust publishers.
Obioha, E. (2009). Climate variability, environmental change and food security nexus in Nigeria. *Journal of Human Ecology, 26*(2), 107–121.
Odjugo, P. A. O. (2010). General overview of climate change impacts in Nigeria. *Journal of Human Ecology, 29*(1), 47–55.
Odjugo, P. A. O., & Ikhuoria, A. I. (2003). The impact of climate change and anthropogenic factors on desertification in the semiarid region of Nigeria. *Global Journal of Environmental Science, 2*(2), 118–126.

Olaniran, O. J. (2002). *Rainfall anomalies in Nigeria: The contemporary understanding*. Paper presented at 55th inaugural lecture, University of Ilorin, Nigeria, 55.
Ologeh, I. O., Akarakiri, J. B., & Adesina, F. A. (2015). Promoting climate smart agriculture through space technology in Nigeria. In *Filho W.L. (2016) Innovation in climate change adaptation* (pp. 99–112). Berlin: Springer.
Oni, K.C. (2008). Agricultural growth and development in Kwara state: The way forward maiden lecture delivered at the fund raising for the building of Agriculture house. Ilorin: Kwara State Council of the Agricultural and Allied Employees' Union of Nigeria (AAEUN).
Parry, M. L, Rosenzweig, C., & Iglesias, A. (2001). *UNEP handbook on methods for climate change impact assessment and adaptation strategies* (pp. 8-1–8-89). UNEP.
Reilly, J. (1996). Climate change, global agriculture and regional vulnerability. A book chapter in *FAO's book—Global climate change and agricultural production. Direct and indirect effects of changing hydrological, pedological and plant physiological processes*. Rome: FAO
Susmita, D., Laplante, B., Meisner C., Wheeler, D., & Yan, J. (2007). The impact of sea level rise on developing countries. A comparative analysis. World Bank Policy Research Working Paper 4136.
United Nations Environment Programme—UNEP. (2008). *Lake Chad: almost gone*. Retrieved May 12, 2015 from http://www.unep.org/dewa/vitalwater/article116.html.
United Nations Framework Convention on Climate Change—UFCCC. (2014). *Land use, land-use change and forestry (LULUCF)*. Bonn: UNFCCC.
United States Environmental Protection Agency—USEPA. (2014). *Climate change impacts and adapting to change*. Retrieved March 31, 2015, from http://www.epa.gov/climatechange/impacts-adaptation/.
United States Global Change Research Program. (2009). Global climate change impacts in the United States. In T. R. Karl, J. M. Melillo, & T. C. Peterson (Eds.), *United States global change research program*. New York: Cambridge University Press.
Wisner, B., Blaikie, P., Cannon, T., & Davis, I. (2004). *At risk: Natural hazards; people's vulnerability and disasters* (2nd ed.). London: Routledge.

Chapter 10
Adapting Agriculture to Climate Change: Suitability of Banana Crop Production to Future Climate Change Over Uganda

Geoffrey Sabiiti, Joseph Mwalichi Ininda, Laban Ayieko Ogallo, Jully Ouma, Guleid Artan, Charles Basalirwa, Franklin Opijah, Alex Nimusiima, Saul Daniel Ddumba, Jasper Batureine Mwesigwa, George Otieno and Jamiat Nanteza

Introduction

Climate change has been identified as a major socioeconomic challenge both globally and regionally (Davidson et al. 2003; Fischer et al. 2005; IPCC 2012). On the global scale, Fischer et al. (2002) observed that the impacts of climate change on agricultural Gross Domestic Product (GDP) are likely to be relatively small ranging between −1.5 and +2.6% by 2080 depending on the scenario. The impacts of climate change will exhibit decreases in agricultural productivity in most developing regions including Uganda due to low adaptive capacity. On the regional scale, climate variability and change is expected to affect rainfall distribution patterns and possibly result in more intense dry spells in East Africa (Hulme et al. 2001; Sabiiti 2008; Shongwe et al. 2011; Washington and Pearce 2012; James et al. 2014). Arnell et al. (2002) and Gregory and Ingram (2000), pointed out that, even with a stabilization of carbon dioxide (CO_2), crop yields in many parts of Africa will still decrease by 2.5–5% by the 2080s. This was illustrated at the national levels

G. Sabiiti (✉) · C. Basalirwa · A. Nimusiima · S. D. Ddumba · J. Nanteza
Department of Geography, Geo-Informatics and Climatic Sciences, Makerere University, P. O. Box 7062, Kampala, Uganda
e-mail: geofsab@gmail.com

J. M. Ininda · L. A. Ogallo · F. Opijah
Department of Meteorology, University of Nairobi, P. O. Box 30197, Nairobi 00100, Kenya

G. Sabiiti · L. A. Ogallo · J. Ouma · G. Artan · J. B. Mwesigwa · G. Otieno
Inter-Governmental Authority on Development's (IGAD) Climate Prediction and Applications Centre (ICPAC), P. O. Box 10304, Nairobi 00100, Kenya

J. B. Mwesigwa
Institute for Climate Change and Adaptation, University of Nairobi, P. O. Box 30197, Nairobi 00100, Kenya

W. Leal Filho and J. Nalau (eds.), *Limits to Climate Change Adaptation*, Climate Change Management, https://doi.org/10.1007/978-3-319-64599-5_10

(Arnell et al. 2002) using models such as DSSAT to assess the impacts of climate change on crop yields.

Banana is one of the major food and income crops in Uganda. Due to its high dependence on rain-fed agriculture and limited adaptation to climate change, the sensitivity of the banana crop to climate change extremes is quite high (Van Asten et al. 2011; USAID 2013; Sabiiti et al. 2016). Several studies (e.g., Van Asten et al. 2004; Wairegi ct al. 2010; German et al. 2015; Sabiiti et al. 2016) have observed that banana productivity (production per hectare) has been declining in the recent years. This has been attributed to climatic and non-climatic factors. Van Asten et al. (2004) reported that actual banana yields in Uganda are low (5–30 t ha^{-1} yr^{-1}) compared to potential yields (70 t ha^{-1} y^{-1}). A number of yield loss factors have been studied by several authors. These studies have investigated the loss in banana productivity associated with deteriorating soil fertility (Bekunda and Woomer 1996; Gold et al. 1999a; Zake et al. 2000; Nyombi 2013), inadequate soil moisture and drought occurrences (Okech et al. 2004; Nyombi 2010; Van Asten et al. 2011; Sabiiti et al. 2016), banana weevils mainly *Cosmopolites Sordidus* (Gold et al. 1999c), banana parasitic nematodes including *Radolpholus Similis* and *Helicotylenchus Multicintus* (Speijer et al. 1999; Speijer and Kajumba 2000; Harper et al. 2004), and banana plant diseases like Black Singatoka, Banana Streak Virus and Banana Wilt (Tushemereirwe 2006). Wairegi et al. (2010) and Van Asten et al. (2010) observed that drought stress was the primary yield constraint in a quarter of studied farmer fields in southwest Uganda. Gold et al. (1999b) also observed shifts in highland cooking banana production that was attributed to a number of factors. This has affected food production, food security (Parry et al. 1999) and farmers' incomes in Uganda exposing them to high vulnerability to climate change impacts.

Nyombi (2010) observed that the banana crop requires mean monthly temperatures of about 25–27 °C for optimal growth and further indicated that high rainfall of 1400 mm per annum was directly proportional to high banana productivity. Nyombi (2010) noted that rainfall is a crucial factor in banana production because the ability of the plant to take up nutrients is related to the soil moisture content that is mainly supplied through rainfall.

Future climate projections based on climate model simulations indicate and increase in temperature levels with some regions likely to receive depressed rainfall. A set of emissions scenarios which are consistent socio-economic assumptions of how the future may evolve have been used in projection of climate change. Two sets of climate scenarios were used in this study. These comprise the new Representative Concentration Pathways (RCPs, IPCC 2014) and Special Report on Emission Scenarios (SRES). The RCPs describe different future world with at different levels of radiative forcing by the year 2100, ranging from 2.6 W/m^2 (best case, very strong mitigation) to 8.5 W/m^2 (worst case, no mitigation). The RCPs have been designed to cover a wide range of possible magnitudes of climate change in models rather than being derived sequentially from storylines of socio-economic futures as had been done in earlier SRES scenarios particularly A1B and A2 also used in this paper. The high mitigation (RCP 2.6, less adaptation), medium mitigation (RCP 4.5 and RCP 6.0, medium adaptation), no mitigation (RCP 8.5, very high adaptation) scenarios, SRES A1B and A2 scenarios represent a spectrum of possible future climate states.

This paper is motivated by the fact that future occurrences of climate change extremes are likely to exacerbate the decline in productivity of bananas and other crops through direct and indirect impacts of climate change (Ingram and Dawson 2005). Scientific information that can guide planning adaptation strategies is vital if farmer communities are to promote agricultural productivity in Uganda (Hisali et al. 2011; Dekens and Bagamba 2014). The purpose of the study was to evaluate the effects of climate change on banana growth and production in Uganda. Banana growth suitability indices based on future rainfall and temperature conditions (best case and worst case scenarios) and breaking points (adaptation limits) for Uganda's agricultural sectors were identified.

Area of Study

This study was conducted over Uganda, a country that lies astride the Equator, between latitudes 4°12′N and 1°29′S and longitudes 29°34′E, and 35°0′E (Fig. 10.1). More than two-thirds of the Country is a plateau, lying between 1000 and 2500 m above mean sea level (MSL). The neighbouring countries provide an opportunity for demand and ready market for the agricultural products, which promotes farming activities and incomes of farmers. Most of the Country experiences an equatorial type of climate with some parts in the northern and northeastern regions experiencing relatively dry conditions particularly during La Niña years. Average temperatures are in the range of 15–30 °C (USAID 2013). Rainfall is fairly reliable, with most parts of the Country experiencing a bimodal rainfall regime with March to May (MAM) as the "long" rain season and October to December (OND) as the "short" rain season. Table 10.1 gives the description of the seven sub regions considered in data extraction.

Data and Methods

Data used in this study included projected rainfall and temperature for the period 2011–2090. This climate data was generated under the fifth phase of the Coupled Model Inter-comparison Project (CMIP5) climate projections (Giorgi et al. 2009; Jones et al. 2011; Hewitson et al. 2012). The study used high resolution ($\sim$1 km) data on combined bioclimatic variables (rainfall and temperature, Hijmans et al. 2005) to map suitability zones of the banana crop while the Providing Regional Climate for Impacts Studies (PRECIS) regional climate model temperature simulations were used to estimate the effect of rising temperature on banana growth assuming other factors constant. The downscaled future climate projections were

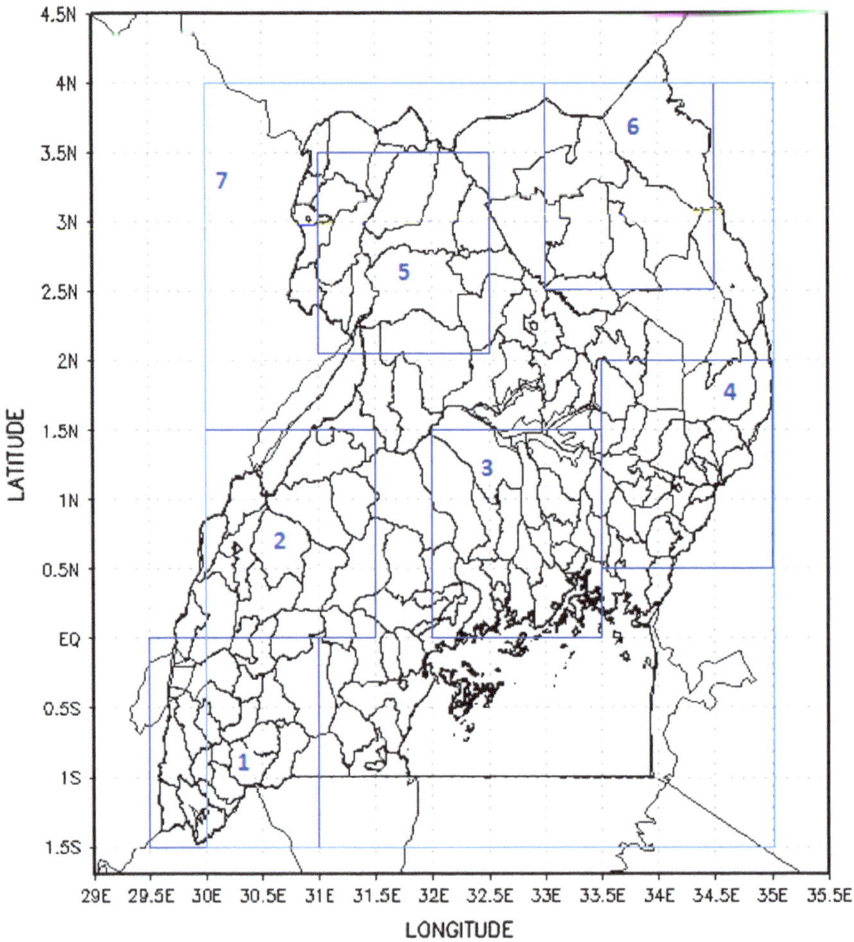

Fig. 10.1 Analysis regions (1–7) over the study area

Table 10.1 Analysis regions (also Fig. 10.1) over Uganda

Region	Latitude interval	Longitude interval	Region description
Region 1	1.5S–0	29.5E–31.0E	Southwestern
Region 2	0–1.5N	30.0E–31.5E	Western
Region 3	0–1.5N	32.0E–32.5E	Central
Region 4	0.5N–2N	33.5E–35.0E	Eastern
Region 5	2.0N–3.5N	31.0E–32.5E	Northwestern
Region 6	2.5N–4.0N	33.0E–34.5E	Northeastern
Region 7	1.5S–4N	30.0E–35.0E	Uganda

based on the Intergovernmental Panel on Climate Change (IPCC) Representative Concentration Pathways (RCPs, 2.6, 4.5, 6.0 and 8.5, Moss et al. 2010; IPCC 2014) and Special Report on Emission Scenarios (SRES, A1B and A2, Nakicenovic and Swart 2000; IPCC 2007) across the period 2011–2090. The approach used sought to understand future trends in banana growth between two periods (2011–2040 and 2061–2090) due to expected changes in rainfall and temperature over selected sub regions of Uganda (Fig. 10.1).

The study determined the potential impacts (Ingram and Dawson 2005) of projected future climate on banana production under different future climate scenarios based on AR4 SRES and AR5 RCPs. The climate change projections were used to examine the potential effects of such climate changes on growth patterns and changes in suitability of production of bananas across Uganda. Apart from comparing the projected climate change scenarios with the specific climate thresholds for banana growth (Table 10.2), the FAO ECO-Crop tool (Ramirez-Villegas et al. 2011b; Van den Bergh et al. 2012) was used to determine and map future suitability of banana growth in Uganda. This tool provided additional information on the thresholds that were used to classify production patterns (suitability) for the four bioclimatic variables under the high mitigation (RCP 2.6), medium mitigation (RCP 4.5 and RCP 6.0) and no mitigation (RCP 8.5) scenarios.

This study identified classes of rainfall and temperature for different banana suitability classes. Banana production indices were determined using a suitability model in the Geographical Information System (GIS) spatial analyst tool. The model was based on the classes of the bioclimatic variables in Table 10.2. Using suitability mapping techniques, Uganda was delineated into six zones for (1) excellent (86–100%), (2) very suitable (71–85%), (3) suitable (56–70%), (4) marginal (41–55%), (5) very marginal (26–40%) and (6) unsuited (less than 25%). The suitability patterns

Table 10.2 Climate thresholds for GIS based banana production suitability mapping

Suitability classes	Annual average temperature (°C)	Temperature of coldest quarter (°C)	Annual rainfall (mm)	Rainfall of driest quarter (mm)
1 (Excellent)	3 (24.1–25)	4 (23.1–24)	6 (>1800)	6 (>320)
2 (Very suitable)	1 (<23)	2 (21.1–22)	5 (1601–1800)	5 (271–320)
3 (Suitable)	2 (23.1–24)	1 (<21)	4 (1401–1600)	4 (221–270)
4 (Marginal suited)	4 (25.1–26)	3 (22.1–23)	3 (1201–1400)	3 (121–220)
5 (Very marginal)	5 (26.1–27)	5 (24.1–25)	2 (1001–1200)	2 (71–120)
6 (Not suited)	6 (>27)	6 (>25)	1 (<1000)	1 (<70)

were constructed based on two periods (2041–2060 and 2061–2080) and also across a full range of AR5 RCPs that included RCP 2.6, RCP 4.5, RCP 6.0, and RCP 8.5 (Hijmans et al. 2005; Moss et al. 2010; IPCC 2014). A similar approach to assess global banana production and suitability under climate change scenarios based on SRES A2 scenario with specific reference on the tropics and subtropics has been used in Bouwmeester et al. (2009) and German et al. (2015).

The non-linear banana temperature regression model (Turner and Lahav 1983; Sastry 1988) was used to assess the impact of future changes in temperature on banana growth. Increase in banana growth translates into increased bunch size and also shorter banana harvest cycles (high banana productivity). The temperature constrained banana growth is given by Eq. 10.1.

$$V_t = 75.35 * e^{-0.0.5 * (t-26.3)^2} \quad (10.1)$$

In Eq. (10.1), V_t is the percentage growth rate of banana crop at a given temperature t (°C), t = 26.3 °C. A temperature of 26.3 °C was identified from experimental studies and used as the optimal surface temperature for banana growth in banana ecological zones (Nyombi 2010; German et al. 2015) over Uganda. The projected temperatures for different climate scenarios were super-imposed on the banana growth-temperature curve for the two periods which provided an indication of the anticipated growth under temperature projections. The major limitation of this research and the results in Section "Results and Discussion" is the assumption that climate factors will be more important than non-climatic factors in future to define banana production zones and growth patterns. In addition, climate projections used might exhibit some uncertainty that might limit the level of accuracy in the results. The results, however, remain useful for planning adaptation options to ameliorate future impacts of climate change in Uganda and neighboring regions (Fig. 10.2).

Results and Discussion

Figures 10.3, 10.4 and 10.6 show results of suitability conditions for the growth of bananas under different projected rainfall and temperature conditions for the RCP 2.6 (Fig. 10.3), RCP 4.5 (Fig. 10.4), RCP 6.0 (Fig. 10.5) and RCP 8.5 (Fig. 10.6) for two periods 2041–2060 and 2061–2080. Under RCP 2.6, it is observed that most of central and western Uganda is likely to remain suitable with banana growth rates of about 56–70% in the 2040–2061 period. The central region (north of Lake Victoria) is projected to be excellent with banana growth rate above 86%. The eastern and southwestern Uganda is projected to be very suitable with banana growth rate of about 71–85%. The northern region (particularly north eastern) is likely to experience very marginal (26–40%) to not suitable (less than 25%) conditions for banana production (Fig. 10.3a). Similar conditions are expected for the period 2061–2080 under the RCP 2.6 (Fig. 10.3b) with a general slight reduction in

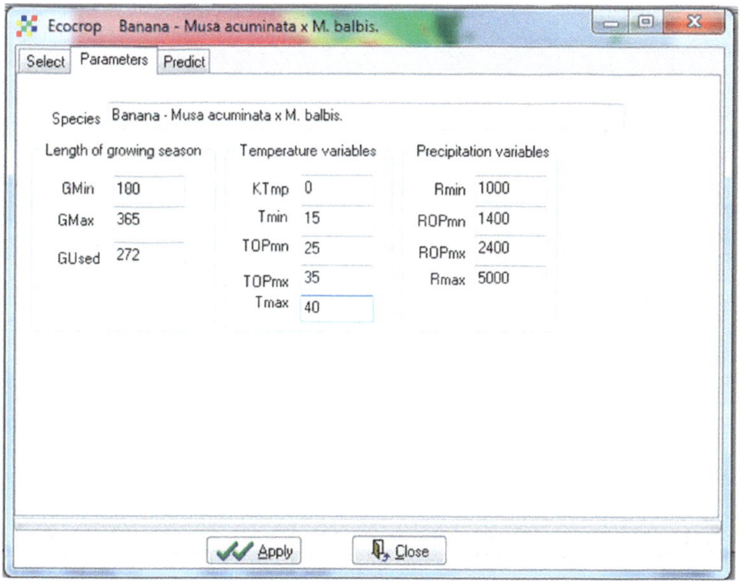

Fig. 10.2 Modelled rainfall and air temperature requirements for banana growth based on the FAO ECOCROP tool

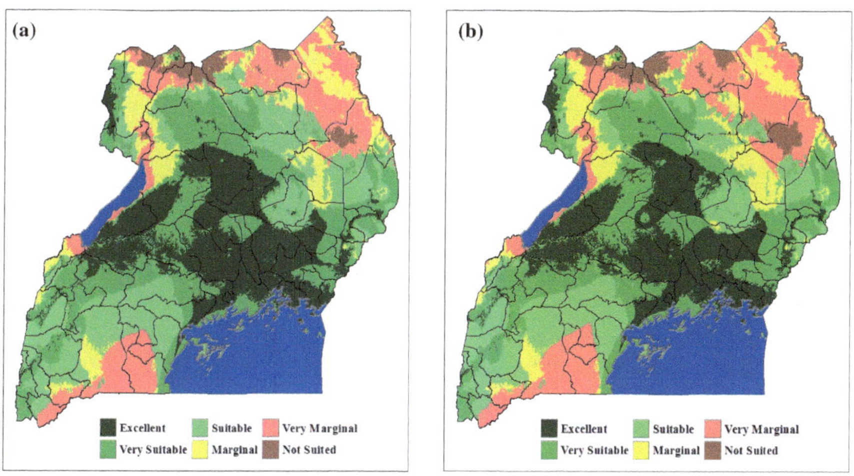

Fig. 10.3 Projected future suitability of banana growth under RCP 2.6 for 2041–2060 (**a**) and 2061–2080 (**b**)

the areas suitable for banana production. This may be attributed to increase in temperature that will be accompanied by slight reductions in annual rainfall over some parts of the Country under the RCP 2.6 between the two periods considered in this study.

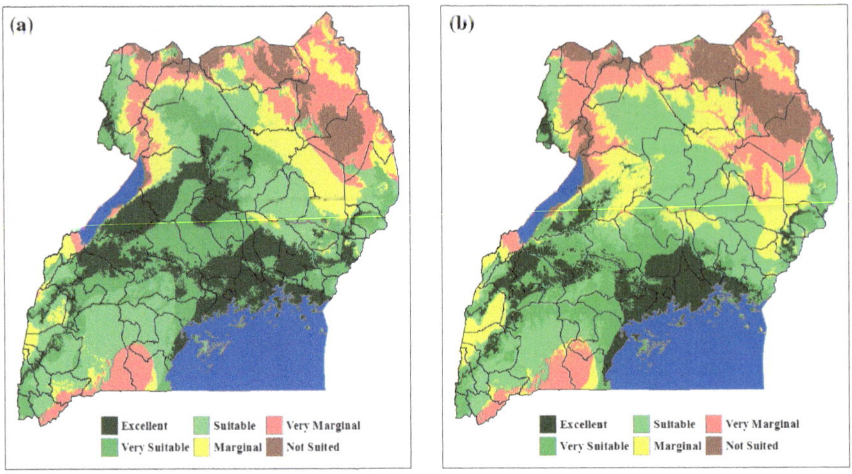

Fig. 10.4 Projected future suitability of banana growth under RCP 4.5 for 2041–2060 (**a**) and 2061–2080 (**b**)

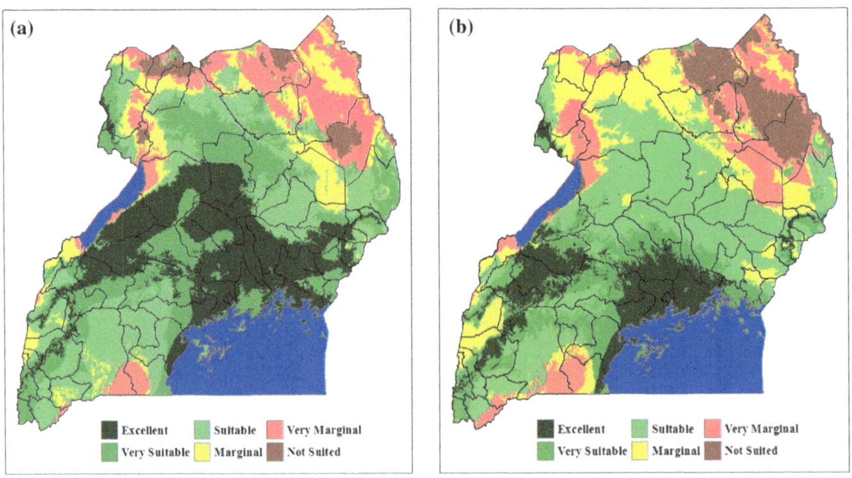

Fig. 10.5 Projected future suitability of banana growth under RCP 6.0 for 2041–2060 (**a**) and 2061–2080 (**b**)

Under the RCP 4.5 (Fig. 10.4), there is a notable reduction in the area suitable for the production of bananas as compared to the RCP 2.6 conditions during the 2041–2060 period. The climatic conditions are expected to become less conducive for the production of the banana crop under the RCP 4.5 relative to RCP 2.6 during the period 2041–2060. The climatic conditions for the future (2061–2080) worsen under the RCP 4.5 scenario with most of the country becoming slightly less suitable for banana growth. Under RCP 4.5 only areas around Lake Victoria show excellent

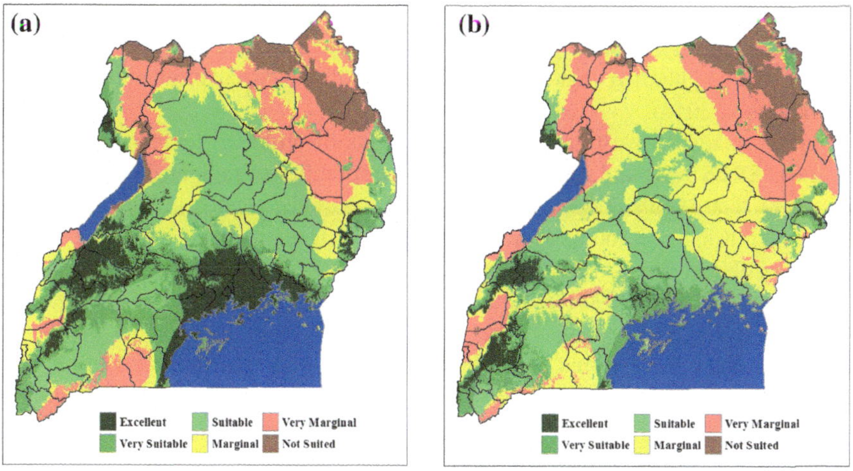

Fig. 10.6 Projected future suitability of banana growth under RCP 8.5 for 2041–2060 (**a**) and 2061–2080 (**b**)

conditions while the northern and northwestern and northeastern remain marginal (26–40%) to not suitable (less than 25%) for production of bananas (Fig. 10.4b).

In the RCP 6.0 scenario (Fig. 10.5) there is high potential for production of bananas (suitable to excellent, 56–100%) during the period 2041–2060 compared to all other scenarios over most of the country. It was further observed that banana production during this period is likely to remain highest compared with all scenarios and periods. The study further observed that, the expected decrease in banana production is significant during 2061–2080 with most of the country indicating suitable to very suitable (56–85%, Fig. 10.5b) under RCP 6.0.

For RCP 8.5, the study observed that during 2041–2060 (Fig. 10.6a), projected rainfall and temperature conditions will favor higher production of bananas over central, western and south western Uganda. On the other hand, the period 2061–2080 is likely to experience climatic conditions that marginally (less that 40%) favor banana production over most parts of Uganda (Fig. 10.6b). The results showed that most parts including central will be marginal to not suitable for production of bananas under this scenario. Generally, the results have observed the worse anticipated condition for banana production in the future under RCP 8.5 due to very high temperatures associated with the scenario.

The best case scenario for production of bananas includes RCP 2.6 and RCP 6.0. The change in production of banana is sudden and high under RCP 6.0 between the two periods while that under RCP 2.6 is small between the two periods. RCP 4.5 projects climatic conditions that favor banana production better than the RCP 8.5 over most of Uganda. For example, the results have indicated that a larger (smaller) area suitable for banana production under RCP 2.6 and RCP 6.0 (RCP 4.5 and RCP 8.5) for the intermediate period 2041–2060 over Uganda. Similar results have been reported in Ramirez-Villegas et al. (2011a) and Van den Bergh et al. (2012), which investigated climate change patterns and impacts on the growing conditions of bananas in the tropics. Effective and integrated adaptation options are required if Uganda and neighboring regions are to continue high bananas production (Dekens and Bagamba 2014).

Further analysis investigated the likely effects of projected temperatures under the IPCC SRES A1B and A2 on banana growth in different regions of Uganda (see Fig. 10.1). The results (Figs. 10.7, 10.8, 10.9) show the change in banana growth due to increasing minimum, average and maximum temperature under the SRES A1B for different region of Uganda. The period 2011–2040 (near future) is being contrasted with 2061–2090 (far future) during a wet (MAM) and dry (JJA) season. The study observed that temperatures (21–22 °C) projected under A1B SRES are still lower than the optimal temperature (26.3 °C) for banana production and increasing temperatures would favor high growth (Fig. 10.7). This is likely to promote high banana growth from lower than 50% to over 60% in these regions under SRES A1B scenario rainfall and other conditions. In contrast, under SRES A2 scenario, temperatures are already close to optimal temperature for banana

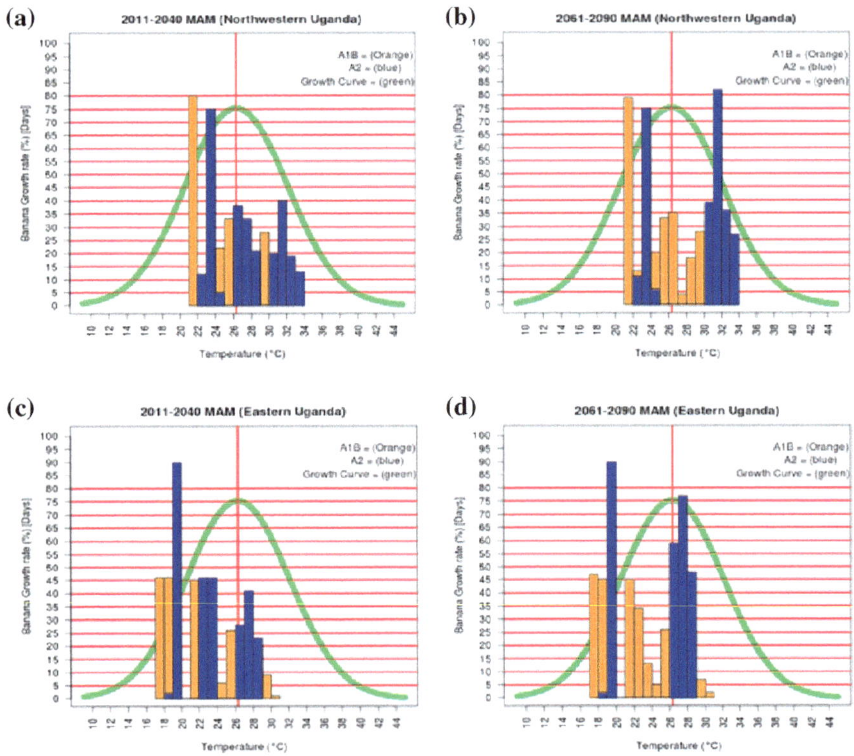

Fig. 10.7 Projected effects of temperature on banana growth during MAM season for northwestern (**a**, **b**) and Eastern (**c**, **d**) region for 2011–2040 and 2061–2090. *Orange bars* A1B, *blue bars* A2, *green curve* temperature-banana growth

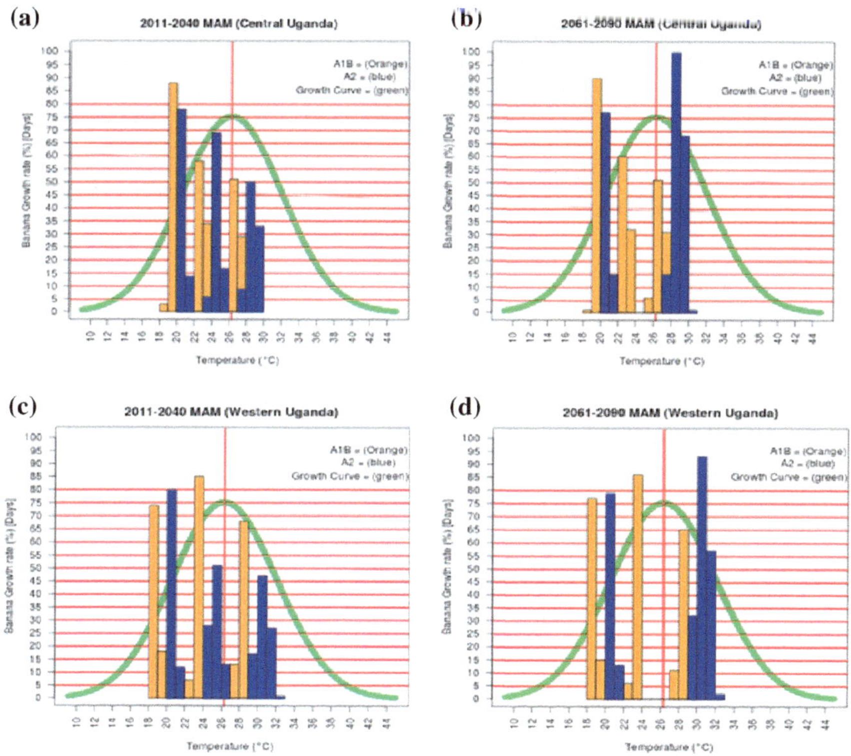

Fig. 10.8 Projected effects temperature on banana growth during MAM for central (**a**, **b**) and western (**c**, **d**) region for 2011–2040 and 2061–2090. *Orange bars* A1B, *blue bars* A2, *green curve* temperature-banana growth

growth. The study observed that any further increases in projected temperatures would retard banana growth from over 60% to about 45% or lower in some areas (Fig. 10.7a, b). Over the eastern region, temperatures projections under both SRES A1B and A2 scenarios are slightly lower than optimal conditions for maximum banana growth. For example, the further increases in projected temperatures under A1B for the period 2061–2090 favor high banana growth from less than 45% to over 65% in the eastern region (Fig. 10.7c, d).

Figure 10.8a–d show results for central and western Uganda respectively. Over the central region, projected temperature increase are expected to promote banana growth from about 45% to over 55% under SRES A1B scenario while the projected increase in temperature under A2 scenario will most likely retard banana growth (Fig. 10.8a, b). Banana growth will retard towards the end of the century under both scenarios.

Figure 10.9a–d presents results for southwestern Uganda during MAM and JJA seasons. Over the southwestern parts of Uganda, temperature changes projected under both SRES A2 and A1B scenario will favor growth and productivity of bananas

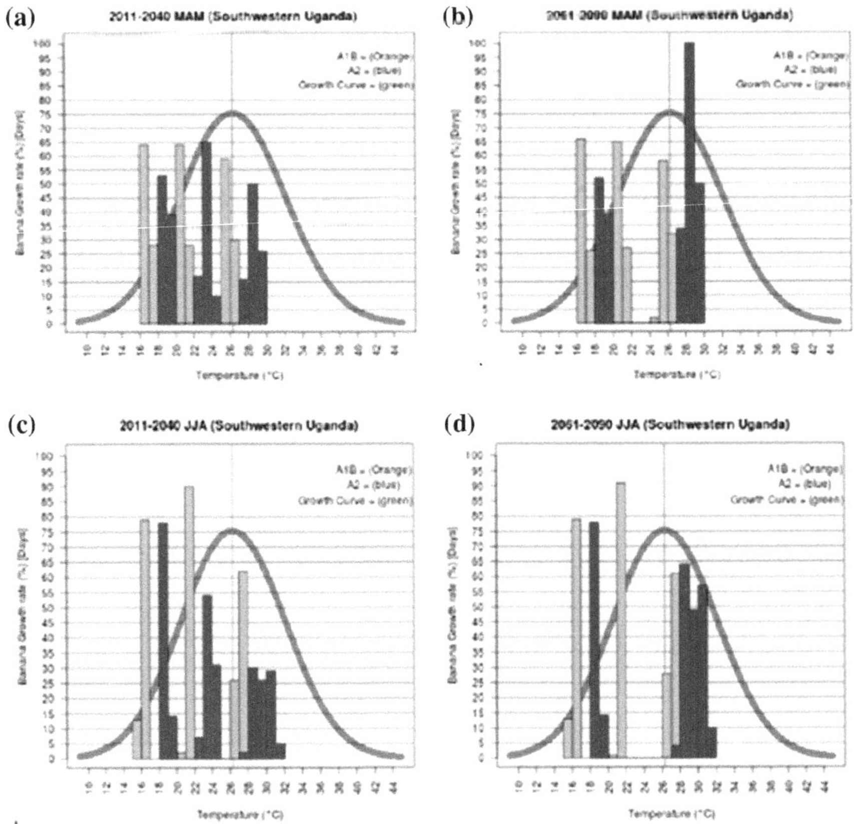

Fig. 10.9 Projected effects temperature on banana growth for MAM (**a**, **b**) and JJA (**c**, **d**) over southwestern region for 2011–2040 and 2061–2090. *Orange bars* A1B, *blue bars* A2, *green curve* temperature-banana growth (Color figure online)

during the MAM season. Due to low temperature projected for southwestern Uganda under SRES A1B, an increase in temperatures (20–22 °C) will favor high banana growth of about 65% in the near future (2011–2040) under this scenario. During the period 2011–2040, projected temperatures in SRES A2 may favor banana growth and negatively affect banana growth in the far future (2061–2090) over the southwestern region (Fig. 10.9a, b). During JJA, temperature increases projected under SRES A1B scenarios favor banana growth with increase from 45% to over 53% while the projected temperatures under SRES A2 will negatively affect banana growth in most area (Fig. 10.9c, d) especially by end of the century.

The results have for the first time shown that the projected temperature increases under SRES A1B will enhance banana growth from about 45% to over 60% while further increases in projected temperatures are likely to retard banana growth under SRES A2 from 55% to lower than 40% in most areas due to the already warmer temperatures and depressed rainfall projected under the SRES A2 scenario. In conclusion, the study observed that favorable conditions for production of banana over Uganda are expected under SRES A1B scenario due to enhanced rainfall and relatively low temperature while A2 scenario may have low rainfall and relatively high temperature that will limit the production of banana to only isolated areas such as those around Lake Victoria basin by end the century. This part of the study has observed that suitability levels of banana production will significantly reduce in the far future period across all the RCPs scenarios. The effects of future climate change on banana production are, however, region specific and vary across climate scenarios. The results have further observed that the northern part of the country remains unsuitable for banana production under all scenarios. In addition, SRES A1B with enhanced rainfall and relatively low temperature increase is likely to favor high future banana production while A2 with decrease in rainfall and high temperature increase will stress banana production over most parts of Uganda. There is therefore a need to promote climate smart agricultural policies (Dekens and Bagamba 2014; Ampaire et al. 2015) and adaptation practices to the impacts of climate change in the country in addition to mitigation of climate change through the reduction of GHGs emission. Hisali et al. (2011) points out that access to reliable off-farm employment could improve the chances of accessing credit during times of need/hardship, which reduces cash constraints and allows farmers to purchase food and inputs such as drought tolerant varieties.

Conclusion

The results revealed unique and distinct banana production suitability and growth patterns for each climate scenarios in the sub-periods. RCP 2.6 and RCP 6.0 are likely to be associated with higher levels of banana production suitability than RCP 4.5 and RCP 8.5. The results further showed that projected temperature increase and enhanced rainfall under SRES A1B will promote banana growth. In contrast, expected increases in temperatures and depressed rainfall under SRES A2 are likely to retard banana growth due to high moisture deficits. The projected patterns of future temperature and rainfall will have varying potential impacts on the production and growth of bananas in Uganda. The effectiveness of adaptation options needed to combat the impacts will be influenced by the magnitude of the expected climatic changes associated with each scenario, the timing of expected climate change extremes and sensitivity of the crop to climate.

In order to offset the projected (likely) negative impacts of increasing temperatures and depressed rainfall under SRES A2 and RCP 8.5, farmers would have to practice moisture conservation farming practices such as mulching, shade systems, agro-forestry and water harvesting to sustain moisture in banana plantations in most parts of Uganda. In addition, application of manure and fertilizers will be necessary to ensure that banana plants can get the nutrients from top layers of the soil. Introduction of drought tolerant banana varieties is necessary for sustainable banana production under the worst case scenarios of climate change. This study has provided critical information that will be useful for planning integrated adaptation practices in the banana farming subsector to promote productivity.

Acknowledgements This work was jointly funded by the Makerere University, IGAD Climate Prediction and Applications Centre (ICPAC) and Department for International Development (DfID) of UK Government. Data/materials were sourced from the Uganda National Meteorological Authority (UNMA), Ministry of Agriculture, Animal Industry and Fisheries (MAAIF), International Institute for Tropical Agriculture and (IITA), http://www.worldclim.org/cmip5_30s. I wish to thank the staff of these institutions for the support to accomplish this work.

References

Ampaire, E. L., Happy, P., Van Asten, P. J., & Radeny, M. (2015). The role of policy in facilitating adoption of climate-smart agriculture in Uganda. CGIAR Research Program on Climate Change, Agriculture and Food Security (CCAFS). Copenhagen, Denmark. www.ccafs.cgiar.org.

Arnell, N., Warren, R., Nicholls, R., Levy, P., & Price, J. (2002). Understanding the regional impacts of climate change; Research report prepared for the Stern Review, Tyndall Centre Working Paper 90. Norwich: Tyndall Centre. Available from http://www.tyndall.ac.uk/publications.

Bekunda, M., & Woomer, P. L. (1996). Organic resource management in banana based cropping systems of the Lake Victoria basin, Uganda. *Agriculture, Ecosystems & Environment, 59,* 171–180.

Bouwmeester, H., Van Asten, P., & Ouma, E. (2009). Mapping key variables of banana based cropping systems in the Great Lakes Region, partial outcomes of the baseline and diagnostic surveys. CIALCA Technical Report 11. http://www.cialca.org/IITA, Ibadan www.cialca.org.

Davidson, O., Halsnaes, K., Huq, S., Kok, M., Metz, B., Sokona, Y., et al. (2003). The development and climate nexus: The case of sub-Saharan Africa. *Climate Policy, 3,* 97–113.

Dekens, J., & Bagamba, F. (2014). Promoting an integrated approach to climate adaptation: Lessons from the coffee value chain. Climate resilient value chains and food systems. Briefing note series. Climate Development Knowledge Network.

Fischer, G., Mahendra, S., & Velthuizen, H. V. (2002). Climate change and agricultural vulnerability. A special report prepared by the International Institute for Applied Systems Analysis as a contribution to the World Summit on Sustainable Development, Johannesburg 2002.

Fischer, G., Shah, M., Tubiello, F. N., & Velhuizen, H. (2005). Socio-economic and climate change impacts on agriculture. *Philosophical Transactions of the Royal Society B, 360,* 2067–2083.

German, C. G., Staver, C., & Siles, P. (2015). An assessment of global banana production and suitability under climate change scenarios. In A. Elbehri (Ed.), *Climate change and food systems: Global assessments and implications for food security and trade*. Rome: Food and Agriculture Organization of the United Nations (FAO).

Giorgi, F., Jones, C., & Asrar, G. R. (2009). Addressing climate information needs at the regional level: The CORDEX framework. *WMO Bulletin, 58*(3), 175–183.

Gold, C. S., Karamura, E. B., Kiggundu, A., Bagamba, F., & Abera, A. M. K. (1999a). Monograph on geographic shifts in highland cooking banana (Musa, group AAA-EA) production in Uganda. *African Crop Science Journal, 7,* 223–298.

Gold, C. S., Karamura, E. B., Kiggundu, A., Bagamba, F., & Abera, A. M. K. (1999b). Geographic shifts in highland cooking banana (Musa spp., group AAA-EA) production in Uganda. *International Journal of Sustainable Agriculture and World Ecology, 6,* 45–59.

Gold, C. S., Rukazambuga, N. D. T. M., Karamura, E. B., Nemeye, P., & Night, G. (1999c). Recent advances in banana weevil biology, population dynamics and pest status with emphasis on East Africa. In E. A. Frison, C. S. Gold, E. B Karamura, & R. A. Sikora, (Eds.),

Proceedings of a workshop on banana IPM held in Nelspruit, South Africa, 23–28 November 1998 (Vol. 59, pp. 35–50, 171–180).

Gregory, P. J., & Ingram, J. S. I. (2000). Global change in food and forest production: Future scientific challenges. *Agriculture, Ecosystems & Environment, 82,* 3–14.

Harper, G., Hart, D., Moult, S., & Hull, R. (2004). Banana streak virus is very diverse in Uganda. *Virus Research, 100,* 51–56.

Hewitson, B., Lennard, C., Nikulin, G., & Jones, C. (2012). CORDEX-Africa: A unique opportunity for science and capacity building. *CLIVAR Exchanges, 17*(3), 6–7.

Hijmans, R. J., Cameron, S. E., Parra, J. L., Jones, P. G., & Jarvis, A. (2005). Very high resolution interpolated climate surfaces for global land areas. *International Journal of Climatology, 25,* 1965–1978.

Hisali, E., Birungi, P., & Buyinza, F. (2011). Adaptation to climate change in Uganda: Evidence from micro level data. *Global Environmental Change, 21,* 1245–1261.

Hulme, M., Doherty, R., Ngara, T., New, M., & Lister, D. (2001). African climate change: 1900–2100. *Climate Research, 17,* 145–168.

Ingram, J. C., & Dawson, T. P. (2005). Climate change impacts and vegetation response on the island of Madagascar. *Philosophical Transactions of the Royal Society of London, Series A, Mathematical Physical and Engineering Sciences, 363*(1862), 55–59.

IPCC. (2007). In M. L. Parry, O. F. Canziani, J. P. Palutikof, P. J. van der Linden, & C. E. Hanson, (Eds.), *Climate change (2007) impacts, adaptation and vulnerability: Contribution of working group II to the fourth assessment report of the intergovernmental panel on climate change* (pp. 433–467). Cambridge: Cambridge University Press.

IPCC. (2012). In C. B. Field, V. Barros, T. F. Stocker, D. Qin, D. J. Dokken, K. L. Ebi, M. D. Mastrandrea, K. J. Mach, G.-K. Plattner, S. K. Allen, M. Tignor, & P. M. Midgley (Eds.), *Managing the risks of extreme events and disasters to advance climate change adaptation. A special report of working groups I and II of the intergovernmental panel on climate change* (p. 582). Cambridge: Cambridge University Press.

IPCC. (2014). In T. F. Stocker, D. Qin, G. K. Plattner, M. Tignor, S. K. Allen, J. Boschung, A. Nauels, Y. Xia, V. Bex, & P. M. Midgley, (Eds.), *Climate change (2014) the physical science basis. Contribution of working group I to the fifth assessment report of the intergovernmental panel on climate change* (p. 1535). Cambridge: Cambridge University Press.

James, R., Washington, R., & Rowell, D. P. (2014). African climate change uncertainty in perturbed physics ensembles: Implications of warming to 4 °C and beyond. *Journal of Climate, 27,* 4677–4692.

Jones, C., Giorgi, F., & Asrar, G. (2011). The coordinated regional downscaling experiment: CORDEX—An international downscaling link to CMIP5. *CLIVAR Exchanges, 16*(2), 34–40.

Moss, R. H., Edmonds, J. A., Hibbard, K. A., Manning, M. R., Rose, S. K., Van Vuuren, D. P., et al. (2010). The next generation of scenarios for climate change research and assessment. *Nature, 463,* 747–756.

Nakicenovic, N., Swart, R. (2000). In IPCC, N. Nakicenovic, & R. Swart, (Eds.), *Special report on emissions scenarios*. Cambridge: Cambridge University Press.

Nyombi, K. (2010). *Understanding growth of East Africa highland banana: Experiments and simulation*. Ph.D. Thesis, Wageningen University, Netherlands.

Nyombi, K. (2013). Towards sustainable highland banana production in Uganda: Opportunities and challenges. *African Journal for Food, Agriculture, Nutrition and Development, 13,* 7544–7561.

Okech, S. H., Asten, P. J. A., Gold, C. S., & Ssali, H. (2004). Effects of potassium deficiency, drought and weevils on banana yield and economic performance in Mbarara, Uganda. *Uganda Journal of Agricultural Sciences, 9,* 511–519.

Parry, M., Rosenzweig, C., Iglesias, A., Fischer, G., & Livermore, M. (1999). Climate change and world food security: A new assessment. *Global Environmental Change, 9,* 51–67.

Ramirez-Villegas, J., Jarvis, A., & Laderach, P. (2011a). Empirical approaches for assessing impacts of climate change on agriculture: The EcoCrop model and a case study with grain sorghum. *Journal of Agricultural and Forest Meteorology, 170*(2013), 67–78.

Ramirez-Villegas, J., Jarvis, A., Van den Bergh, I., Staver, C., & Turner, D. (2011b). Changing climates. Effects on growing conditions for banana and plantain (*Musa* spp.) and possible responses. In S. S. Yadav, R. Redden, H. Lotze-Campen, & A. Hall (Eds.), *Crop adaptation to climate change* (pp. 426–438). Oxford: Wiley-Blackwell.

Sabiiti, G. (2008). *Simulation of climate scenarios using the PRECIS Regional Climate Model over the Lake Victoria basin*. M.Sc. Dissertation, Department of Meteorology, University of Nairobi, Kenya.

Sabiiti, G., Ininda, J. M., Ogallo, L. A., Opijah, F., Nimusiima, A., Otieno, G., et al. (2016). Empirical relationship between banana yields and climate variability over Uganda. *International Journal of Environmental and Agricultural Sciences, 7,* 03–13.

Sastry, P. S. N. (1988). *Agrometeorology of the banana crop*. Geneva: World Meteorological Organization.

Shongwe, M. E., Van Oldenborgh, G. J., Van den Hurk, B., & Van Aalst, M. (2011). Projected changes in mean and extreme precipitation in Africa under global warming. Part II: East Africa. *Journal of Climate, 24*(14), 3718–3733.

Speijer, P. R., & Kajumba, C. (2000). Yield loss from plant parasitic nematodes in East African highland banana (*Musa* spp. AAA). In K. Craenen, R. Ortiz, B. E. Karamura, & D. R. Vuylsteke (Eds.), *Proceedings of the first international conference on banana and plantain for Africa, 14–18 October, 1996* (Vol. 540, pp. 453–459). Kampala, Uganda: Acta Horticulture.

Speijer, P. R., Kajumba, C., & Sango, F. (1999). East African Highland banana production as influenced by nematodes and crop management in Uganda. *International Journal of Pest Management, 45,* 41–49.

Turner, D. W., & Lahav, E. (1983). The growth of banana plants in relation to temperature. *Australian Journal of Plant Physiology, 10,* 43–54.

Tushemereirwe, W. K. (2006). Experiences with banana bacterial wilts: The national strategy for the management of BXW in Uganda. In E. B. Karamura, M. Osiru, G. Blomme, C. Lusty, & C. Picq (Eds.), *Proceedings of the Banana Xanthomonas Wilt regional preparedness and strategy development workshop. Kampala, Uganda, 14–18 February 2005*. Montpellier: International Plant Genetic Resources Institute, 13–16.

USAID. (2013). Uganda climate change vulnerability assessment report. A report of the African and Latin American resilience to climate change project (pp. 1–78). Available online at: http// www.community.eldis.org.

Van Asten, P. J., Fermont, A. M., & Taulya, G. (2011). Drought is a major yield loss factor for rain-fed East African highland banana. *Agricultural Water Management, 98,* 541–552.

Van Asten, P. J., Nikolaou, G. N., & Grone, A. (2010). The occurrence of cpb2-toxigenic Clostridium perfringens and the possible role of the β2-toxin in enteric disease of domestic animals, wild animals and humans. *The Veterinary Journal, 183*(2), 135–140.

Van Asten, P. J., Verhaest, I., Lamzira, S., Hernandez-Aguado, I., Zangerle, R., Boufassa, F., et al. (2004). Spread of hepatitis C virus among European injection drug users infected with HIV: A phylogenetic analysis. *Journal of Infectious Diseases, 189*(2), 292–302.

Van den Bergh, I., Ramirez, J., Staver, C., Turner, D., Jarvis, A., & Brown, D. (2012). Climate change in the subtropics: The impacts of projected averages and variability on banana productivity. *Acta Horticulturae, 928,* 89–99.

Wairegi, L. W. I., Van Asten, P. J., Tenywa, A., & Bekunda, M. (2010). Abiotic constraints override biotic constraints in East African highland banana systems. *Field Crops Research, 117,* 146–153.

Washington, R., & Pearce, H. (2012). Climate change in East African agriculture: Recent trends, current projections, crop-climate suitability, and prospects for improved climate model information. CGIAR Research Program on Climate Change, Agriculture and Food Security (CCAFS), Copenhagen, Denmark. www.ccafs.cgiar.org.

Zake, Y. K., Bwamiki, D. P., & Nkwiine, C. (2000). Soil management requirements for banana production on the heavy soils around Lake Victoria in Uganda. *Acta Horticulturae, 540,* 285–292.

Chapter 11
Local Adaptation to Climate Extremes in Domboshawa: Opportunities and Limitations

Vincent Itai Tanyanyiwa and Rejoice Madobi

Introduction

The livelihoods of many people in developing countries are depended on climate sensitive resources and rural livelihoods of particularly many people in Zimbabwe including those residing in Domboshawa rely heavily on rain fed agriculture. Thus climate change is one of the greatest challenges facing developing countries. Climate related disasters such as droughts, famines, floods and heat waves are on the increase globally and developing countries are the most affected (Brooks and Adger 2004). Development and poverty alleviation efforts are being deterred and livelihoods are being threatened by climate change impacts. The existing non-climate stressors in developing countries worsens the vulnerability to climate change, hence the imperative need for communities in such countries to adapt.

Adaptation to climate change is not new as communities have always coped with climate extremes but as Fujikura and Kawanishi (2011) posit, climate change is expected to increase risks and to inundate copying mechanisms especially in Africa, where direct dependence upon the natural environment for livelihood support exists alongside a lack of infrastructure and high levels of poverty thereby increasing vulnerability to environmental change hence the need for a contextualised study of climate change adaptation practices in Domboshawa, a rural area in Zimbabwe.

The need to adapt has been reiterated locally and internationally. Adaptation is defined as initiatives and measures to reduce the vulnerability of natural and human systems against actual or expected climate change effects (IPCC 2007). It is important to note that adaptation does not take place in a vacuum; hence various

V. I. Tanyanyiwa (✉) · R. Madobi
Department of Geography and Environmental Studies, Faculty of Science and Technology, Zimbabwe Open University, P.O. Box MP1119 Mt. Pleasant, Harare, Zimbabwe
e-mail: tanyanyiwavi@yahoo.com

R. Madobi
e-mail: elshyd@gmail.com

W. Leal Filho and J. Nalau (eds.), *Limits to Climate Change Adaptation*,
Climate Change Management, https://doi.org/10.1007/978-3-319-64599-5_11

factors influence the adaptive capacity of a society making the capacity to adapt to differ among groups within a given geographical locality. Vulnerability is dynamic and so is adaptation and as such adaptation is an on-going process in a range of stressors and is inherent of social-ecological systems (Fujikura and Kawanishi 2011). Adaptation is affected by a number of factors that include economic and natural resources, social networks, entitlements, institutions, governance, human resources and technology. These factors vary from one community to the other and this also means that the barriers or limits to climate change adaptation differ from one community to the other hence the ways in which Domboshawa Community adapts or fails to adapt to climate change forms the basis of this study. The purpose of this study was to assess the adaptation strategies that are currently practised in Domboshawa and the factors influencing the adaptation mechanisms.

Background of Domboshawa: The Historical and Geographical Setting

The research was carried out in Domboshawa is comprised of about 40,000 households with an average household size of 5 people and with more than 75% of these people relying on farming (Mashonaland East Provincial Census Report 2012). There are 5 wards in Domboshawa. The population has since increased due to natural increase and net migration. Domboshawa has a land area of 35.312 ha and is situated in natural farming regions 2 and 3 on the Highveld of Zimbabwe at 1200 m.a.s.l. The average rainfall ranges from 800–1000 mm per annum (Unganai 1996). In drought years however the rainfall tends to decrease e.g. in the 1991/92 drought the rainfall was 405 mm. Rainfall is seasonal with approximately 90% falling from October to March (Vincent and Thomas 1962). Temperatures in Domboshawa are generally warm in the summer months ranging from 22 to 30 °C (ZMO 2012). However winters can be quite cold with temperatures dropping down to freezing levels. Since the source of livelihoods for the majority of the people is horticultural crops, they need adequate protection in winter especially in dry areas so that their crops are not damaged by frost. The soils of Domboshawa are generally course grained sands/sandy loams which are naturally deprived in principal plant nutrients (the paraferralitic group of soils). Paraferralitic soils absorb and lose water easily. The topography of the area is quite varied with some undulating areas but generally the area is rugged and rocky hence its neglect by early colonial settlers (GRDC 1996). The Ngomakurira Hills (Mountain of Drums) and the Domboshawa Hill (The Red Rock) are some of the spectacular topographic features which attract tourists. Vegetation is mainly Miombo woodland dominated by muzhanje (Uapaca kirkiana), muhacha (Parinali curatefolia), munhondo (Julbennadia globifora), musasa (Brachystegia spiciformis), mupfuti (Brachystegia boehmii), mususu (Terminalia sericia) and muunze (Brachystegia globifora). The Miombo vegetation is declining due to uncontrolled cutting down of trees by urban, peri-urban and communal residents. The few remaining woodlands form part of the grazing lands and is restricted to the rugged terrain which cannot be used for crop production. Varying climate and vegetation characteristics depicted in each Ward were captured. Households for each ward were randomly selected. The questionnaire was pilot tested before it was used for the larger sample. Varied perspectives of households on LTK on drought forecasting dry spells and droughts were obtained with the use of this structured questionnaire, aspects such as demographic features of the population and drought history; response to droughts; drought adaptation

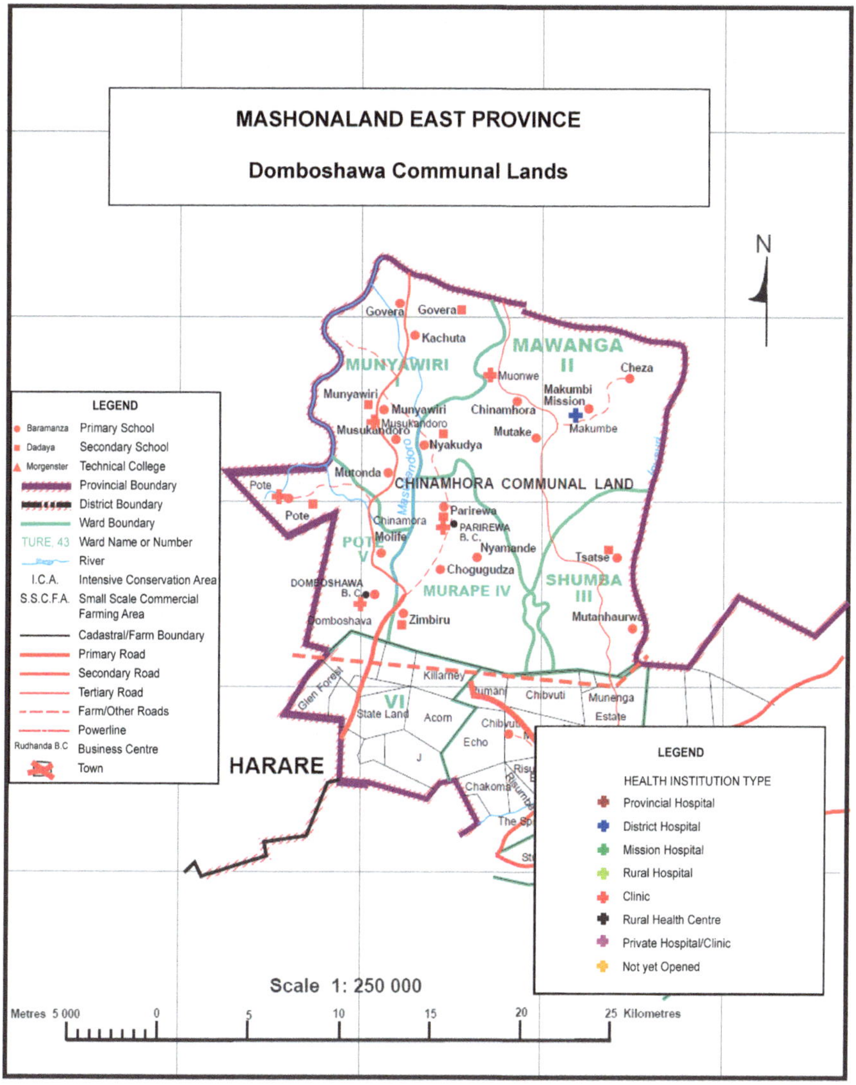

Fig. 11.1 Domboshawa communal lands. *Source* Surveyor General: June 2014

strategies; local traditional indicators as well as their perception and knowledge of meteorological drought forecasting methods were taken on board (Fig. 11.1).

Research Problem

Historically communities have always been exposed to climate extremes and have over time developed adaptation and coping strategies. With climate change, current adaptation strategies may not be sufficient to address new forms of dynamic risks and vulnerability. Adaptation to climate change is more than knowing current and future climate change impacts but how the changes in climate are affecting the local people and how they are responding to these changes. The majority of the farming communities in Domboshawa are smallholder farmers dependent on rain-fed agriculture and climate sensitive resources. Because of Domboshawa's proximity to Harare, vegetable gardening plays an important role in household food security. However, recent evidence has shown the area is already experiencing the effects of climate change, notably rainfall variability and extreme events. Vegetable gardening has the potential to reduce the effects of prolonged dry spells, recurrent droughts and an early end to the rain season. There are also some questions on whether farmers are adapting or failing to adapt to climate change. There is insufficient information on adaptation/maladaptation to climate change at household level. This study is an attempt to address this gap by assessing the status of adaptation to climate change in Domboshawa. No research has been carried out in the study area on the factors that influence a household's choice of adaptation strategies at household level. Such limitations undermine the mainstreaming of climate change adaptation plans at local level.

Research Questions

In light of the adverse impacts of climate change under rain fed agriculture being experienced in Domboshawa, the following research questions arise

- What are the major classes of adaptation practices used by the rural poor and how do rural institutions assist households in pursuing these practices?
- What kind of support do external interventions provide to assist rural institutions, communities, and households and how do individuals within communities respond to the prospect of changes to their lives?
- What lessons emerge for those working to secure lives and livelihoods in the face of climate change?
- How are the perceptions of vulnerability affecting adaptation actions among the vulnerable populations? And what are the barriers arising from such perceptions?

Literature Review

Determinants of Adaptive Capacity

IPCC (2001) defines adaptive capacity as the capability of a system to adapt to climate stimuli or their effects or impacts. It is the ability of a system to adjust to climate variability and change, including climate extremes, to moderate potential damages, to take advantage of the opportunities, or to cope with the consequences. There are many stressors that drive vulnerability; vulnerability in itself is differential and dynamic and so is climate change adaptation. Adaptation to climate change takes place in a dynamic environment (social, economic, technological, biophysical and political) that varies over time, location and sector (Fujikura and Kawanishi 2011). Factors that determine adaptive capacity range from economic resources to information and skills. The factors when intertwined give a clear picture of why the global south is more vulnerable to climate change than the global north.

Economic Resources

Economic resources (whether expressed in terms of economic assets, capital resources, financial means, wealth or poverty) are determinants of adaptive capacity (Muzari and Mutambara 2014). Developing countries are therefore more vulnerable to climate change because of the high poverty levels that hinder the ability of communities to adapt. Muzari and Mutambara (2014) point out that there is a link between poverty and vulnerability though according to Dow (1992) poverty is not synonymous with vulnerability but it is an indicator of the ability to cope. Moreover, people in developing countries are more dependent on climate sensitive resources which makes them more unable to adapt and therefore more vulnerable. IPCC (2007) points out that a person's vulnerability depends in part on gender roles and relations (gendered vulnerability). Poor women in both rural and urban areas in developing countries have a low adaptive capacity compared to their male counter parts. Developed or wealthier nations on the other are more able to adapt to and cope with the impacts of climate variability and change because they have developed technology and infrastructure. Therefore countries that have the ability to develop technology are better able to adapt to climate change. Therefore countries that have the ability to develop technology are better able to adapt to climate change.

Technology

Technology is needed in the management of water resources, modeling climate change, development and construction of flood control structures, improving the agricultural sector (for example development of drought tolerant crops, better crop varieties, animal breeds, etc.) and designing of buildings. Collection of reliable climate data and early warning systems require relevant and reliable technology. Lack of technology thus has the potential to lower a community's adaptive capacity. Technologies can become obsolete hence the need for a community to be able to develop new technologies as Muzari and Mutambara (2014) point out that a community's current technology and its ability to develop new technologies influences its adaptive capacity.

Information, Knowledge and Skills

The majority of underprivileged and vulnerable people in developing countries lack access to climate change information. Yet, Fankhauser and Tol (1997) argue that for adaptation to be successful there is need for a recognition of the need to adapt, knowledge of the available options, the ability to assess them and the ability to implement the appropriate ones. Climate change information dissemination has not been significant in many developing countries (Tawodzera 2012). Climate change debates and policies lack relevance to the poor people as they (poor people) "fail to make a clear link between the long term impact of climate change and their immediate needs and challenges" (Tawodzera 2012: 69).

Communities are aware that the weather patterns are changing but they are not always sure of what exactly has changed. Where rainfall patterns and temperatures have changed communities may be able to observe only rainfall patterns because they (rainfall patterns) are obvious (Muzari and Mutambara 2014). For communities to be able to adapt, there is need for awareness raising and to build on what the communities already know about climate change. Without knowledge, poor people particularly those in developing countries continue to rely on unsustainable and damaging coping strategies in the hope that things will improve in the near future (Muzari and Mutambara 2014).

Making information available depends on the generation of data and information and this requires governments to invest in the generation of that information. The adoption of climate change information in Africa's agricultural sector has been hampered by lack of a comprehensive baseline data and lack of adequate skills for climate change modeling (Fujikura and Kawanishi 2011). Lack of resources has also left many meteorological services departments in developing countries poorly equipped to collect reliable and adequate data for climate change monitoring and adaptation purposes. More so, developing countries have limited capacity to run climate models and where the information is available access to the users is also limited (Fujikura and Kawanishi 2011).

Infrastructure

Adaptive capacity is also determined by the availability of social infrastructure. Poor housing for example exposes the poor communities to climate extremes. The unavailability or shortage of infrastructure such as roads, drainage systems and flood control structures reduces the capacity of a community to respond to climate extremes.

Institutions

Adequate and well-functioning institutions (civic, public and private) are a requirement for successful adaptation. Countries with well developed institutions have a greater adaptive capacity than those with less efficient institutional arrangements. The ability of households to adapt depends on the "ways institutions regulate and structure their interactions both among themselves and with external actors" (Mearns and Norton 2010). This is so because institutions regulate access and exposure to resources (Mearns and Norton 2010). Institutional constraints limit entitlements and access to resources and therefore increase poverty and vulnerability. The existing institutions in developing countries are poorly coordinated (Fujikura and Kawanishi 2011); there is poor coordination within and between civil society, NGOs and governments in addressing adaptation to climate change. Governance structures therefore influence the adaptive capacity of a community and as such there is a link between institutions, adaptation and livelihoods.

Equity (in Access to Resources)

Equitable access to resources varies across space, time, sectors, and social groupings yet equity in the distribution and allocation of resources is necessary for climate change adaptation. Adaptive capacity is not only a function of the availability of resources but also of access to those resources by decision makers and vulnerable sectors of the population (Adger and Kelly 1999). Social inequalities and lack of access to resources in developing countries' communities have enhanced the vulnerability of poor communities to climate extremes.

Limits to Adaptation

Most studies of specific adaptation plans and actions argue that there are likely to be both limits and barriers to adaptation as a response to climate change (IPCC 2007). Limits are defined as the conditions that render adaptation ineffective as a response to climate change and are basically insurmountable. These limits to adaptation are closely linked to the rate and magnitude of climate change, as well as associated key vulnerabilities. An adaptation limit as espoused by Klein et al. (2014) is more restrictive and means that there are no adaptation measures that can be implemented over a given period of time to achieve management objectives, maintain values or sustain natural systems. There are hard and soft adaptation limits (Klein et al. 2014). Hard adaptation limits are those which no adaptive actions are possible to avoid intolerable risks and soft adaptation limit are those options that are currently not available to avoid intolerable risks through adaptive action (Klein et al. 2014).

Klein et al. (2014) distinguishes constraints or barriers from limits. The adaptation constraints thus according Klein et al. (2014) pose limits to the ability of actors to adapt and they are factors that make it more difficult to adapt. However, Barnett et al (2015) argue that the distinction between barriers and limits to adaptation is ambiguous. Many of the theorized limits to adaptation are social constructions; they arise because social processes expose groups to climate change risks, constrain their adaptive capacities, or impede adaptation responses. Some barriers can become limits, so some limits to adaptation can potentially be overcome and transformed into barriers with new cultures, values, technologies, or governance systems. Therefore limits are subjective and dependent upon the values of diverse groups. Vulnerability to climate change effects and adaptation strategies vary temporally and spatially and so do the perceived limits to adaptation.

Therefore, there is significant if seldom acknowledged interplay between barriers and limits to adaptation. Time and path dependence can change some perceived barriers into limits as it becomes apparent that they will not be overcome. Conversely, transformations in institutions, values, and technologies may render some factors perceived to cause limits into barriers, because the shared meanings of loss change or the capacity to avoid these losses improves (Barnett et al, 2015).

Types of Limits to Adaptation

The discourse around limits to adaptation is frequently constructed around three dimensions; ecological and physical limits, economic limits, and technological limits (Adger et al. 2009). Ecological and physical limits or thresholds are the limits beyond which ecosystem degradation can be avoided. Economic limits take into account the costs of adaptation against the costs of impacts to be reduced or prevented. Cost-benefit analysis approaches are therefore applied (Adger et al. 2009). Technological limits are the thresholds beyond which available technologies can avoid or reduce climate change impacts. Various communities are affected differently by climate change and their interpretation of vulnerability and their response (adaptation) to climate change effects. This brings about social limits to climate change adaptation. According to Adger et al. (2009) limits (to adaptation) "are endogenous and emerge from 'inside' society" e.g. the use of recycled water is viewed differently by different communities particularly in developing countries.

Maladaptation to Climate Change

Adaptation to climate change can be successful or it can lead to maladaptation. Maladaptation is a process that results in increased vulnerability to climate variability and change, directly or indirectly, and/or significantly undermines capacities or opportunities for present and future adaptation (Magnan 2014). According to Barnett and O'Neill (2010) in Magnan (2014) the Pathway Framework gives five characteristics of maladaptation which include;

- disproportionately burdening the most vulnerable
- high opportunity costs
- reduce incentives to adapt
- path dependency (relative to alternatives, they set paths that limit future choices)
- increasing emissions of greenhouse gases.

Materials and Methods

To understand existing climate change adaptation initiatives in Zimbabwe in general and Domboshawa in particular, data collection process was structured in two major components: (1) desk review of adaptation policies, including institutional analysis at both national and local level, and (2) case study based primary data collection from sampled households using a questionnaire. In addition, informal and formal interviews and focused group discussions were conducted with relevant stakeholders. A case study was used to get a detailed insight of people's engagements with different climate change adaptation strategies at household level. Data for this study was collected qualitatively and quantitatively. Specifically, for the quantitative approach the survey was used whilst participatory rural appraisal (PRA), in particular, resource mapping, historical trend lines, seasonal and daily activity calendars were used to collect qualitative data (Creswell 2003). Questions sought to characterize the farming households in terms of their education, asset base, and access to five capital resources (physical, human, natural, social and financial) and how they relate to their land, livelihoods and food security. Focus group discussions and household case-story. The study used purposive or judgment sampling since informants were selected on the basis of their social position, roles in the community, knowledge of the community and critical information rather than on representative consideration. Selected relevant professionals of Domboshawa such as Agritex Officers were interviewed. Domboshawa is made up of five wards: Munyawiri, Mawanga, Shumba, Murape and Pote. Population registers for all were obtained from the respective village heads and compared with that obtained from Goromonzi Rural District Council (GRDC), the local authority. Data generated was mainly qualitativc in naturc and relied on descriptive statistics for analysis. Data gathered through PRA techniques, FGDs and historical timelines were analysed in a thematic approach. This entailed coding of this data according to the themes that had already been developed for the study based on the objectives of the study, relevant literature reviewed and those emerging from the data collection process. Survey data was entered and analysed through descriptive statistics to enable categorization of behavioural data, for purposes of classification, summarization and tabulation.

There were challenges in controlling expectations as farmers thought we were donors hence in some cases farmers tended to under report assets they had. Triangulation with other methods served to address this bias. Farmers were fearful of potential repercussions if they openly gave their thoughts, an anomaly catered for as we advised that this study was purely academic. In-depth case studies were used to supplement data whilst review of literature was used to validate collected data which in our view may have been falsified.

Results and Discussion

Current Information

Farmers should also have regular information on current issues related to climate change and agriculture. This can be achieved through the strengthening of the nation's extension services perhaps by devolving the bulk of the services down to the local councils, which is closer to the farmers, and encouraging farmers to form farmer groups for enhanced capacity through group efforts. This may help them take advantage of the internet. However despite the fact that there is almost 100% penetration in the use of cell phones, the cell phone is mainly used by the farmers as for voice calls, very few if any use the cell phone to gain insights that are given by the service provider. all there is increasing use of solar for charging cell phones the majority of residents rely on electricity grid for charging their cell phones, this tend to be a disadvantage especially for those who may not want to receive information, this also applies to those who rely on electricity for television sets. At the moment the Zimbabwe meteorological services are not well decentralised and thereof contacts with the farmers is very limited. In the same vein Domboshawa is located about 25 km from Harare and relies on weather information that applies to Harare yet in reality the characteristics of these two places are geographically different.

The use of community weather stations may go a long way in solving this problem. Farmers may work with extension officers so that weather information is permanently recorded, very important because this is will be specific to Domboshawa and therefore relevant to the farmers. All extension workers that deal with farmers have to work concurrently with farmers. At present the situation on the ground is different, synergies between government departments is nil or rather slime Environmental Management Agency (EMA), Agriculture and technical extension service (Agritex) and the Forestry Commission all belong to different Ministries and therefore have different budgets although they work with farmers daily on the ground.

Indigenous Knowledge Systems and Culture

The high climate unpredictability that characterizes the Zimbabwean context presupposes that people have developed thriving indigenous adaptation strategies. In reality this is not the case in Domboshawa. Some of the strategies that people use to forecast the weather have also been affected by climate change such that these are no longer useful in weather prediction. For example derivation of the weather forecasts from vegetation has been disturbed by natural fires, deforestation and the clearance of land for agriculture. The prevailing characteristics of specific vegetation, used to derive a weather forecast. Wild fruits and abundance of fruits like *Parinari curatellifolia* (hacha) and *Uapaca kirkiana* (mazhanje) indicate a good rainfall season but this cannot be the case in areas where vegetation has been disturbed and may not be a good measure of adaptation It is therefore advocated that indigenous knowledge and practices should be integrated into formal climate change mitigation and adaptation strategies at the local level and in the school curriculum. School pupils, especially those in secondary schools, and are day scholars are presumably supposed to disseminate the information they would have acquired in schools back home. There is need to localise climate change discourse so that that villagers may know what it is. There is very little literature on climate change in Zimbabwe; as a result, studies are dominated by examples from Western countries which may not necessarily be relevant to the Zimbabwean situation. In Zimbabwe, there is little documentation of this knowledge yet it is a viable alternative knowledge base. Policy should be put in place through provision of human and financial a resource to support IKS within communities, no policy has been created in Zimbabwe specifically on IKS, extinction of indigenous knowledge is imminent In the same vein, in Domboshawa, due to increasing variability in rainfall and temperatures farmers are always advised to grow drought tolerant crops such as rapoko and finger millet. Farmers, however continue to grow maize, the staple food crop of Zimbabwe. Their reasons for growing maize is because it tastes better, yields more per hectare, easy and cheaper to grind, easy to hoe and harvest as compared to the two crops mentioned above.

Limited Investment by Government

The explanation and analysis of climate change adaptation programmes in the global south hardly ever examine the political contexts or the way in which climate change programmes shape local and national politics. Investment on improved agricultural technology by government and other stakeholders are very necessary for agriculture to be able to cope with climate change. At the moment, irrigation facilities on commercial farms were vandalized and in the rural areas not much was done by the government to build dams and therefore improve irrigation facilities, however Domboshawa has always been producing horticultural products for Harare and this has been done through simple irrigation facilities developed y the small scale farmers. The only noticeable government intervention is the deployment of extension workers who assist the farmers as and when the need arises. The extension workers also organize green shows in the area and this improves efficiency and productivity in farming. The Zimbabwe government has suffered from donor fatigue as from around 1999 when the government embarked on the farm grabs exercises. This has negatively affected farmer development programmers such as the training in Master Farmer and hence acquisition of new knowledge has not transcended to the local people and hence their limited adaptation knowledge. The 5 A's of scientific responses may encompass mutable barriers or de facto limits to adaptation and these are availability, accessibility, affordability, advisability and acceptability. In Zimbabwe climate change has the following governance structures; (i) National Climate Change Office based in the Ministry of Environment, water and climate and funded by the UNDP. Its mandate is to engage with UNFCCC and for production of the National Communications (ii) National Climate Change Task Team (NCCTT) which is coordinated by the Office of the President and Cabinet (OPC). The NCCTT has a Permanent Secretary and its small budget is from the central government. The NCCTT is comprised of permanent secretaries from relevant ministries (environment, water, and climate agriculture, and economic planning and finance). This entity ensures that the National Climate Change Response Strategy is produced and, then, the National Climate Change Plan (iii) National Climate Change Steering Committee is made up of government and civil society representatives and is anticipated to guide the production of the National Climate Change Response Strategy. In spite of the existence of these entities to address climate change concerns, there has been very inadequate allocation of resources by the government to address the issue due to extremely limited budget available for climate change adaptation.

Limited High Yielding Crops

With the increasing rate of erratic rainfall patterns, drought and desertification, drought resistant and short duration high yielding crops should be developed through research efforts and made available to farmers. The Zimbabwean government should take a bold step to establish better-equipped weather stations as against the scanty and ill-equipped ones we currently have in Zimbabwe. With this, accurate weather forecast and predictions will be possible and this will help to prevent weather-related disasters through early warning and effective response/adaptation system. In addition, efforts need to be made towards tackling the dilapidated infrastructure in the country.

Professionalisation of Agriculture

There is a need for a radical shift to irrigation rather than over reliance on rain-fed food production. For that reason there is need for sufficient provision of irrigation and drainage infrastructures which is critical for climate change adaptation. Agriculture needs to be fully professionalized i.e. people should view agriculture as a business or commercial venture as opposed to being a part time activity, this is one of the reasons why people fail to adapt because they will consider other livelihood options instead of concentrating on how they can adapt in the agriculture sector which is their core business (Thomsen et al. 2012). With educational training incentives, and development of human capital rural communities can easily adapt because knowledge empowers in the direction of crop and livestock production. A better educated farmer would for instance be able to absorb new information faster than one who is not educated, the reason why in Domboshawa there is continuous training development by government and other stakeholders such NGOs in training farmers.

Government Role

The Zimbabwe governments' poverty and Hiv/aids programs should be decentralised so that they become in structure for effectiveness and not for tokenism. Hence in the same vein agricultural adaptation to climate change should be mainstreamed into government's poverty alleviation programme. At the national level a comprehensive agricultural research policy framework and decentralised research funding into districts and provinces where the farmers at the local level can actively benefit. This is mainly done to provide a favourable environment for stability and efficiency in agricultural programmes. The government also plays a key role in determining the rate at which farmers adapt to climate and for a community such as the Domboshawa the following questions came to the fore; in which arena are adaptation decisions to be made? Who initiates or holds responsibility for developing adaptation options? Who has (potentially) decision-making and implementation authority? What influences adaptation decisions? What are the outcomes of a decision, once made, and how do the decision-makers and affected stakeholders live with them?

Links Between Institutions and Coping Strategies

In Domboshawa adaptation to climate change is affected by a number of factors, there are many rural institutions that that play apart in determining adaptation at the household level; the aspects include information, communication and technology (ICTs) such as the cell phone, radio and television as they are the basis of dissemination weather information. Technology is not only limited to ICTs, but also include tractors and other modern equipment such as rain gauge. Financial resources are also play a significant role in Domboshawa, those farmers that have been able to farm profitably in the past have invested heavily in irrigation hence they produce crops throughout the year. Financial resources are also important for fencing and buying other equipment important in farming sector. Apart from that, leadership is very important of which in Domboshawa Agritex extension officers play a part in advising farmers the types of crops to grow and when they should be grown. This is quite pertinent because more than 50% of farmers rely on rainfed farming. The factors highlighted above play a critical role in enabling mobility of inputs, produce and the farmers themselves. Storage of harvested crops and inputs play s a significant role in the adaptation practises because some vegetables such as rape, cabbage and chomolier are dried for later consumption should the season be poor or as a matter of preference for dried vegetables. This also applies to cereals such as maize and finger millet. In the process farmers have also diversified so that if on crop failure they will be smoothened by another which is more drought tolerant e.g. finger millet.

Communal pooling is common in Domboshawa; in the local Shona language this is called the *zunde ramambo* concept (The chief's granary). Zunde ramambo is a customary social security understanding designed to address food shortages in a drought or famine (Dhemba et al. 2002). This practise existed before colonisation of Zimbabwe and was meant to show the powers of a chief in promoting the welfare of his/her people. Traditionally, a chief designated land for growing food crops as protection against food insecurity through a common land; the *Zunde*. Labour was voluntarily provided by locals even if they did not benefit from produce. The harvest was/is stored in granaries at the chief's homestead as food reserves, only distributed to the community when food was/is in short supply. Local Chief Chinamhora controls this whole process. As a suitable community response to food insecurity, the effectiveness of the chief's granary is compromised by a shortage of land and unaffordable costs of inputs. The harsh economic environment has made volunteering a futile endeavor as members struggle to meet their own basic needs. In spite of this anomaly, the chiefs granary has is improved the welfare of orphans, other vulnerable children and older persons in Domboshawa. Exchange of commodities is also important adaptation measure in Domboshawa for example one household reported to have exchanged a 750 ml of paraffin (worth USD1) with two bundles of rape. Exchange takes place amongst a variety of goods and services.

Both coping (short term, flexible) and adaptation (longer term and more permanent) are influences by rural institutions that are inherent in Domboshawa and are affected by mobility, storage, diversification, communal pooling and exchange, all these factors are affected by exchange in information, technology, financial investments and the type of leadership in the area. In the case of Domboshawa the bureaucratic agencies include the ministries of agriculture, environment water and climate as well as local government, the Goromonzi Rural District Council, membership organisations such as the church, farmer group's burial groups among others in Zimbabwe. However, there is a polarised political landscape with considerable levels of mistrust, to some extent climate change plays a convenient role as an issue around which partnership and teamwork can be practiced because things need "hold things together. Numerous issues are highly contested in Zimbabwe to the extent that there is a value for both government and civil society agencies alike to have areas in which they can cooperate. Without or limited, social compact between citizens, civil society, the state would be more tenuous than it is at present. In other words, does engagement with climate change adaptation build institutional practices that offer broader application to other development linked challenges and with the intention of building a consciousness that is likely to lead to broader transformations in climate change adaptation? There is growing debate on outcomes that emanate from adaptation activities and whether these contribute simply to "coping" with complex circumstances or add to more "transformative" change that reduces long-term exposure to an extensive variety of shocks and stresses which are climate change related and others which are not (O'Brien 2012). Moser and Boykoff (2013, 14) reveal "successful adaptation" as operating along a continuum from "maladaptation" to "building something different or better". However, there is need for "transformative" approaches which are financially, socially, politically, technologically, and or culturally strong for standard adaptation approaches to be met in Domboshawa, a feat difficult to achieve given the economic, and socio-political problems bedeviling Zimbabwe since early 2000. Adaptation takes place within this broad macro political framework.

Main adaptation strategies used by farmers in Domboshawa

Adaptation strategy	Explanation	Reason(s) for such strategy (ies)
Potholing	Inputs such as fertilizers/manure, seed, water and lime are all concentrated in the prepared hole	To conserve moisture and inputs
Crop rotation	To maintain soil nutrients	To have variety in crops grown conservation of the soil
Buying food	To augment food supplies in both good and bad farming seasons	To have adequate food
Off farm work	To supplement income from the farm	To get income so as to purchase other goods

(continued)

(continued)

Adaptation strategy	Explanation	Reason(s) for such strategy (ies)
Migration	To find alternative livelihoods/farms	Remittances are sent back home
Hiring labour	To increase labour force	Enhance their yields regardless of a good or bad season
Ripping	Deep cultivation (greater than 20 cm)	Soils with poor structure deep ripping can loosen and 'soften' the oil, giving plants access to deeper soil
Market gardening	Growing horticultural crops throughout the year	To get income throughout the year
Irrigation	Growing horticultural crops throughout the year	To supplement rainfed agriculture and to get income throughout the year
Collection of wild fruits	To supplement home food	To ensure food security

Source Field data November, 2014

From the data gathered it appeared adaptation is influenced by age, educational level, and access to weather information, training and group membership and location among others and all this happens through the interaction of rural institutions that operate in Domboshawa and the farmers themselves. In reality adaptation has not been easy because the vulnerability index keeps on increasing due to some of the factors highlighted above. As a result farmers fail to adapt and found themselves in a vicious cycle of vulnerability which is difficult to escape. This is evident in Domboshawa through farmers' failure to pay fees for their school going children and one meal per day per household. Farmers express grief over the deterioration in their standard of living since early 2000 to the present day.

Conclusion

- In Domboshawa, drought and climate change have merely accelerated what was already occurring especially on the economic landscape. Zimbabwe has been facing serious socio-economic and political problems which have led to deterioration in living standards. These effects are comprehensive and complex.
- Existing support services to enable adaptation in the farming sector are being stretched to their limits. Solutions will require multiple support strategies by working with the locals and proactive approaches in envisioning, planning for and adapting to diverse climatic and socio-economic futures e.g. what did the

locals do to deal with previous extreme climate extremes episodes and/or droughts? What worked in the coping/adaptation process and what did not?
- Selection of coping and adaptation strategies to climate change is influence by factors such as demography, access to information and assets and vulnerability levels. There is therefore need to reorganize agriculture and for the rural inhabitants to turn off farm activities such as brick moulding.
- Farmers are confronted with multiple stressors which have an effect on responses to both coping and adaptation, and these are affected by the general economic and political context

The study recommends the following:

- Strengthen the capacity of farmers and institutions for to enable the m to cope and adapt to climate change including an integral component of multiple stressors that confront farmers.
- There is need to provide support for suitable agricultural research and innovations of new livelihood option in agricultural extension and meteorology to name but a few.
- Educate farmers so that farming is seen/viewed as a real livelihood option rather than complement to formal work. Once this is done farmers are more likely to find innovative ways of coping and adapting to the effects of climate change.
- Rural communities need to be well-supported throughout the processes of change, in ways that are respectful and invigorating for people who are tired from seemingly chronic climatic extremes in addition to non-climatic pressures such as political intolerance.
- The required proactive approach means that drought and climate change adaptation need no longer be the centrepiece of government policy but should be part of a comprehensive policy towards poverty alleviation especially at a local level. The problems rural communities encounter are more than just drought or climate change but are multifaceted strategic planning that deals with ongoing climate change and other future uncertainties is a must for government and local communities need.

References

Adger, W. N., Dessai, S., Goulden, M., Hulme, M., Lorenzoni, I., Nelson D, R., et al. (2009). Are there social limits to adaptation to climate change? *Climatic Change, 93*(3–4), 335–354.

Adger, W. N., & Kelly, P. M. (1999). Social vulnerability to climate change and the architecture of entitlements. *Mitigation and Adaptation Strategies for Global Change, 4*(3–4), 253–266.

Barnett, J. L., Evans, L. S., Gross, C., Kiem, A. S., Kingsford, R. T., Palutiko, J. P., et al. (2015). From barriers to limits to climate change adaptation: Path dependency and the speed of change. *Ecology and Society, 20*(3), 5. doi:10.5751/ES-07698-200305.

Barnett, J., & O'Neill, S. (2010). Maladaptation. *Global Environmental Change-Human and Policy Dimensions, 20,* 211–213.

Brooks, N., & Adger, W. N. (2004). Technical paper 7: Assessing and enhancing adaptive capacity. In B. Lim & E. Spanger-Siegfried (Eds.), *Adaptation policy frameworks for climate change: Developing strategies, policies and measures*. Cambridge: Cambridge University Press.

Creswell, J. (2003). *Designing and conducting mixed methods research* (2nd ed.). Thousand Oaks: Sage.

Dhemba, J., Gumbo, P., & Nyamusara, J. (2002). Social security in Zimbabwe. *Journal of Social Development in Africa, 17*(12), 132–156.

Dow, K. (1992). Exploring the differences in our common future (s): The meaning of vulnerability to global environmental change. *Geoforum, 23,* 417–436.

Fankhauser, S., & Tol, R. S. J. (1997). The social costs of climate change: The second IPCC assessment report and beyond. *Mitigation and Adaptation Strategies for Global Change, 1,* 385–403.

Fujikura, R., & Kawanishi, M. (2011). *Climate change adaptation and international development: Making development cooperation more effective*. London: Japan International Cooperation Agency Research Institute.

Goromonzi Rural District Council. (1996). *Goromonzi rural district council economic survey report*. Goromonzi: Ruwa.

IPCC. (2001). *Climate change 2001: Impacts, adaptation and vulnerability*. Geneva: IPCC.

IPCC. (2007). Summary of policymakers. In *Climate change 2007: The physical science basis. Contribution of working group 1 to the fourth assessment of report of the IPCC*. Cambridge: Cambridge University Press.

Klein, J. R. T., Midgley, G. F., Preston, B. L., Alam, F. G. H., Berkhout, K., Dow, K., et al. (2014). Adaptation opportunities, constraints and limits. In C. B. Field, V. R. Barros, D. J. Dokken, K. J. Mach, M. D. Mastrandrea, T. E. Bilir, et al. (Eds.), *Climate change 2014: Impacts, adaptation, and vulnerability. Part A: Global and scrotal aspects. Contribution of working group II to the fifth assessment report of the intergovernmental panel on climate change* (pp. 899–943). Cambridge: Cambridge University Press.

Magnan, A. (2014). Avoiding maladaptation to climate change: Towards guiding principles. *SAPIENS, 7*(1).

Mashonaland East Provincial Census Report. (2012). Harare, Zimbabwe.

Mearns, R., & Norton, A. (2010). *Social dimensions of climate change: Equity and vulnerability in a warming world*. Washington, DC: World Bank.

Moser, S., & Boykoff, M. (2013). Climate change and adaptation success: The scope of the challenge. In S. Moser & M. Boykoff (Eds.), *Successful adaptation to climate change: Linking science and policy in a rapidly changing world* (pp. 1–34). London: Routledge.

Muzari, W., & Mutambara, J. (2014). *Climate change impacts and response options among rural communities: Issues and strategies*. ASARE.

O'Brien, K. (2012). Global environmental change II: From adaptation to deliberate transformation. *Progress in Human Geography, 36*(5), 667–676.

Tawodzera, G. (2012). *Urban household vulnerability to food security and climate change experiences from urban areas of Zimbabwe*. New York, NY: Earthscan.

Thomsen, D. C., Smith, T. F., & Keys, N. (2012). Adaptation or manipulation? Unpacking climate change response strategies. *Ecology and Society, 17*(3), 20.

Unganai, L. S. (1996). Historic and future climate in Zimbabwe. *Climate Research, 6,* 137–145.

Vincent, V., & Thomas, R. G. (1962). *An agricultural survey of southern Rhodesia: Part I: Agro-ecological survey*. Salisbury: Government Printers.

Zimbabwe Meteorological Organisation (ZMO). (2012). Seasonal Weather Forecasts. Harare, Zimbabwe

Part III
Limits to Climate Change Adaptation in Australia, North-America and Europe

Chapter 12
The Limits of Imagination

Liese Coulter

Introduction

Visions of likely future climate patterns are influenced in part by what is known about climate change, the way that information is shared, and by differing human capacities to imagine the future. Based on those visions, adaptation plans are being developed to moderate harm by adjusting policy and practice to consider actual or expected climate change and, possibly, to also exploit opportunities (IPCC 2014a). Determining when, and if, people and systems need to adapt is often influenced by the physical and social factors used to evaluate risks. These assessments consider both exposure and sensitivity to likely climate change hazards and capacity to deal with consequent impacts (IPCC 2014a). However, the processes of human cognition and perceptions are also important drivers in decision-making for adaptation (Grothmann and Patt 2005; Lee et al. 2015). Mental rehearsals that build on an understanding of climate change risks and hazards are a necessary part of the future thinking required to prepare for expected activities, events, and changes. This future thinking, or prospection, is linked to memories of both personal lives, *episodic thinking*, and broader experiences of the world and society, *semantic thinking* (Tulving 1985). The cognitive process of linking future expectations to memories of the past poses a significant complication in mentally preparing for uncertain, and possibly abrupt, effects of climate change. The perceived need to adapt is affected by the extent to which people and communities imagine themselves as being vulnerable to climate change, sensitive to climate impacts, and able to manage resulting risks. A lack of perceived need can limit the capacity to adapt in the future when a failure of imagination may lead to choices that reduce the flexibility of future options, or to investments in infrastructure that will too quickly become inappropriate or inadequate to meet their intended purpose.

L. Coulter (✉)
Griffith University, 170 Kessels Rd, Nathan, QLD 4111, Australia
e-mail: l.coulter@griffith.edu.au

W. Leal Filho and J. Nalau (eds.), *Limits to Climate Change Adaptation*,
Climate Change Management, https://doi.org/10.1007/978-3-319-64599-5_12

In many places, entrenched cultural, economic, and practical expectations of ongoing prosperity lend support to plans that explicitly aim to maintain or increase stability (Gardner and Assadourian 2004). As the climate system becomes increasingly dynamic (Rockstrom et al. 2009), assumptions of ongoing prosperity and stability may become limiting factors in planning to adapt to climate change. Senior decision-makers draw on personal experience and episodic memory, as well as their professional expertise and semantic memory, to inform the choices that affect so many others. Therefore, imagination plays an important role in whether expected climate impacts are reflected in current decisions, particularly those that initiate or approve investments with multi-decadal lifespans.

This chapter addressed three main aims to better understand the roles of imagination in adaptation planning. First, it examined the perceived need for adaptation to climate change over the next twenty years in Australia and Canada, to provide context for conceptual and cultural factors that influence imagination in adaptation planning. Second, it drew on empirical evidence from a study of Australian and Canadian professionals who develop, communicate, or apply climate information in their work. Based on interview data, this study considered the relative roles of climate change knowledge, narrative communication, and future thinking in personal adaptation planning. Third, a thematic analysis of participant interviews explored the role of imagination in discussing and planning adaptation in relation to perceived vulnerability to climate change. Participants provided useful insights into how imagined futures are influencing current decision-making and adaptation planning. This chapter argues that, in addition to physical limitations, a deficit in perceived vulnerability and, therefore, a reluctance to imagine adaptation, can be a limiting factor in planning for climate change impacts in developed nations.

Adaptation Challenges

Near-Term Adaptation

Highly credible and widely accepted information and projections about expected climate change are freely available through the Intergovernmental Panel on Climate Change (IPCC) reports. The most recent IPCC report into physical science is the first to include a chapter on climate change in the near-term, using the timeframe 2016–2035 (Kirtman et al. 2013). The report shows that even over the next two decades, global increases in average temperatures will be larger and faster in the atmosphere over land than over the oceans. The poles and equator will see larger temperature increases than in the mid-latitudes. On average, globally, there will be fewer cold days and nights during this period and the runs of very hot days will be hotter, last longer, and cover larger areas. Rainfall will become less regular, with some areas experiencing longer dry periods interrupted by intense rain events. Storms are not expected to be more frequent however, they are likely to be larger when they occur. In addition, locally higher temperatures will increase the likelihood that air pollution will become more detrimental. It is a challenge to reflect this knowledge in specific plans for families and communities over the next few decades.

These climate change impacts and their effects will not be evenly distributed around the globe. For Australia, in Australasia, key climate challenges where adaptation options may prove effective have been identified in the changing coral reef systems, increasing urban flood damage, and the growing risks to coastal and low-lying infrastructure and ecosystems (IPCC 2014b). For Canada, in North America, key challenges associated with significant adaptation opportunities relate to managing wildfire damage, reducing the growing risk of heat-related deaths and minimising urban damage from coastal and river flooding (IPCC 2014b).

While these challenges are based on physical changes directly affected by the climate system, the adaptation options are further defined by social and institutional factors (Brown and Westaway 2011). Despite a relatively high adaptive capacity, adaptation in both Australia and Canada is practically constrained by issues of leadership, institutional change, lack of coordination, and legal challenges (Reisinger 2014; Romero-Lankao 2014). Agreeing where the responsibility lies in managing adaptation is fundamental to address these challenges.

As stories of our expected future are developed and shared, constructions of meaning about our world, from our past to the present, and into the future, are exchanged. Especially through geographically dispersed and socially disparate communities, these shared stories become cognitive and conceptual meeting places where people recognise those with aligned viewpoints (Oyserman and James 2011). Including possible climate change impacts in a personal future vision begins the process of adapting on cognitive and psychological levels (Reser and Swim 2011) and influences the cultural changes that will allow individuals and communities to orient themselves toward the future (Bateson 2007). Although these future visions may be flawed in their foundations, they provide a starting place from which to develop a collective narrative of adaptation to climate change (Roncoli et al. 2009). This research seeks to better understand emerging views of how people may each adapt to climate change, and subsequently share these imaginative narratives to support broader involvement in their evolution.

Benefits and Responsibilities of Adaptive Capacity in Australia and Canada

Assessments and assumptions about adaptive capacity reflect past actions and benefits; however, they may not adequately account for the implications of expected climate impacts. Global negotiations and media reports employ two common frames of reference for ethical conceptions of responsibility for climate change. These principles of *sharing burdens* and *avoiding harm* (Caney 2014) are applied to benefit vulnerable nations exposed to the negative effects of climate change. These frame place responsibility on developed nations who contributed substantially to historical emissions; benefited significantly from fossil fuel use; and currently have the greatest capacity to pay to adapt and mitigate. Both these benefits and obligations are inferred from past actions and advantages, and a currently high adaptive capacity. While these nations are seen by many to have a significant obligation in climate negotiations, the current benefits gained from a high adaptive capacity may diminish with repeated impacts from unprecedented and extreme weather events. In negotiations under the United Nations Framework Convention for Climate Change (UNFCCC), a loose umbrella group of non-European developed countries, including Australia and Canada, accounts for approximately 30% of past global greenhouse gas emission (Hurlbert 2011). With a public focus on climate change mitigation, many of the adaptation efforts in Australia and Canada so far have been in planning and awareness raising, rather than implementation (Serrao-Neumann et al. 2014; Warren and Lemmen 2014). Managing the expected impacts from climate change is not given a high priority in government and communities compared to issues that are seen as more urgent and immediate, such as jobs and the economy.

Adaptation and Imagination

How climate change projections are imagined will influence personal visions of the future because, to some extent, the future must be imagined to be discussed. The human brain uses stored information to imagine, and so simulate and possibly predict, future events (Schacter et al. 2007). These mental simulations allow sharing of future imaginings, which can then be reflected in discussions to influence planning, policy, and actions (Szpunar 2010). On a personal level, past experiences are incorporated and memories are shared to include current understandings of climate change in order to make sense of likely and possible futures. In this way, imagination creates a space that allows the formation of new connections between objects and subjects, and forms a bridge between perceived external environments and more private interior worlds (Fincher et al. 2014). Imagined scenarios that represent plausible futures are useful for estimating risks and informing adaptation planning (Atance and O'Neill 2001). When planned responses to climate change raise additional barriers and create limitations to adaptation, it is usually considered maladaptation (Barnett and O'Neill 2010). However, this can be seen rather as a failure to adapt when a lack of action, based on inadequately imagined vulnerability and consequent adaptation options, has negative consequences (Juhola et al. 2016). Critically engaging with these imagined futures can increase a capacity to manage social, cultural, and political implications of changes (Wright et al. 2013).

Assuming the high adaptive capacity of the past when imagining a future affected by climate change may limit the usefulness of adaptation planning. To develop effective adaptation, new information will need to be reflected in basic assumptions, which will differ from those of the past. Over the coming decades, both societies and climate patterns will be changing, so realistically imagined scenarios will need to reflect future climates as environments for future societies. Past societies, which were not affected by rapid climate change, will not be found in the climate of the future. Only a future society can realistically be imagined in a future climate (Feenstra et al. 1998).

Conceptual and Cultural Limits

Across any nation, people have different initial levels of vulnerability, exposure, and capacity to manage external threats, which create different profiles of risk to any hazard, not just to climate change. Ethics, knowledge, risk, and culture have been suggested as fundamental social considerations in adaptation planning (Adger et al. 2009). How the ethical consideration of who has the power to decide is navigated may determine for example, which parts of a river or coastline are flooded, and whose property and livelihood may be sacrificed for the greater good (Serrao-Neumann et al. 2014). The ways that new information is acknowledged as true and credible has broadened from traditional cultural and local knowledge pools, and from scientific proofs delivered through learned journals and education, to encompass mass media and digital communities (Moser 2010). Perceived vulnerability acts as a filter in assessing risk, which is balanced against the sense of urgency and scale of likely damage, to affect how and when actions take place (Lee et al. 2015). Imagining a future world affected by climate change forces a re-examination of how culture and nature fit together as the focus shifts between local and global issues affecting individuals and their neighbours (Hulme 2010). The people, things, and ideas that are valued, and what might be done to protect these, are at the core of adaptation decisions. The key question this research addressed is "How is knowledge of climate change projections reflected in personal visions, discussions, and decisions about the future?"

Methods

Case Study Boundaries

This research aims to better understand individual imaginations and subjective views of likely futures. Therefore, qualitative methodologies were chosen that suit an interpretivist approach. This interpretivist perspective includes recognising imaginatively constructed futures as valid subjects for research, and acknowledging influence from individual cultural backgrounds and idiosyncratic histories, desires, needs, and perceived opportunities and challenges (Reed 2008). Due to this focus on individual narrative, a qualitative research approach was selected using semi-structured interviews for primary data collection (Savin-Baden and Niekerk 2007). The study employed a Future Climate Narrative (FCN) typology as an analytic framework to assess participants' relative focus on factual knowledge about climate changes, experiences of climate change communication, and more imaginative thoughts about the future.

This study involved 31 participants from Australia and Canada who have been publicly recognised as working in climate-related areas. Recruitment was based on the rationale that this population would be more experienced and knowledgeable in discussing climate change than non-professionals. In addition, there are fundamental similarities between the two countries. Counted among the highest greenhouse gas-producing nations per capita, Australia and Canada both feature resource-based economies, have similar approaches to federalism in regulatory frameworks to deal with climate change, and have vulnerable urban areas leading the way on policy development for both climate change mitigation and adaptation (Jones 2012). In both countries, public discourse has been influenced by conservative governments' communication and policy initiatives (Moser 2010). The interviews were conducted in Australia (June to December 2014) while Tony Abbot was Prime Minister, and in Canada (January to April 2015) while Stephen Harper was Prime Minister. There is evidence to suggest that during the 2010s, Canadian perceptions of the threat of climate change has remained relatively constant while climate threat perception in Australia has declined somewhat (Capstick et al. 2015).

Interviews were designed to encourage participants to construct narratives discursively, building detail into the telling and allowing for contradiction and expectations that might be considered unreal. The key concern for the study was the practice of imagining a personal relationship to the future that included climate change. Participants were prompted to iteratively situate climate change impacts within the narratives and reflect on proposed adaptation strategies. Many participants were practiced in discussing climate-related energy and carbon issues, and in relating to social issues that are distant in time and place. Therefore, during the interviews it was frequently necessary to draw the conversation back to personal vulnerabilities to climate impacts and possible adaptation plans. For the participants who said they had never before discussed personal adaptation to climate change, this was especially challenging.

Participants were chosen from professionals who had dealt with climate information in their work in research organisations, governments, extension activities, or non-government organisations. Few people have extensive knowledge or a deep understanding of the complexities of climate change. This information deficit is often cited as one barrier to discussing adaptation options (O'Neill and Hulme 2009; Patchen 2006; Stuiver et al. 2013). While this is no guarantee of extensive climate knowledge, all of the participants had access to current climate projections, many had consciously applied these in their work, and some had contributed directly to the production of the information. In addition, selecting people who were publicly recognised as working with climate change established a participant group who were prepared to discuss climate issues, at least on a professional level. To encourage frank discussions of personal perspectives on areas of professional expertise, all interview material is de-identified and participant identity is kept confidential. Hence, selected participants had a high potential to understand, compare, and discuss factors that may influence personal imaginings related to climate change.

In public discourse, the impacts from climate change have been largely discussed as distant, both physically in space and time, and psychologically through uncertainty and by expecting climate impacts to affect 'someone else' (Spence et al. 2012, p. 958). The reference time for this study is the period approximately twenty years from initial data collection: 2015–2035. While twenty years is quite distant in one's lifetime, in terms of even near-term climate change where a decade is often counted as one unit of measure, this is very immediate. Many participants noted that this was a much shorter reference period than was common in their ongoing discussions of expected climate change.

Future Climate Narrative Typology

Participant interviews were analysed in relation to the key thematic areas of first, future thinking; second, climate change knowledge; and third, narrative communication; as well as expected fourth, change over time. As described in Table 12.1, these themes form the foundation of the Future Climate Narrative (FCN) typology, which was employed to guide thematic analysis. The future thinking theme is the most personal. It represents thoughts that may be little shared but which play an important role in individual simulation, prediction, intention, and planning for the future (Szpunar et al. 2014). The second theme is climate change knowledge. This relates to the external environment, and knowledge constructed through global research and government networks as embodied by the work of the IPCC (IPCC 2014a). Narrative communication, the third theme, connects personal insight and external situations. It operates on multiple social levels where information is exchanged professionally and personally to develop shared understandings (Dilling and Moser 2007). Change over time is an overarching theme that was more or less present, depending on participant' responses that referenced time as dynamic.

Table 12.1 Future climate narrative (FCN) typology related to themes

Type	Unfolding futures [change over time (C)]					
	Dangerous futures (n/c)		Imagined futures (f/n)		Unfamiliar futures (c/f)	
Theme	Narrative communication (n)	Climate change knowledge (c)	Future thinking (f)	Narrative communication (n)	Climate change knowledge (c)	Future thinking (f)
Key expressions	How climate change is talked about		How the future is talked about		How the future is imagined with climate change	

Although on some level climate change inherently implies change over time, this is not always reflected in mental models and discussions about climate change (Chen 2011).

The FCN typology was designed to identify and contextualise how the expected future as affected by climate change is imagined and shared. The following describes the relationship of the FCN types and how they were applied to the data. Change over time was labelled Unfolding futures, which is related to both environmental and societal changes, as well as to the changes that come from personal growth and aging. Passages that made direct reference to past or future temporal changes were coded to the overarching Unfolding futures type, in addition to another theme reflecting the main focus of the content. Passages focused on the intersection between narrative communication and climate change knowledge were coded under the Dangerous futures type. Passages focused on both future thinking and narrative communication were assigned to the Imagined futures type. Passages focused on the overlap of climate change knowledge and future thinking were coded to the Unfamiliar futures type. In this way the interview data was categorized according to both the content and the overarching dynamic temporal context. This was useful to identify within the data inherent participant awareness of both climate and change.

This research was limited by the lack of consistent and ongoing conversations about adaptation to climate change reported by participants. While this led to many participants sharing more private thoughts on the vulnerability and exposure to climate change of their families and community, they could not often discuss how others viewed their concerns. In addition, the recruitment process in this study was limited to participants who engage with climate information and communication, and was not controlled for sensitivity to future thinking. More interventionist and longitudinal approaches could have minimised some of these issues by prompting repeated adaptation conversations and allowing participants more time for reflection.

Results

Dangerous Futures (N/C)

Participants were generally comfortable communicating climate change knowledge that conveyed the main messages at the core of their professional practice; whether it was to explain climate data, support carbon mitigation practices, or demonstrate the need for more sustainable behaviours. When pressed by the interviewer to make responses personally relevant, and framed in terms of the next twenty years, there were generally longer pauses, more false starts and slower phrasing of answers. Many participants commented that it was either unusual, or in some cases, the first time, that someone had asked them to talk about their own adaptation thoughts and plans. As shown in Table 12.2, when participants focused on communicating about

Table 12.2 Dangerous futures: key expressions and exemplary quotes coded to both narrative communication (n) and climate change knowledge (c)

Expression	Exemplary quote
Minimising threat and immediacy	"The urgency or the extreme nature is what's not coming across because we put things out in a scientific way, in a conservative sort of way. But then if we say 'By the way, I bought a place down in Tasmania because this is real', that's not what's coming across." Practitioner 1 (Aus)[a], 40–60
Creating personal distance	"... we're wealthy people in an extremely wealthy nation and I just don't see, I don't feel that personal vulnerability, I think for that reason." Researcher 9 (Aus), 40–60
Expressing fear	"People often, when I tell them what I do, the first thing they say is' Oh, is it terrible? Are you scared every day?' And I really don't think about it that way." Researcher 10 (Can), under 40
Avoiding discussions of climate issues	"I think we are not willing to have the conversation about where our own vulnerabilities... Where those pockets are." Researcher 8 (Can), under 40

[a]Participants were identified by their role and age which was categorised as under 40, 40–60, or over 60 years

climate change knowledge, key expressions involved (1) minimising threat and immediacy; (2) creating personal distance; (3) expressing fear; and (4) avoiding discussions of climate issues.

Imagined Futures (F/N)

Participants who reported that they did not usually focus on the future expressed difficulty in discussing the future in detail, especially as it became more personally relevant. When the discussion was more distant in time or focused on larger social issues and problems, these people spoke more freely. This study did not concentrate on the relationship between age and future thinking; however, remaining life expectancy was frequently cited as a factor in how a personal future was imagined. In the 40–60 age bracket, the most common frame for personal futures related to retirement planning. Interestingly, imagined retirement rarely included consideration of likely climate change impacts even when work duties included communicating those impacts. As shown in Table 12.3, when participants focused on communication related to personal thoughts about the future, key expressions involved one, expecting others to be worse off than themselves; two, uncertain scale and time frame for possible losses; three, lacking basic necessities; and four, those responsible failing to adequately manage complex problems.

Table 12.3 Imagined futures: key expressions and exemplary quotes coded to both future thinking (f) and narrative communication (n)

Expression	Exemplary quote
Expecting others to be worse off	(C) "… things are going to be bad, it's going to be bad for other people and so we should be doing something about it for that reason, but we'll cope." Practitioner 1 (Aus), 40–60
Uncertain scale and time frame for possible losses	(C) "What is going to become unviable that we've been investing in, developing? That's scary. … But I don't know if that's in the next 20 years or not." Practitioner 3 (Aus), under 40
Lacking basic necessities	(C) "Obviously, the type of things they are growing may change, and that may not be over 20 years, that may be over 50 years. Global food shortages will certainly affect prices. They already have." Practitioner 6 (Can), 40–60
Those responsible failing to adequately manage	"I'm pessimistic. Little of my own professional career has indicated to me there's some kinds of risks that governments have learned to be reasonably good at managing." Policy-maker 11 (Can), over 60

Unfamiliar Futures (C/F)

Almost all participants reported that combining future thinking and climate change knowledge was rare for them, even on a professional basis. Aside from the few participants who actually generated climate projections, this information was prepared by others and accepted as a given, so that questioning the meaning and implications was outside of their expertise. Many participants became overtly emotional when discussing future climate impacts as personally relevant, especially for those who were responsible for other family members such as children. In addition, most participants reported not discussing these future possibilities with their families to avoid bringing negative emotions into social situations. As shown in Table 12.4, when participants focused on thinking about the future while reflecting what they knew about climate change, key expressions involved one, feeling emotionally overwhelmed; two, trying to imagine beyond their own memories; three, personal experience of change; and four, needing to make transformational changes.

Unfolding Futures (C)

Participants often made reference to changing expectations and perspectives over time, in addition to commonly relating responses to the present. As it explicitly considers change, this typology is useful for highlighting ongoing shifts in what is known about climate change, the effect of new narratives on frames of understanding, and the influence of altered patterns in thinking about the future. For

Table 12.4 Unfamiliar futures: key expressions and exemplary quotes coded to both future thinking (f) and climate change knowledge (c)

Expression	Exemplary quote
Feeling emotionally overwhelmed	(C) "You can imagine being in government going "What are we going to do about 3 million people who might not have enough water tomorrow?" Researcher 1 (Aus), over 60
Trying to imagine beyond memory	(C) "… we had the massive Black Saturday[a] bush fires here in Victoria … Well it's almost hard to imagine it being worse than that…" Researcher 2 (Aus), 40–60
Personal experience of change	(C) "We are losing glaciers and we are seeing changes in patterns … and more people are becoming aware of it, more people are starting to understand it. So I think in the next 20 years it is possible to start engaging in that. Will we see absolutely profound changes in the next 20 years? No. Will we see them in 80 years? Probably." Policy-maker 7 (Can), over 60
Needing to make transformational changes	(C) "I can foresee climate and weather changes affecting my life in that I may have to consider moving from the house that I'm currently in because it borders on a small floodplain, it's not in a floodplain, but it's pretty close and in the extreme storm I may be flooded." Policy-maker 9 (Can), over 60

[a]Bushfires that swept through parts of the Australian state of Victoria in late January and February 2009 were focused on and around Saturday, 7 February, which is referred to as Black Saturday

example, a participant who had worked on climate change issues for decades reported experiencing changes in the relative emphasis on mitigation and adaptation during policy planning, as recent observations have made it clear that some adaptation is unavoidable. In addition to external and environmental change, participants also acknowledged changes in personal expectations and opportunities as they moved through different life stages. As shown in Table 12.5, all instances coded to Unfolding futures are also included in another type; therefore, the main expressions have been captured in that context.

Beyond Imagination

This interview data provides an opportunity to investigate how people who work with climate information imagine and discuss likely futures over the next two decades. By situating the emphasis of participant responses relative to future thinking, narrative communication, and climate change knowledge, the study offers insight into personal perspectives on the expected future. It is useful that situating those responses in relation to references of change over time creates a framework that can be applied iteratively. Noting explicit references to ongoing change helps to identify patterns in the quality and characteristics of participants' engagement in future thinking. In contrast to much work focused on the physical limits of climate change adaptation, this research indicates that individual hopes and fears play a

Table 12.5 Unfolding futures; key expressions and exemplary quotes coded to change over time (C) and related content type: (n/c); (f/n); (c/f)

Expression	Exemplary quote
Unfolding futures (C) and Dangerous futures (n/c)	
Minimizing threat and immediacy	(C) "…we shy away from talking about the possible positives of climate change because of the typical reaction that we've got for the last five or 10 years from Canadians when we raise climate change and that is, I'm looking forward to that, less winter, less snow to shovel." Policy-maker 9 (Can), over 60
Unfolding futures (C) and Imagined futures (f/n)	
Those responsible failing to adequately manage	(C) "I think what a lot of people shy away from, is once you start to imagine something, then you might actually be able to do something. And then you might actually be responsible for something." Practitioner 2 (Aus), 40–60 [S]
Unfolding futures (C) and Unfamiliar futures (c/f)	
Trying to imagine beyond memory	(C) "I think it's hard to imagine the society as a whole, including government, going to the end of the century on a pathway that has that many impacts from climate change." Practitioner 3 (Aus), under 40

significant role in whether [or not] climate change knowledge is considered, shared, or applied by professionals in the field. In line with current debate, psychological distancing has been shown as a common mechanism to manage emotional stress and fear responses aroused by imagining climate change impacts (Reser and Swim 2011). This research expands on previous research by clearly linking future thinking, or prospection, to climate change discussions.

As seen in Table 12.6, participants' key expressions coded to Dangerous futures (n/c) showed narratives of climate change to be mostly framed to create distance and reduce fear, with little evidence of imagining the likely effects of climate change. It was interesting to note that this type held few direct references to change over time (C), past or future, and was generally spoken about from the participants' perspective in the present. Where thinking about the future was explicitly discussed in the Imagined futures (f/n) type, many, but not all participants reflected change

Table 12.6 Key expressions by type

Dangerous futures (n/c)	Imagined futures (f/n)	Unfamiliar futures (c/f)
Minimising threat and immediacy	Expecting others to be worse off	Feeling emotionally overwhelmed
Creating personal distance	Uncertain scale and time frame for possible losses	Trying to imagine beyond memory
Expressing fear	Lacking basic necessities	Personal experience of change
Avoiding discussions of climate issues	Those responsible failing to adequately manage	Needing to make transformational changes

over time (C). Key expressions in this type addressed uncertainties, possible losses, and issues of dependence. Here there was more evidence of imagined futures, mainly based on past or present perspectives. Unfamiliar futures (c/f) linked future thinking with what the participants knew about climate change. This type held the highest level of imagination and all expressions coded here also referenced change over time (C). Here key expressions reflected feeling overwhelmed and expecting change to go beyond what could be remembered from past experiences.

Conclusion

Among those who work with climate knowledge in Australia and Canada, immediacy and threat from climate change is often minimised based on perceived high adaptive capacity and low exposure. Even for those in the upper 40–60 age bracket, there is little personal exposure to the more extreme impacts that will be experienced in 30–40 years; relatively high levels of personal financial security and education support a high adaptive capacity; and citizenship in such stable and well-governed nations have traditionally offered the multiple safety nets needed to deal with both direct and indirect impacts from climate change. Professional discussions of climate change often focus on risk management strategies where uncertain risks seem less immediate or likely, and where other nations are generally seen as more vulnerable. Fears were generally focused on concerns that those who managed traditional environmental issues will not be equipped to deal with climate change; and that certain losses will make basic necessities such as food, water, and energy contested commodities. However, in trying to imagine how climate change will affect the next few decades, most people felt overwhelmed and avoided discussions centred on that future, instead focusing on familiar problems and solutions from the past.

This study suggests that thought leaders and decision-makers in Australia and Canada have limited capacity to discuss and imagine themselves as vulnerable to climate change over the next two decades. Emotional responses to uncertainty, fear, and lack of clear reference points made it difficult for them to engage with imagining the future affected by climate change. This lack of imagination among the most knowledgeable may place significant limits on the adaptive effectiveness of current investments with long-term consequences for the wider population. The large contingent of civil society who lack scientific understanding or a high adaptive capacity has little help to imagine their own adaptation needs and options. This lack of guidance to imagine the future may in turn reduce the real and perceived agency of socially and economically marginalised people to influence forward planning in their favour.

To better understand the roles of imagination in climate change research and communication will require research that is both broader and deeper than the present study. Recruiting a larger sample of professional participants and controlling for future orientation could yield results more closely focused on the role of imaginative future thinking in climate change adaptation plans and decisions. It would also be valuable to interview other national, social and cultural groups to identify what factors influence how future climate change is imagined and discussed. Interdisciplinary synthesis drawing together social, cognitive, and climate sciences is needed to better understand what personal and imaginative constraints affect swift and effective adaptation.

Acknowledgements Thank you to the researchers, policy-makers, and climate change practitioners whose frank comments and willingness to discuss a difficult topic was invaluable. The research was supported by Griffith University School of Environment through Ph.D. funding, and an International Experience Incentive Scheme (IEIS) grant. A Griffith University/Simon Fraser University Collaborative Grant supported Canadian data collection. Special thanks to my Canadian hosts, The Pacific Institute for Climate Solutions (Tom Pedersen) through their member centres at the University of Victoria (Francis Zwiers); the University of British Columbia (John Robinson); and Simon Fraser University (Mark Roseland). Thanks also to Ann-Maree Dowd at CSIRO for transcription assistance. The mentoring of Silvia Serrao-Neumann, Eddo Coiacetto, and Peter Hegedus has increased the depth and focus of this research. This work would not be possible without the ongoing support of Malcolm Coulter and Lucas Mahrt.

References

Adger, W. N., Dessai, S., Goulden, M., Hulme, M., Lorenzoni, I., Nelson, D. R., et al. (2009). Are there social limits to adaptation to climate change? *Climatic Change, 93,* 335–354. doi:10.1007/s10584-008-9520-z.

Atance, C. M., & O'Neill, D. K. (2001). Episodic future thinking. *Trends in Cognitive Sciences, 5* (12), 533–539. doi:10.1016/S1364-6613(00)01804-0.

Barnett, J., & O'Neill, S. (2010). Maladaptation. *Global Environmental Change, 20*(2), 211–213. doi:10.1016/j.gloenvcha.2009.11.004.

Bateson, M. (2007). Narrative, adaptation, and change. *Interchange, 38*(3), 213–222. doi:10.1007/s10780-007-9030-3.

Brown, K., & Westaway, E. (2011). Agency, capacity, and resilience to environmental change: Lessons from human development, well-being, and disasters. *Annual Review of Environment and Resources, 36*(1), 321–342. doi:10.1146/annurev-environ-052610-092905.

Caney, S. (2014). Two kinds of climate justice: Avoiding harm and sharing burdens. *Journal of Political Philosophy, 22*(2), 125–149. doi:10.1111/jopp.12030.

Capstick, S., Whitmarsh, L., Poortinga, W., Pidgeon, N., & Upham, P. (2015). International trends in public perceptions of climate change over the past quarter century. *Wiley Interdisciplinary Reviews: Climate Change, 6*(1), 35–61. doi:10.1002/wcc.321.

Chen, X. (2011). Why do people misunderstand climate change? Heuristics, mental models and ontological assumptions. *Climatic Change, 108*(1–2), 31–46. doi:10.1007/s10584-010-0013-5.

Dilling, L., & Moser, S. C. (2007). *Creating a climate for change: Communicating climate change and facilitating social change.* Cambridge: Cambridge University Press.

Feenstra, J. F., Burton, I., Smith, J. B., & Tol, R. S. J. (1998). *Handbook on methods for climate change impact assessment and adaptation strategies.* Retrieved from http://www.ivm.vu.nl/en/Images/UNEPhandbookEBA2ED27-994E-4538-B0F0C424C6F619FE_tcm53-102683.pdf.

Fincher, R., Barnett, J., Graham, S., & Hurlimann, A. (2014). Time stories: Making sense of futures in anticipation of sea-level rise. *Geoforum, 56,* 201–210. doi:10.1016/j.geoforum.2014.07.010.

Gardner, G., & Assadourian, E. (2004). *Rethinking the good life state of the world 2004; a Worldwatch Institute report on progress toward a sustainable society, special focus: The consumer society* (pp. 164–179). New York and London: WW Norton.

Grothmann, T., & Patt, A. (2005). Adaptive capacity and human cognition: The process of individual adaptation to climate change. *Global Environmental Change, 15*(3), 199–213. doi:10.1016/j.gloenvcha.2005.01.002.

Hulme, M. (2010). Cosmopolitan climates: Hybridity, foresight and meaning. *Theory, Culture & Society, 27*(2–3), 267–276. doi:10.1177/0263276409358730.

Hurlbert, M. A. (2011). Evaluating climate justice—attitudes and opinions of individual stakeholders in the United Nations Framework Climate Change Convention Conference of the Parties. *Journal of Integrative Environmental Sciences, 8*(4), 267–286. doi:10.1080/1943815X.2011.599812.

IPCC. (2014a). *Climate change 2014: Impacts, adaptation, and vulnerability. Working group II summary for policymakers*. Cambridge: Cambridge University Press.

IPCC. (2014b). *Climate change 2014: Synthesis report. Contribution of working groups I, II and III to the fifth assessment report of the intergovernmental panel on climate change*. Geneva: IPCC.

Jones, S. (2012). A tale of two cities: Climate change policies in Vancouver and Melbourne—Barometers of cooperative federalism? *International Journal of Urban and Regional Research, 36*(6), 1242–1267. doi:10.1111/j.1468-2427.2011.01083.x.

Juhola, S., Glaas, E., Linnér, B.-O., & Neset, T.-S. (2016). Redefining maladaptation. *Environmental Science & Policy, 55,* 135–140. doi:10.1016/j.envsci.2015.09.014.

Kirtman, B., Power, S. B., Adedoyin, J. A., Boer, G. J., Bojariu, R., Camilloni, I., et al. (2013). Near-term climate change: Projections and predictability. In *Climate change 2013: The physical science basis* (pp. 953–1028). Cambridge: Cambridge University Press.

Lee, T. M., Markowitz, E. M., Howe, P. D., Ko, C.-Y., & Leiserowitz, A. A. (2015). Predictors of public climate change awareness and risk perception around the world. *Nature Climate Change, 5*(11), 1014–1020. doi:10.1038/nclimate2728.

Moser, S. C. (2010). Communicating climate change: History, challenges, process and future directions. *Wiley Interdisciplinary Reviews: Climate Change, 1*(1), 31–53. doi:10.1002/wcc.011.

O'Neill, S. J., & Hulme, M. (2009). An iconic approach for representing climate change. *Global Environmental Change, 19*(4), 402–410. doi:10.1016/j.gloenvcha.2009.07.004.

Oyserman, D., & James, L. (2011). Possible identities. In S. Schwartz, K. Luyckx, & V. L. Vignoles (Eds.), *Handbook of identity: Theory and research* (Vol. 1, pp. 117–145). New York: Springer.

Patchen, M. (2006). *Public attitudes and behavior about climate change: What shapes them and how to influence them.* Retrieved from http://www.purdue.edu/DP/climate/assets/pdfs/PatchenOP0601.pdf.

Reed, I. (2008). Justifying sociological knowledge: From realism to interpretation. *Sociological Theory, 26*(2), 101–130. doi:10.1111/j.1467-9558.2008.00321.x.

Reisinger, A., Kitching, R. L., Chiew, F., Hughes, L., Newton, P. C. D., Schuster, S. S., et al. (2014). Australasia. In V. R. Barros, C. B. Field, D. J. Dokken, M. D. Mastrandrea, K. J. Mach, T. E. Bilir, M. Chatterjee, K. L. Ebi, Y. O. Estrada, R. C. Genova, B. Girma, E. S. Kissel, A. N. Levy, S. MacCracken, P. R. Mastrandrea, & L. L. White (Eds.), *Climate change 2014: Impacts, adaptation, and vulnerability. Part B: Regional aspects. Contribution of working group II to the fifth assessment report of the intergovernmental panel on climate change* (pp. 1371–1438). Cambridge: Cambridge University Press.

Reser, J., & Swim, J. K. (2011). Adapting to and coping with the threat and impacts of climate change. *The American Psychologist, 66*(4), 277–289. doi:10.1037/a0023412.

Rockstrom, J., Steffen, W., Noone, K., Persson, A., Chapin, F. S., Lambin, E. F., et al. (2009). A safe operating space for humanity. *Nature, 461*(7263), 472–475. doi:10.1038/461472a.

Romero-Lankao, P., Smith, J. B., Davidson, D. J., Diffenbaugh, N. S., Kinney, P. L., Kirshen, P., et al. (2014). North America. In V. R. Barros, C. B. Field, D. J. Dokken, M. D. Mastrandrea, K.

J. Mach, T. E. Bilir, M. Chatterjee, K. L. Ebi, Y. O. Estrada, R. C. Genova, B. Girma, E. S. Kissel, A. N. Levy, S. MacCracken, P. R. Mastrandrea, & L. L. White (Eds.), *Climate change 2014: Impacts, adaptation, and vulnerability. Part B: Regional aspects. Contribution of working group II to the fifth assessment report of the intergovernmental panel on climate change* (pp. 1439–1498). Cambridge: Cambridge University Press.

Roncoli, C., Crane, T., & Orlove, B. (2009). Fielding climate change in cultural anthropology. In M. Nutall & S. Crate (Eds.), *Anthropology and climate change: From encounters to actions* (pp. 87–115). San Francisco: Left Coast Press.

Savin-Baden, M., & Niekerk, L. V. (2007). Narrative inquiry: Theory and practice. *Journal of Geography in Higher Education, 31*(3), 459–472. doi:10.1080/03098260601071324.

Schacter, D. L., Addis, D. R., & Buckner, R. L. (2007). Remembering the past to imagine the future: The prospective brain. *Nature Reviews Neuroscience, 8*(9), 657–661. doi:10.1038/nrn2213.

Serrao-Neumann, S., Harman, B., Leitch, A., & Low Choy, D. (2014). Public engagement and climate adaptation: Insights from three local governments in Australia. *Journal of Environmental Planning and Management, 58*(7), 1196–1216. doi:10.1080/09640568.2014.920306.

Spence, A., Poortinga, W., & Pidgeon, N. (2012). The psychological distance of climate change. *Risk Analysis, 32*(6), 957–972. doi:10.1111/j.1539-6924.2011.01695.x.

Stuiver, M., Leeuwis, C., Klostermann, J. E. M., Turnhout, E., & Harms, B. (2013). New roles of science in society: Different repertoires of knowledge brokering. *Science and Public Policy, 40*(3), 354–365. doi:10.1093/scipol/scs114.

Szpunar, K. K. (2010). Evidence for an implicit influence of memory on future thinking. *Memory & Cognition, 38*(5), 531–540. doi:10.3758/MC.38.5.531.

Szpunar, K. K., Spreng, R. N., & Schacter, D. L. (2014). A taxonomy of prospection: Introducing an organizational framework for future-oriented cognition. *Proceedings of the National Academy of Sciences USA, 111*(52), 18414–18421. doi:10.1073/pnas.1417144111.

Tulving, E. (1985). *Memory and Conciousness Canadian Psychology/Psychologie canadienne, 26*(1), 1–12. doi:10.1037/h0080017.

Warren, F. J., & Lemmen, D. S. (Eds.). (2014). *Canada in a changing climate: Sector perspectives on impacts and adaptation*. Ottawa, ON: Government of Canada.

Wright, C., Nyberg, D., De Cock, C., & Whiteman, G. (2013). Future imaginings: Organizing in response to climate change. *Organization, 20*(5), 647–658. doi:10.1177/1350508413489821.

Author Biography

Ms Liese Coulter is a climate change adaptation specialist who holds a Master's Degree in Science Communication from the Australian National University. Liese is completing a Ph.D. at the Cities Research Center, Griffith University where she also teaches Science, Technology and Society subjects. She has served as national Knowledge Communication Manager for both the CSIRO Climate Change Adaptation Flagship and the National Climate Change Adaptation Research Facility (NCCARF). In 2016, Liese was awarded a research travel grant by the Max Planck Institute to contribute to the Future Earth cross community workshop, Extreme Events and Environments: from Climate to Society. "How to project climate extremes that really matter?"

Chapter 13
Climate Change Adaptation in Coastal East Arctic Ecosystems: Complexity and Challenges of Monitoring and Evaluation

Moktar Lamari, Line Poulin-Larivière and Johann L. Jacob

Introduction

The Arctic ecosystem is warming up at a faster pace than the northern hemisphere average, and associated changes will have major impacts on the global climate system (Larsen et al. 2014; Ford et al. 2010; McBean et al. 2005; Weller et al. 2004). Located north of the 64th parallel, Nunavut is largely characterized by an extremely low demographic density and Inuit indigenous group predominance (Statistics Canada 2013). The territory has an estimated ocean coastline of 104,000 km and the majority of the population lives in coastal areas (Public Works and Government Services Canada 2014). Local mitigation opportunities are limited as the territory is responsible for a small proportion of Canada's greenhouse gas (GHG) emissions; therefore, adaptation measures appear to be the main path to follow (Klein et al. 2014; Bowron and Davidson 2011).

A first step toward adaptation implies mainstreaming or integrating into planning and ongoing sectoral decision-making policies and initiatives expected to enhance adaptive capacity and to reduce the vulnerability and sensitivity of natural and human systems to the effects of actual or anticipated climate change (Anisuzzaman and Jennings 2012; Klein et al. 2007). Knowledge development activities play an active role in this process. Knowledge about the various dimensions of adaptation may be obtained through the assessment of risks and vulnerabilities, the documentation of local climate knowledge and practices, the measurement of the progress of adaptation activities via a set of performance indicators, the identification of what works, for whom, when, using monitoring and evaluation (M&E) activities, etc. It then serves as a crutch for governmental organizations (national/territorial/local), non-governmental organizations and local stakeholders. This knowledge is

M. Lamari (✉) · L. Poulin-Larivière · J. L. Jacob
École National d'administration Publique, University of Quebec, 555 Blvd. Charest Est, Quebec G1K 9E5, Canada
e-mail: moktar.lamari@enap.ca

W. Leal Filho and J. Nalau (eds.), *Limits to Climate Change Adaptation*, Climate Change Management, https://doi.org/10.1007/978-3-319-64599-5_13

critical for identifying options for adaptation, designing and implementing policies, programs and plans, and integrating adaptation. Conversely, knowledge, through its multiple forms, in the process of adaptation mainstreaming is a prerequisite for an adaptive governance of climate change (Nursey-Bray 2010).

As is often the case when dealing with serious policy problems, such as situations with important uncertainties, limited experience to rely on, costs and benefits unevenly distributed among constituents, the production of knowledge can help in reducing uncertainties and identifying acceptable adaptation options. Production of knowledge, especially via information gathered through robust M&E indicators, is critical to document the progress of adaptation activities, disseminate the realities of an environment to the greatest number, identify innovations, advocate for the integration of the identified responses to climate change into government policies and plans, and support government (local/national) and non-governmental organizations to design and implement policies, programs, plans, which integrate adaptation to climate change. That is why difficulties in the production of knowledge about climate risks, elements of vulnerability and sensitivity, operational effectiveness and efficiency of adaptation options, as well as their impacts, could represent important limits to adaptation (Nursey-Bray 2010). Such is the situation in Nunavut. Over the last decade, several government and community-based adaptation initiatives have been implemented (Ford et al. 2010, 2014; Healey et al. 2011), in addition to important efforts to produce knowledge on anticipated climatic threats and risks, climate scenarios, vulnerability maps, elements of adaptive capacity (Lamari et al. 2016b; Bunce et al. 2016; Ford et al. 2013). Still, difficulties remain linked to the absence of standardized metrics and definitions in the M&E process, as well as a lack of documentary evidence of formal adaptation evaluation (Ford et al. 2014).

For adaptation in general, despite numerous frameworks for measuring and evaluating climate change adaptation, the institutionalized utilization of common, reliable and robust adaptation indicators is still limited (Lamari et al. 2016b; Adger et al. 2005; Kates et al. 2012; Quay 2010). Beyond the persistent vagueness concerning concepts related to adaptation (Hinkel 2011; Hedger et al. 2008; Thywissen 2006), high uncertainties still surrounding climate change impacts, in a context of complex and dynamic socio-economic and environmental processes, prevent establishing baselines from which evaluative criteria and, performance indicators for adaptation interventions could be derived (Fisher et al. 2015).

Because adaptation efforts are rooted in local settings, a plurality of contextual elements may impact the implementation of an adaptation project (Duerden 2004). Adaptation produces adjustments in human systems, occurring at different scales, through different actors, with successes likely to be perceived differently by individuals (Berrang-Ford et al. 2015). Therefore, available M&E guidelines from grey literature will be of limited use because their indicators cannot be used without prior efforts to identify and adapt the few relevant ones for the particular setting in which the adaptation initiative is implemented. For instance, in the Canadian Arctic, differences exist regarding how Western and indigenous knowledge is integrated in assessing the impact of climate change adaptation (Pennesi et al. 2012; Dowsley et al. 2010). Communities' understanding of projected climate change and potential vulnerabilities is also often limited (Ford et al. 2015; Reid et al. 2009), which may complicate the identification of shared and accepted adaptation indicators. Far from being value-neutral, indicators must be grounded within social constructions and infused with political and social values and ethical power (Lamari et al. 2016b).

The M&E of adaptation to climate change also faces other important methodological issues, notably dissonances related to the measurement of tangible (versus intangible) effects, measurable (versus diffuse and latent) effects, specific (versus combined) effects, long- and medium-term (versus short-term) effects. Methodological issues are also linked to the dynamics by which the impacts associated with adaptation to climate change interventions are materialized. Evaluating adaptation in the light of a single theory or paradigm proves to be complicated, as is finding indicators that capture the "invisible" and that reliably quantify all derived benefits. Most indicators will tackle the "visible part of the iceberg", but will have limited relevance when one looks at the underlying dynamics through which various adaptation interventions produce their effects (Lamari et al. 2016b; Lamari and Jacob 2015).

The design and utilization of adaptation indicators also appears dissonant and fragmented based on the advocated approaches and disciplines of their promoters. Unlike with mitigation, there is no unique and accepted adaptation indicator as multiple points of view can be used to evaluate adaptation actions (Noble et al. 2014; Gagnon-Lebrun and Agrawala 2007; Perkins et al. 2007; Füssel 2008). Instead, adaptation outcomes are mostly tagged with scalar indicators and indexes aggregating and combining different concepts derived from multiple theoretical backgrounds, and different methods are mobilized to document them. Strategic weighting is thus needed to determine, which inputs, outputs, outcomes and impacts should be measured and considered first. Collective judgments are also at the heart of the valuation process (see this chapter of IPCC fifth Assessment Report (Noble et al. 2014) for a review of the subject). Alone, many indicators presented in available guidelines are of limited usefulness because adaptation results will be evaluated by experts via a subjective assessment based on the triangulation of elements of information on key parameters.

Our contribution presents the main challenges to M&E faced in Nunavut as related by local stakeholders concerned by climate change adaptation (CCA), as well as the indicators commonly used in that context. Our paper will specifically answer three questions: (1) What are examples of data, statistics and indicators linked to dimensions of CCA currently used in the context of Iqaluit and what can be said about their attributes? (2) How does the lack of robust monitoring and evaluation activities and indicators influence adaptation limits in the region? and (3) What are the current insights emerging on this topic and what do these limits mean for local communities? The latter will explore how Western and Inuit *Qaujimajatuqangit* is integrated in assessing the impact of adaptation activities in the Iqaluit context. The following chapters will allow us to briefly introduce the methodology used, to present the results obtained, to discuss the main challenges and limits identified and, finally, to conclude with key findings and relevant opportunities for future investigation.

Theory and Method

Ecosystems in the Arctic are facing rapid transformations, mainly because of the socio-economic mutations and political competition for field and common pool resources appropriation, which are all having important consequences on addressing climate change issues (Ford et al. 2010). Collective action theory has inspired our methods and research approaches. Ostrom (2010) operationalized the collective action theory for climate change and demonstrated that climate change adaptation is a lengthy process involving (1) actors' attributes and motivations, (2) physical and natural attributes of ecosystems exposed to climate change externalities (public resources, private resources, common pool resources, rivalry, etc.) and (3) the capability of communities to draft new rules, institutions and values, in order to "acclimatize" collective action. Ostrom (2010) claims that if individuals are informed about, and sensitized to, the challenges they are facing, they should be able to craft new arrangements enhancing social capital (confidence, reciprocity, cost sharing, etc.), without asking for external and exogenous powers. Knowledge and capabilities are therefore required, among other factors, as a precondition for the implementation of actions whose long-term benefits are maximized for the community.

Ostrom's theoretical concepts were helpful because of their capacity to structure and identify tools and incentives adapted to the complexity associated with the climate change context. Our research particularly aimed at identifying and describing knowledge development activities and tools in the Iqaluit context, in the form of adaptation indicators used by local actors as well as public agencies, and understanding the factors constraining their development and use. Indicators are snapshots of data and information that describe what is going on in a large context at a certain time. They are dependent on the availability of a select set of data (statistics). Indicators are metrics that can be dedicated to helping decision making, mainly by monitoring changes and progress over time, or by mixing different information and knowledge to highlight a specific issue sometime, somewhere. To provide helpful, valid, and understandable insight for decision makers, indicators must be specific, reliable, valid, complete and accessible.

In order to identify and describe adaptation indicators used in the context of Iqaluit, our exploratory research followed a two-phased methodology: (1) a review of recently published open access data (statistics/reports); (2) a data collection in Iqaluit-Nunavut. As a first step, we searched open source databases to identify existing data, statistics, and indicators linked to dimensions pertaining to climate change adaptation in Canada's Arctic region. Robust and official statistics were extracted from the following databases: Canadian Coast Guard, Statistics Canada, Transport Canada, Environment and Climate Change Canada, Natural Resources

Canada and the Nunavut Bureau of Statistics. Indicators and statistics retrieved were linked to dimensions such as adaptation, vulnerability, exposure, sensitivity, and adaptive capacity. We have gathered and synthesized 62 documents dealing with climate change in Nunavut. About 23 scientific articles, published between 2005 and 2015, were examined in order to extract and delimit the monitoring and evaluation dimensions associated with climate change adaptation in the Nunavut territory.

In this study, we have chosen for these concepts the definitions presented in Smit and Wandel (2006). According to their analysis of available definitions of these interrelated concepts, *adaptation* refers to manifestations of *adaptive capacity*, as well as ways of reducing *vulnerability* of a system. A concept leading to other concepts, *vulnerability* of a system (at any scale) is reflective of (or a function of) the *exposure* and *sensitivity* of that system to hazardous conditions and the ability or capacity or resilience of the system to cope, adapt or recover from the effects of those conditions. "The *exposure* and *sensitivity* of a system to an environmental change risk reflect the likelihood of the system experiencing the particular conditions and the occupancies and livelihood characteristics of the system which influence its sensitivity to such exposure" (Smit and Wandel 2006, p. 286). Finally, *adaptive capacity* refers to the forces (determinants) that influence the ability of a system to adapt. Such determinants could refer to factors like managerial ability, access to financial, technological and information resources, infrastructure, the institutional environment within which adaptations occur, political influence, kinship networks, etc. (Smit and Wandel 2006, p. 287).

This first step was followed by a field trip to Iqaluit, Nunavut, during the summer of 2015 to conduct seven semi-structured interviews with key local stakeholders (representatives from Nunavut government agencies, the city of Iqaluit, NGO, civil society organizations involved with issues related to climate change). Respondents were selected by a purposeful sampling based on criteria such as the mandates of the organizations of the respondents, their links with climate change adaptation, as well as the availability of the respondents at the time of our visit. In addition to the elements derived from Ostrom's theory of collective actions, semi-structured interviews explored climate change experienced in the Iqaluit region, as well as the current state of practice of adaptation M&E and factors influencing the development and use of adaptation indicators. The meetings with these respondents also allowed us to question them about possible documents related to climate change indicators that we should consult. Upon our return, we completed our review of relevant documentation by analyzing the documents provided or suggested by respondents met in Iqaluit, which led us to update the information initially gathered.

Our team identified four main limitations to the research: geographical coverage during data collection due to long distances and absence of road infrastructure, language as the documentation consulted excludes material only available in Inuktitut, the limited quantity and quality of data specific to coastal areas available and the lack of information shared between organizations. De facto, data are greatly affected by the collection parameters heterogeneity and the contextual monitoring constraints including remoteness, access to sites and performance of technological tools in extreme cold conditions (Forbes 2011). Therefore, caution is advised in the interpretation of our results, as our research is exploratory.

Results

Our analysis started with examples of available data, statistics and indicators linked to dimensions of CCA currently used in the Iqaluit context. Before a thorough presentation of the data retrieved via our research, some general comments on the attributes of the identified indicators follow. Our analysis suggests that the deficit of indicators dealing with climate change in the Iqaluit region is tremendous. In trying to collect our data and empirical information, we have discovered the scarcity of indicators and reliable data for the Iqaluit region and communities. In spite of the multiplication of documents and information about Nunavut, very little of the available information can be used as an indicator for measuring climate change adaptation and mitigation.

The prevailing lack of robust monitoring and evaluation is observed locally through the irregular definitions of associated key concepts and the aggregated and inconsistent inventory of CCA indicators. In line with the literature (Hofmann et al. 2011; Sanahuja 2011), our study confirmed that multi-sectoral, associated concepts are fragmented and expressions remain polysemous. The examination of publications issued by government, community and academic sectors fails to provide a uniform definition of CCA. Compared to IPCC's definition of CCA, the publications issued in Nunavut give particular attention to vulnerability. The definition of adaptive capacity emphasizes individuals within their community (Government of Nunavut 2011; Nunavut Climate Change Center n.d.; Lewis and Miller 2010; Government of Nunavut, Department of Culture, Language, Elders and Youth and Nunavut Tunngavik Incorporated 2005). As an illustration, Nunavut's Government (2011) mentioned that adaptive capacity is a region or community's ability to manage the impacts and risks of climate change.

Our analysis also shows conceptual discrepancies, characterized by a holistic vision of climate change by local indigenous communities. A review of numerous translations of the definitions of key adaptation concepts shows that important efforts have been made in the translation of these concepts. Indeed, the emergence of neologism in Inuktitut is visible in Nunavut official publications as part of building a process of ownership (Government of Nunavut 2011; Nunavut Climate Change Center n.d.; Lewis and Miller 2010).

A list of current identified adaptation indicators could be arranged in three categories: socioeconomic data, vulnerability (exposure and sensitivity) indicators and indicators more generally associated with locally-based initiatives. Over the last decade, federal and territorial authorities have tried to collect relevant data to measure local community resilience and adaptation capacity through socioeconomic indicators. Available data draw a mixed portrait of the population.

Socioeconomic Data

The concepts of resilience and adaptive capacity, while being distinct, are quite similar and synonymous. Resilience refers to the capacity of individuals or groups to implement many proactive adaptive actions over a time period (McDaniel et al. 2008), while adaptive capacity encompasses the ability of individuals and communities to adapt to adversity and stressful conditions and reorganize into networks and institutions that learn and solve problems (Marshall 2010), cited in Sovacool and Linnér (2015, p. 10). Since these concepts are not always directly measurable, socioeconomic data can be used as proxies, in the same way as with social capital.

A 2015 food price survey confirms the inflated price of groceries as the inhabitants pay, on average, twice as much as in the rest of the country (Nunavut Bureau of Statistics 2010). The situation is mainly related to the inexistence of local agriculture, the air delivery of perishable food and boat transportation of any other merchandise. Additionally, the increased supply from the South is likely to continue due to the consumption habits of the growing number of residents born outside the territory (Statistics Canada 2015).

Infrastructure and housing data confirm the precariousness and deficient quality of installations: the 2009–2010 Nunavut Housing Needs Survey indicated that the majority of the housing is public and is below housing standards (Nunavut Bureau of Statistics 2010). The survey also highlights that 18% of occupied Iqaluit dwellings require major repairs (defective plumbing, electrical wiring, leaking oil or sewage tank, or a broken hot water tank) (Nunavut Bureau of Statistics 2010). The National Household Survey labor force indicators also demonstrate regional precariousness as the unemployment rate reached 17.9% and the average individual income was estimated at $43,305 (Statistics Canada 2015).

Additional pressures are documented through demographic growth (8.3% between 2006 and 2011) as nearly 40% of the population is under 18 years of age and that density remains below zero (Statistics Canada 2015). Nearly half the residents were born outside the territory and one-third of the 610 migrants living in Nunavut arrived between 2001 and 2011 (Statistics Canada 2015). The two main countries of origin are the United Kingdom and the Philippines (Statistics Canada 2015).

Data specific to Iqaluit also demonstrate interconnected issues such as remoteness, unemployment, poverty and food insecurity (Statistics Canada 2013). The capital is in a particular situation due to its population density of 127.6/km^2 and its young population, as only 2.5% is over 65 years of age and the median age is 30.1 years (Statistics Canada 2013).

Vulnerability (Exposure and Sensitivity) Indicators

"Vulnerability is not only exposure or sensitivity to disasters, but also the capacity of an individual or community to be resilient to these disasters, to evolve in order to accommodate the hazards, or change positively to cope with it" (Adger 2006, cited in Sovacool and Linnér 2015, p. 10). The glossary of IPCC fifth assessment report (IPCC 2014) defines exposure as "the presence of people, livelihoods, species or ecosystems, environmental functions, services, and resources, infrastructure, or economic, social, or cultural assets in places and settings that could be adversely affected", while sensitivity is "the degree to which a system or species is affected, either adversely or beneficially, by climate variability or change".

At a regional level, inter-state initiatives were put into place in order to promote M&E collaboration. A recent evaluation on the impacts of climate change on snow, water, ice and permafrost included numerous indicators measuring vulnerability in the Arctic and highlights the urgency to integrate climate change adaptation at all levels (AMAP 2012). Other initiatives suggest establishing a standardized protocol in order to facilitate comparison between regions (Ramesh et al. 2015).

The Circumpolar Biodiversity Monitoring Program (CBMP) of the Conservation of Arctic Flora and Fauna (CAFF) working group of the Arctic Council also developed indicators that provide objective information to orient decision making (CAFF 2016). Recent work constitutes a coproduction of data that does not only integrate traditional knowledge but also selects indicators based on specific criteria: scientific validity, facility to understand and communicate, response to change, relevance (to the circumpolar region, organizational mandate, other programs, local and external population and policy makers), ecological relevance, likelihood of being subject to targets and thresholds, long-term commitments and finally practicality (is cost effective, relies on accessible data, is technically feasible to measure and is representative of multiple species, ecosystems and habitats) (CAFF 2016).

Regional M&E indicators specific to coastal zones have been incontestably initiated but Canadian initiatives remain limited. Based on Gornitz's vulnerability index, Shaw et al. (1998) developed a coastal sensitivity of Canada's coasts to sea level rise caused by global warming. The sensitivity index scores range from 0.8 to 56.7 and are grouped together into three categories: low 0–4.9, moderate 5–14.9 and high >15. Based on this index, the Frobisher Bay coastal sensitivity was considered moderate. We were unable to obtain an updated score regarding this sensitivity.

The evolution of the Canadian climate is increasingly documented and is mainly characterized by the diminution of sea-ice coverage (Environment Canada 2013). The perspective of growing traffic in the Arctic raises several challenges: water pollution, law enforcement in a vast territory, disruption of traditional lifestyles, etc. For the time being, the Canadian Coast Guard mainly monitors and coordinates maritime traffic from its Iqaluit office and gathers data related to the number of vessel voyages (Office of the Auditor General of Canada 2014), the maritime route schedule, new routes facilitating ore transportation to Europe and new types of travelers such as adventurers crossing the Northwest Passage.

Promoting capacity building and mainstreaming, the Iqaluit CCA action plan also highlights the technicality of data, the retention of information within organizations and the lack of data regarding climate change impacts and risk mapping (Lewis and Miller 2010). The rapid urban growth has caused the multiplication of local actors and affected the adaptive capacity due to workforce training and regular arrival of professionals (Lewis and Miller 2010). Using technologies such as Light Detection and Ranging (LiDAR) and the Global Positioning System (GPS), the coastal hazard evaluation completed by Hatcher et al. (2011) identified the zones at risk and concluded that main changes to the coastal zones of the city are related to annual sea ice in Frobisher Bay.

Local interviews enabled us to identify certain informal indicators utilized by the population to measure their perception of climate change impacts and also local adaptation processes. Inhabitants have developed indicators taken from daily life that are context specific, non-transferable and descriptive of new phenomena: the number of polar bears seen or killed in Iqaluit and Apex, the frequency of the incidents, changes in snowmobile routes due to ice thinning, changes in snow cover and the number of accidental drownings during traditional activities such as hunting (Table 13.1).

Table 13.1 Indicator overview

Indicator	Concept measured	Source
Percentage of dwellings below housing standards	Exposure	Nunavut Bureau of Statistics (2010)
Percentage of population under 18 years of age	Exposure	Statistics Canada (2012)
Number of vessel voyages in Canadian arctic	Exposure	Office of the Auditor General of Canada (2014)
Number of days with snow on the ground (duration)	Sensitivity	AMAP (2012)
Area of land covered by snow at a given time (extent)	Sensitivity	AMAP (2012)
Actual amount of snow (depth)	Sensitivity	AMAP (2012)
Water quality index	Sensitivity	CAFF (2016)
Arctic human well-being index	Sensitivity	CAFF (2016)
Arctic land cover change index	Sensitivity	CAFF (2016)
Use of traditional knowledge in research, monitoring and management	Adaptive capacity	CAFF (2016)
Mean annual air temperature	Sensitivity	Sladen (2012)
Mean annual ground temperature	Sensitivity	Sladen (2012)

Locally-Based Adaptation Initiatives

Without formalized M&E practice, we have however documented the use of CCA indicators in the main local initiatives. First, the CCA toolkit for Nunavut suggests a 5-step process that includes a monitoring and review phase and key indicators and milestones (Bowron and Davidson 2011). Also, at the municipal level, the 2014 annual progress report of the Iqaluit sustainable community plan includes four indicators associated with the measurement of CCA progress (Campbell 2014). Last, a recent initiative organized by the Qaujigiartiit Health Research Center explores health-related indicators of climate change and is based on the consultation of local communities (Healey 2015) (Table 13.2).

Discussions and Implications

Our investigation results and interviews in local communities highlight two main questions: How does the lack of robust monitoring and evaluation influence adaptation limits in the region? What are the current insights emerging on this topic and what do these limits mean for local communities?

Q1) How does the lack of robust monitoring and evaluation reduce climate change adaptation in the Iqaluit landscape?

Table 13.2 Examples of local-based indicators

Indicator	Concept measured	Source
Promote awareness of climate change adaptation in iqaluit	Adaptation	Campbell (2014)
Improve wastewater infrastructure	Adaptation	Campbell (2014)
Improve housing quality	Adaptation	Campbell (2014)
Share information on local research initiatives related to housing infrastructure	Adaptive capacity	Campbell (2014)
Sea level rise vulnerability (population by community within 5 km of coast with "very high" vulnerability to sea level rise)	Exposure	Healey (2015)
Flooding vulnerability (the elderly, those living in poverty, infants, and the disabled living in 100- and 500-year flood zones)	Exposure	Healey (2015)
Food security vulnerability (elderly, those living in poverty, infants, and the disabled)	Exposure	Healey (2015)
Mental Health (reports of depression and anxiety due to climate change)	Sensitivity	Healey (2015)
Number of injuries or deaths from sea-ice instability	Sensitivity	Healey (2015)

As mentioned by Klein et al. (2014): "Multiple constraints can significantly reduce the range of adaptation options and opportunities available to actors and therefore may pose fundamental limits to adaptation" (p. 12). Based on the typology developed by the IPCC, our research identifies key findings associated with the five categories of constraints (Klein et al. 2014, pp. 12–22). These constraints have to be documented with regard to adaptation efforts and collective action dealing with climate change. Our investigations have characterized five major constraints involving different dimensions and parameters.

1. Knowledge and technology gaps. The first and main constraint is related to knowledge and technology gaps. In these communities, traditional knowledge is declining, weakened by different socio-economic evolution and mutations. Our investigations suggest that traditional knowledge is almost tacit knowledge and orally communicated by local dialects. At the same time, our observations indicate the emergence of growing research on CCA in the Arctic region. These research programs are producing new knowledge, mainly dedicated to the government decision-making process. Endogenous local absorptive capacity knowledge seems to be limited, mainly because of the low level of literacy among local community leaders and stakeholders. Also, the rapidity and complexity of climate change may surpass local absorptive capacity, and more has to be done in order to enhance knowledge advancement and knowledge utilization. Knowledge utilization is still embryonic, mainly regarding how to measure progress and changes, and how to conduct CCA monitoring and evaluation (AMAP 2012). Our investigations demonstrate several local capacity deficiencies. Local experts are needed to build new knowledge, enhance awareness and facilitate the implantation of required new technologies.
2. Physical and biological vulnerabilities. Partial or total ignorance about these vulnerabilities is not helpful for CCA monitoring and evaluation. The Iqaluit region is facing intense, rapid physical climate change. In the city of Iqaluit, research on permafrost has confirmed temperature rise since the early 1990s (Leblanc et al. 2012). Permafrost is about soil, rock, or sediment that remains at or below a temperature of 0 °C for at least two successive years. Many reports provide reviews of the linkages between permafrost conditions, hydrology, soil quality and vegetation including the implications of warming and thawing permafrost for aquatic and terrestrial ecosystems. In the Iqaluit region permafrost is continuous and may be several hundred meters thick and have temperatures colder than −5 °C.
3. Economic and infrastructure constraints. Our interlocutors were aware of climate change and the imperative for public action towards mitigation and adaptation. As a local leader mentioned: "Climate change is affecting a number of things, including buildings, roads, water supply, treatment of wastewater, excessive presence of waste in the community, etc." The most common comments were related to snow and sea ice. In recent years, ice has usually formed later than usual, melted or broken up earlier, and been thinner; snowfalls and accumulations have been less abundant, spring thaw has occurred earlier and the

open water season has been extended. Some interviewees mentioned they had observed changes in wind patterns in terms of frequency, direction and strength. Regarding wildlife, recent changes have introduced a greater number of seals and harps, while fish spawned at different times because of natural factors such as increased rainfall. Birds and ducks seem to adapt their habits as they stay longer during summer. New species have also been spotted.

4. Human capital and health. Our investigation tends to confirm that many diseases are associated with the polluted water in the city of Iqaluit. Recent media reports have revealed high rates of enteric illness (diarrhea, vomiting, etc.) caused by contaminated drinking water (i.e. waterborne diseases), contaminated food (i.e. foodborne disease), or contaminated soil.

During the last century, a high level of resilience has been observed within local indigenous communities as they overcame rapid transformations that included a decline in the traditional subsistence lifestyle, a sedentarization process, public programs imposed by southern institutions newly implemented in the Arctic region, and the exploitation of natural resources.

Recent efforts to elaborate vulnerability indicators have led to the undertaking of an inclusive dialogue and may attenuate current discrepancies, as well as provide a better regional understanding of the climatic risks and measures to implement in Arctic coastal zones. In short, identified climate vulnerability and risks are related to food security, water quality, mental wellbeing, new hazards, pollution concerns, and public services delivery (health, education, social services, transportation, etc.).

5. Acclimating governance and institutions. Adaptation has played an important role in northern climatic policy since the early 2000s (Ford et al. 2014). Initiatives and publications are numerous and include the implementation of an interregional strategy, the creation of the Nunavut Climate Change Center and workshops to consult local communities. Government authorities focus on responding to the basic social and economic needs of the population. In that context, M&E is not a priority. The implementation of official adapted and practical indicators to measure CCA in coastal zones is greatly limited by the lack of historical data and specific data on coastal areas. In fact, coastal management is a shared responsibility between federal, territorial and municipal authorities (Environment Canada 2013), which increases the number of stakeholders potentially involved in CCA M&E decision-making.

In this context, governance is facing competing values. On the one hand, our investigations and interviews suggest that local communities are facing multiple scarcities of essential goods, thus impacting population well-being. Therefore, climate change adaptation is one among various issues to address: illness, poverty, suicide, etc. As an illustration, the mortality rate by suicide for Nunavut Inuit over the last few years is 111.4 per 100,000 population, 10 times the global Canadian rate (Statistics Canada's CANSIM Table 102-0552). On the other hand, the federal government is trying to improve the calibration of its development strategies so that they would better reflect local priorities and survival ambitions.

Local communities are facing cross-scale dynamics involving multiple, strong stressors and determinants of climate change on Eastern Arctic coastal zones: the development of new maritime routes, the intensification of the exploitation of natural resources, the growing flow of visitors and integration into the wage economy. Over the last decade, Iqaluit has experienced a profound transformation and is now the territory's hub hosting local, territorial and federal government points of service.

During our observations, we came across areas of concern such as safe drinking water, increased air pollution caused by waste disposal management policies, historical bay deep-freezing inhibiting transportation preventing receiving supplies from the South, and homeowners' obligation to install water tanks and to use the related services. We also observed recent broader phenomena such as cyclical migratory flows, the influx of previously unavailable products, a growing concern about the effects of permafrost melting on infrastructure, and a growing collective awareness of changing vulnerabilities. In light of all these complex issues, one main conclusion that could be drawn is that the current local government's priorities are not focused on climate change but rather on the short-term supply of essential goods, services and infrastructure.

Q2) What are the current insights emerging on this topic and what do these limits mean for local communities?

As confirmed by our field observations, dissonance between how local communities and governmental authorities understand projected climate change impacts and potential vulnerabilities is visible. Adapting to climate change is directly affecting many Nunavut inhabitants' deep beliefs because it would mean giving up activities associated with a traditional lifestyle. Climate change is one of many issues in a global fight for survival by Iqaluit community members. It is competing with issues seen as more urgent, such as the preservation of hunting, fishing and trapping activities, which are at the center of a traditional subsistence economy. Public health concerns as well as access to essential goods are also important sources of concern for local authorities.

This difficulty is also exacerbated by a "polluter-polluted" paradox, where local communities are directly affected by climate change, without being responsible for the experienced negative consequences. This situation helps to anchor the perception that climate change is a difficult issue to influence at the local level, which in turn leads to its relegation to an already long list of issues. Efforts are thus necessary to persuade local community members to enroll in adaptation initiatives, as adaptive management requires establishing strong partnerships and strategic alliances with local stakeholder groups (Nursey-Bray 2010).

The Qaujigiartiit Health Research Centre undertook a project to help identify a series of Nunavut-specific health-related indicators of climate change (Healey 2015). In the course of this project, stakeholders at the community, regional, and territorial levels in Nunavut gathered for a 1-day workshop to identify a subset of indicators specific to Nunavut and the perspectives of Nunavummiut. Potential data sources were also identified for the 20 identified indicators (among others, the Government of

Canada, Environment Canada, Canadian Ice Service, Aboriginal Affairs and Northern Development, Department of Fisheries and Ocean, Statistics Canada, Government of Nunavut, Department of Health).

Such an initiative involving local stakeholders in adaption M&E is promising, even though the documentation of all identified indicators is likely to be challenging, as many interviewees pointed out the scarcity of reliable and regularly updated data. As an illustration of the kind of difficulties encountered, the Nunavut Bureau of Statistics conducted the Nunavut Housing Needs Survey in 2009 and 2010, which is a survey of households in 25 communities in Nunavut (Nunavut Bureau of Statistics 2010). The survey's primary objective was to collect census data on the housing needs in all 25 communities of Nunavut. However, mid-way through collection, difficulties in hiring interviewers and a low response rate led researchers to settle for a sample survey of residential dwellings in Iqaluit and a census of residential dwellings in the other 24 communities.

Knowledge is critical for identifying options of adaption, designing and implementing policies, programs and plans integrating adaptation. Conversely, difficulties associated with the gathering of evidence to document indicators are jeopardizing the identification of acceptable adaptation options, and, as such, represent important limits to adaptation.

Conclusion

On the local scale, we have distinguished multiple and dissimilar definitions of associated key concepts such as vulnerability, resilience and adaptive capacity. We also acknowledge the additional dissimilarity associated with their meaning in the local language, Inuktitut. We have identified a series of indicators related to cultural diversity, political empowerment, and access to new technologies and resources for better welfare. Reviewing past assessment initiatives, our analysis shows that resilience can be measured through cultural, ethnic, economic and political dimensions. We have also observed that factors, such as individual incentives, institutional strength and data collection difficulty, greatly influence the development and use of metrics by local actors such as territorial officials and elected municipal representatives. We determine that in the region, climate change is not only a serious problem characterized by its complexity but also a cross-cutting issue which jeopardizes the survival of local indigenous peoples and their traditions.

References

Adger, W. N., Arnell, N. W., & Tompkins, E. L. (2005). Successful adaptation to climate change across scales. *Global Environmental Change, 15*(2), 77–86.

Anisuzzaman, M., & Jennings, P. (2012). A framework for mainstreaming climate change adaptation within local government operations.*International Journal of Climate Change: Impacts & Responses, 3*(3), 31–46.

Arctic Monitoring and Assessment Programme—AMAP. (2012). *Arctic climate issues 2011: Changes in arctic snow, water, ice and permafrost, SWIPA 2011*, Oslo, Norway, 97 pp. http://www.amap.no/documents/doc/arctic-climate-issues-2011-changes-in-arctic-snow-water-ice-and-permafrost/129.

Berrang-Ford, L., Pearce, T., & Ford, J. D. (2015). Systematic review approaches for climate change adaptation research. *Regional Environmental Change, 15*(5), 755–769.

Bowron, B., & Davidson, G. (2011). *Climate change adaptation planning: A Nunavut toolkit*. Atuliqtuq: Action and adaptation in Nunavut, 29 pp. https://www.cip-icu.ca/Files/Resources/NUNAVUT-TOOLKIT-FINAL.

Bunce, A., Ford, J., Harper, S., Edge, V., & IHACC Research Team. (2016). Vulnerability and adaptive capacity of Inuit women to climate change: A case study from Iqaluit, Nunavut. *Natural Hazards, 83*(3), 1419–1441.

Campbell, R. (2014). *Iqaluit sustainable community plan, annual progress report 2014: The first annual progress report of the Iqaluit sustainable community plan*, 32 pp. https://sustainableiqaluit1.files.wordpress.com/2015/08/2014-annual-progress-report-final-eng-aug-2015.pdf.

Conservation of Arctic Flora and Fauna—CAFF. (2016). *Indices and indicators*. http://www.caff.is/indices-and-indicators3.

Dowsley, M., Gearheard, S., Johnson, N., & Inksetter, J. (2010). Should we turn the tent? Inuit women and climate change.*Études/Inuit/Studies, 34*(1), 151–165.

Department of Environment. (n.d.). *Upagiaqtavut setting the course: Climate change impacts and adaptation in Nunavut*, 30 pp.

Duerden, F. (2004). Translating climate change impacts at the community level. *Arctic, 57*(2), 204–212.

Environment Canada. (2013). *Canada's sixth national report on climate change 2014: Actions to meet commitments under the United Nations Framework Convention on Climate Change*, 279 pp. http://publications.gc.ca/collections/collection_2014/ec/En4-73-2013-eng.pdf.

Fisher, S., Dinshaw, A., McGray, H., Rai, N., & Schaar, J. (2015). Evaluating climate change adaptation: Learning from methods in international development. *New Directions for Evaluation, 2015*(147), 13–35.

Forbes, D. L. (2011). *State of the Arctic coast 2010: Scientific review and outlook, international Arctic science committee, land-ocean interactions in the coastal zone*. Arctic Monitoring and Assessment Programme et International Permafrost Association, Geesthacht, Germany, 178 p. http://ipa.arcticportal.org/files/sac/state%20of%20the%20arctic%20rept-high%20res.pdf.

Ford, J. D., Stephenson, E., Cunsolo Willox, A., Edge, V., Farahbakhsh, K., Furgal, C., & Austin, S. (2016). Community-based adaptationresearch in the Canadian Arctic. *Wiley Interdisciplinary Reviews: Climate Change, 7*(2), 175–191.

Ford, J. D., Stephenson, E., Cunsolo Willox, A., Edge, V., Farahbakhsh, K., Furgal, C., et al. (2015). Community-based adaptation research in the Canadian Arctic. *Wiley Interdisciplinary Reviews: Climate Change*.

Ford, J., McDowell, G., & Jones, J. (2014). The state of climate change adaptation in the Arctic. *Environmental Research Letters, 9*(10). http://iopscience.iop.org/article/10.1088/1748-9326/9/10/104005.

Ford, J. D., McDowell, G., Shirley, J., Pitre, M., Siewierski, R., Gough, W., et al. (2013). The dynamic multiscale nature of climate change vulnerability: An Inuit harvesting example. *Annals of the Association of American Geographers, 103*(5), 1193–1211.

Ford, J. D., Pearce, T., Duerden, F., Furgal, C., & Smit, B. (2010). Climate change policy responses for Canada's Inuit population: The importance of and opportunities for adaptation. *Global Environmental Change, 20*(1), 177–191.

Füssel, H. M. (2008). Assessing adaptation to the health risks of climate change: What guidance can existing frameworks provide? *International Journal of Environmental Health Research, 18* (1), 37–63.

Gagnon-Lebrun, F., & Agrawala, S. (2007). Implementing adaptation in developed countries: An analysis of progress and trends. *Climate Policy, 7*(5), 392–408.

Government of Nunavut, Department of Culture, Language, Elders and Youth and Nunavut Tunngavik Incorporated. (2005). *Terminology on climate change*. http://www.tunngavik.com/documents/publications/2008-00-00%20Terminology%20on%20Climate%20Change.pdf.

Government of Nunavut. (2011). Upagiaqtavut Setting The Course: Climate Change Impacts and Adaptation in Nunavut. *Iqaluit, Nunavut*, 30 pp. http://www.climatechangenunavut.ca/sites/default/files/3154-315_climate_english_reduced_size_1_0.pdf.

Hatcher, S. V., Forbes, D. L., & Manson, G. K. (2011). *C-change working paper: Coastal hazard assessment for adaptation planning in an expanding arctic municipality*. http://ww.coastalchange.ca/images/stories/Documents_Tab/workingpaper23_hatcherforbes_manson_2011.pdf.

Healey, G. K. (2015). *Exploring health-related indicators of climate change in Nunavut*. Iqaluit: Qaujigiartiit Health Research Center.

Healey, G. K., Magner, K. M., Ritter, R., Kamookak, R., Aningmiuq, A., Issaluk, B., et al. (2011). Community perspectives on the impact of climate change on health in Nunavut, Canada. *Arctic, 64,* 89–97.

Hedger, M. M., Mitchell, T., Leavy, J., Greeley, M., & Downie, A. (2008). Desk review: Evaluation of adaptation to climate change from a development perspective. Brighton, Institute of Deveopment Studies. Lisa Horrocks, AEA Group. https://www.researchgate.net/profile/Martin_Greeley/publication/238113590_Desk_Review_of_Adaptation_to_Climate_Change_from_a_Development_Perspective/links/00b7d53c53b3179880000000.pdf.

Hinkel, J. (2011). "Indicators of vulnerability and adaptive capacity": Towards a clarification of the science–policy interface. *Global Environmental Change, 21*(1), 198–208. doi:10.1016/j.gloenvcha.2010.08.002.

Hofmann, M. E., et al. (2011). Classifying knowledge on climate change impacts, adaptation, and vulnerability in Europe for informing adaptation research and decision-making: A conceptual meta-analysis. *Global Environmental Change, 21*(3), 1106–1116.

IPCC (2014) Annex II: glossary Agard, J., Schipper, E. L. F., Birkmann, J., Campos, M., Dubeux, C., Nojiri, Y., Olsson, L., Osman-Elasha, B., Pelling, M., Prather, M. J., Rivera-Ferre, M. G., Ruppel, O. C., Sallenger, A., Smith, K. R., St Clair, A. L., Mach, K. J., Mastrandrea, M. D., and Bilir, T. E. (Eds) In: Barros, V. R., Field, C. B., Dokken, D. J., Mastrandrea, M. D., Mach, K. J., Bilir, T. E., Chatterjee. M., Ebi. K. L., Estrada. Y. O., Genova, R. C., Girma, B., Kissel, E. S., Levy, A. N., MacCracken, S., Mastrandrea, P. R., White, L. L. (Eds) Climate change 2014: impacts, adaptation, and vulnerability. Part B: regional aspects. Contribution of Working Group II to the Fifth Assessment Report of the Intergovernmental Panel on Climate Change. Cambridge University Press, Cambridge, 1757–1776.

Kates, R. W., Travis, W. R., & Wilbanks, T. J. (2012). Transformational adaptation when incremental adaptations to climate change are insufficient. *Proceedings of the National Academy of Sciences, 109*(19), 7156–7161.

Klein, R. J. T., Midgley, G. F., Preston, B. L., Alam, M., Berkhout, F. G. H., Dow, K., et al. (2014). Adaptation opportunities, constraints, and limits. In C. B. Field, V. R. Barros, D. J. Dokken, K. J. Mach, M. D. Mastrandrea, T. E. Bilir, et al. (Eds.), *Climate change 2014: Impacts, adaptation, and vulnerability. Part A: Global and sectoral aspects*. Contribution of working group II to the fifth assessment report of the intergovernmental panel on climate change. Cambridge, and New York, NY: Cambridge University Press, pp. 899–943.

Klein, R. J., Eriksen, S. E., Næss, L. O., Hammill, A., Tanner, T. M., Robledo, C., et al. (2007). Portfolio screening to support the mainstreaming of adaptation to climate change into development assistance. *Climatic Change, 84*(1), 23–44.

Lamari, M., & Jacob, J. L. (2015). *Adaptation aux changements climatiques en zones côtières: politiques publiques et indicateurs de suivi des progrès dans sept pays occidentaux*. Quebec: Presses de l'Université du Québec.

Lamari, M., Bouchard, J., Jacob, J., & Poulin-Lariviere, L. (2016b). Monitoring and evaluation of climate change adaptation in coastal zones: Overview of the indicators in use. Filho, W. L., Musa, H., Cavan, G. & O'Hare, P. (dir), *Climate Change Adaptation, Resilience and Hazards*, Springer International Publishing, 3–20.

Larsen, J. N., Anisimov, O. A., Constable, A., Hollowed, A. B., Maynard, N., Prestrud, P., et al. (2014). Polar regions. In V. R. Barros, C. B. Field, D. J. Dokken, M. D. Mastrandrea, K. J. Mach, T. E. Bilir, et al. (Eds.), *Climate change 2014: Impacts, adaptation, and vulnerability. Part B: Regional aspects*. Contribution of working group II to the fifth assessment report of the intergovernmental panel on climate change (pp. 1567–1612). Cambridge and New York, NY: Cambridge University Press.

LeBlanc, A.-M., Allard, M., Mathon-Dufour, V., Doyon-Robitaille, J., l'Hérault, E., Oldenborger, G., et al. (2012). Mapping of surficial geology and permafrost characterization in Iqaluit. *Changing Times: Climate Change Impacts and Adaptation in Nunavut* (Summer). http://www.climatechangenunavut.ca/sites/default/files/summer_2012_newsletter.pdf.

Lewis, J., & Miller, K. (2010). *Climate change adaptation action plan for Iqaluit, Atuliqtuq: Action and adaptation in Nunavut*, 31 pp. https://www.cip-icu.ca/Files/Resources/IQALUIT_REPORT_E.

Marshall, N. A. (2010). Understanding social resilience to climate variability in primary enterprises and industries. *Global Environmental Change, 20*(1), 36–43.

McBean, G., Alekseev, G., Chen, D., Foerland, E., Fyfe, J., Groisman, P. Y., et al. (2005). *Chapter 2: Arctic climate: Past and present*. Arctic Climate Impact Assessment [np]. http://www.acia.uaf.edu/PDFs/ACIA_Science_Chapters_Final/ACIA_Ch02_Final.pdf.

McDaniels, T., Chang, S., Cole, D., Mikawoz, J., & Longstaff, H. (2008). Fostering resilience to extreme events within infrastructure systems: Characterizing decision contexts for mitigation and adaptation. *Global Environmental Change, 18*(2), 310–318.

Noble, I., Huq, S., Anokhin, Y., Carmin, J., Goudou, D., Lansigan, F., et al. (2014). *Chapter 14: Adaptation needs and options*. Working group II of the intergovernmental panel on climate change fifth assessment report.

Nunavut Bureau of Statistics. (2010). *Overview and methodology 2009/2010 Nunavut housing needs survey*. http://www.stats.gov.nu.ca/Publications/Housing/NHNS%20Pubs/NHNS%20Overview%20and%20Methodology,%202009-2010.pdf.

Nunavut Climate Change Center. (n.d.). *Climate change adaptation*. http://climatechangenunavut.ca/en/understanding-climate-change/climate-change-adaptation.

Nursey-Bray, M. (2010). Local governance for local governments: A framework for addressing climate change. *Commonwealth Journal of Local Governance, 7*, 168–186.

Office of the Auditor General of Canada. (2014). *Fall Report of the Commissioner of the Environment and Sustainable Development: Chapter 3—Marine Navigation in the Canadian Arctic*, 40pp. http://www.oag-bvg.gc.ca/internet/English/parl_cesd_201410_03_e_39850.html.

Ostrom, E. (2010). Polycentric systems for coping with collective action and global environmental change. *Global Environmental Change, 20*(4), 550–557.

Pennesi, K., Arokium, J., & McBean, G. (2012). Integrating local and scientific weather knowledge as a strategy for adaptation to climate change in the Arctic. *Mitigation and Adaptation Strategies for Global Change, 17*(8), 897–922.

Perkins, S. E., Pitman, A. J., Holbrook, N. J., & McAneney, J. (2007). Evaluation of the AR4 climate models' simulated daily maximum temperature, minimum temperature, and precipitation over Australia using probability density functions. *Journal of climate, 20*(17), 4356–4376.

Public Works and Government Services Canada. (2014). *Real property institute of Canada: Marine infrastructure national workshop*. http://www.rpic-ibic.ca/documents/2014_Marine_Workshop/RPIC_Presentation_Pangnirtung_Harbour_Development__2014_01_14.pdf.

Quay, R. (2010). Anticipatory Governance : A tool for climate change adaptation. *Journal of the American Planning Association 76*(4), 496–511.

Ramesh, R., Chen, Z., Cummins, V., Day, J., D'Elia, C., Dennison, B., et al. (2015). Land–ocean interactions in the coastal zone: Past, present & future. *Anthropocene, 12*, 85–98.

Reid, H., Alam, M., Berger, R., Cannon, T., Huq, S., & Milligan, A. (2009). Community-based adaptation to climate change: An overview. *Participatory Learning and Action, 60*(1), 11–33.
Sanahuja, H. E. (2011). *A framework for monitoring and evaluating adaptation to climate change*. Washington, DC: Community of Practice, Global Environment Facility.
Shaw, J., Taylor, R. B., Forbes, D. L., Ruz, M. H., & Solomon, S. (1998). *Sensitivity of the coasts of Canada to sea-level rise*, 114. Ottawa:Geological Survey of Canada. https://www.researchgate.net/profile/John_Shaw8/publication/235241275_The_Sensitivity_of_the_Coasts_of_Canada_to_Sea_Level_Rise/links/0fcfd510977c6dfc8f000000/The-Sensitivity-of-the-Coasts-of-Canada-to-Sea-Level-Rise.pdf.
Sladen, W. (2012). Permafrost facts. *Changing Time: Climate Change Impacts and Adaptation in Nunavut* (Summer), 2–4.
Smit, B., & Wandel, J. (2006). Adaptation, adaptive capacity and vulnerability. *Global Environmental Change, 16*(3), 282–292.
Sovacool, B., & Linnér, B.-O. (2015). *The political economy of climate change adaptation*. Berlin: Springer.
Statistics Canada. (2013). Iqaluit, CY, Nunavut (Code 6204003) (table). National Household Survey (NHS) Profile. 2011 Census. Statistics Canada Catalogue no. 99-004-XWE. Ottawa. http://www12.statcan.gc.ca/nhs-enm/2011/dp-pd/prof/index.cfm?Lang=E.
Statistics Canada. (2015). *National Household Survey*. http://www12.statcan.gc.ca/nhs-enm/2011/dp-pd/prof/search-recherche/frm_res.cfm?Lang=F&TABID=1&Geo1=PR&Code1=01&SearchType=Begins&SearchText=Nunavut&PR=62.
Thywissen, K. (2006). *Components of risk: A comparative glossary*. Bonn: UNU-EHS.
Weller, G., Symon, C., Arris, L., & Hill, B. (2004). *Arctic climate impact assessment (ACIA): Summary and synthesis of the ACIA, Chap. 18* (pp 989–1020). University Press.

Chapter 14
Limits to Adaptation on Climate Change in Bosnia and Herzegovina: Insights and Experiences

Goran Trbic, Davorin Bajic, Vladimir Djurdjevic, Vladan Ducic, Raduska Cupac, Đorđe Markez, Goran Vukmir, Radoslav Dekić and Tatjana Popov

Introduction

Bosnia and Herzegovina has a total surface area of 51.2092 km^2, composed of 51.197 km^2 of land and 12.2 km^2 of sea. Of total land area, 5% is lowlands, 24% hills, 42% mountains, and 29% karst regions (Avdic et al. 2013). Based on its geographical position on the Balkans Peninsula, it belongs to the Adriatic basin and the Black Sea Basin (Fig. 14.1).

According to the preliminary results of 2013 population census, Bosnia and Herzegovina has a total population of 3,791,622 (Bosnia and Herzegovina Statistics Agency 2015). Compared to the last census, conducted in 1991, the population has decreased by 585,411 (the population in 1991 was 4.377.033). The reasons for the population decline include armed conflicts, migration and a marked decrease in the birth rate. Preliminary results of the 2013 population census indicate that the total number of residents counted in the Republic of Srpska is 1,326,991 (The Republic of Srpska Statistics Agency 2015) (35%), in the Federation of Bosnia and Herzegovina 2,371,603 (The Federation of Bosnia and Herzegovina Statistics Agency 2015 (63%) and in Brčko District 93,028 (2%).

Bosnia and Herzegovina is located in the region of the West Balkans which is quite threatened by the climate change. Almost all countries in the region have similar problems caused by climate change (Spasov et al. 2008). Unfortunately, the earlier estimates of the Intergovernmental Panel on Climate Change (IPCC) have proved to be true, more precisely, the increase in temperature, decrease in rainfall during the summer months and increase in the intensity and frequency of extreme climate events (Cramer et al. 2001). The increase in mean annual air temperature is

G. Trbic (✉) · D. Bajic · V. Djurdjevic · V. Ducic · R. Cupac · Đ. Markez · G. Vukmir
R. Dekić · T. Popov
Faculty of Science, University of Banja Luka, Banja Luka, Bosnia and Herzegovina
e-mail: goran.trbic@pmf.unibl.org; trbicgoran@yahoo.com
URL: http://www.pmfbl.org

W. Leal Filho and J. Nalau (eds.), *Limits to Climate Change Adaptation*,
Climate Change Management, https://doi.org/10.1007/978-3-319-64599-5_14

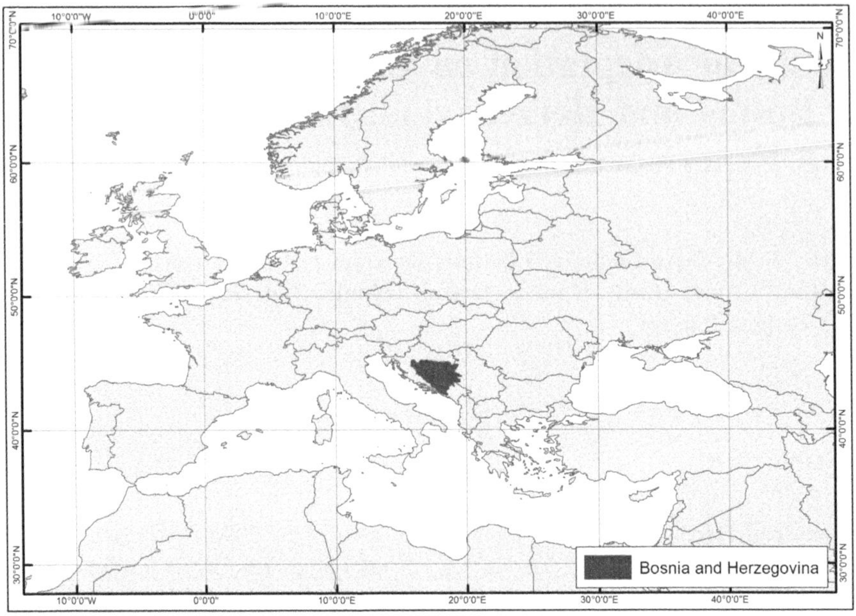

Fig. 14.1 Geographical location of Bosnia and Herzegovina in Europe

evident on the whole territory of Bosnia and Herzegovina. The largest increase goes up to 1.3° annually. Changed distribution of precipitation, and the largest deficit during the delivery period of the summer are also evident (Trbic et al. 2013).

Extreme climate events in Bosnia and Herzegovina have become increasingly frequent. Of the last 12 years, six were very dry to extremely dry (2003, 2007, 2008, 2011, 2012 and 2013) In addition, years with large to catastrophic floods have been very frequent (2001, 2002, 2009, 2010 and 2014) (Hydro Meteorological Institute of the Republika Srpska and the Federation of Bosnia and Herzegovina 2015). Extreme climate events have been particularly pronounced over the past 5 years, with great floods in 2009 and 2010; severe droughts in 2001, 2012 and 2013; heat waves/tropical temperatures in early 2012; a wave of intense cold in early 2012; and gale-force winds in mid-2012. In April and May 2014, the northern part of the country experienced record rainfall (more than 420 mm), which caused catastrophic flooding in the Vrbas and Bosna catchment areas and the Semberija region (European Commission 2014).

For the needs of the second and the third national communication, climate models and scenarios were developed on the basis of EBU-POM climate model and the downscaling projected for the space of Bosnia and Herzegovina by the end of the 21st century (Marsland et al. 2003; Roeckner et al. 2003; Djurdjevic and Rajkovic 2008). For the needs of this paper, Prof. Djurdjevic (a co-author of the paper) has outlined the possible changes of air temperature and the quantity of rainfall according to climate scenarios A1B and A2. The expected increase in

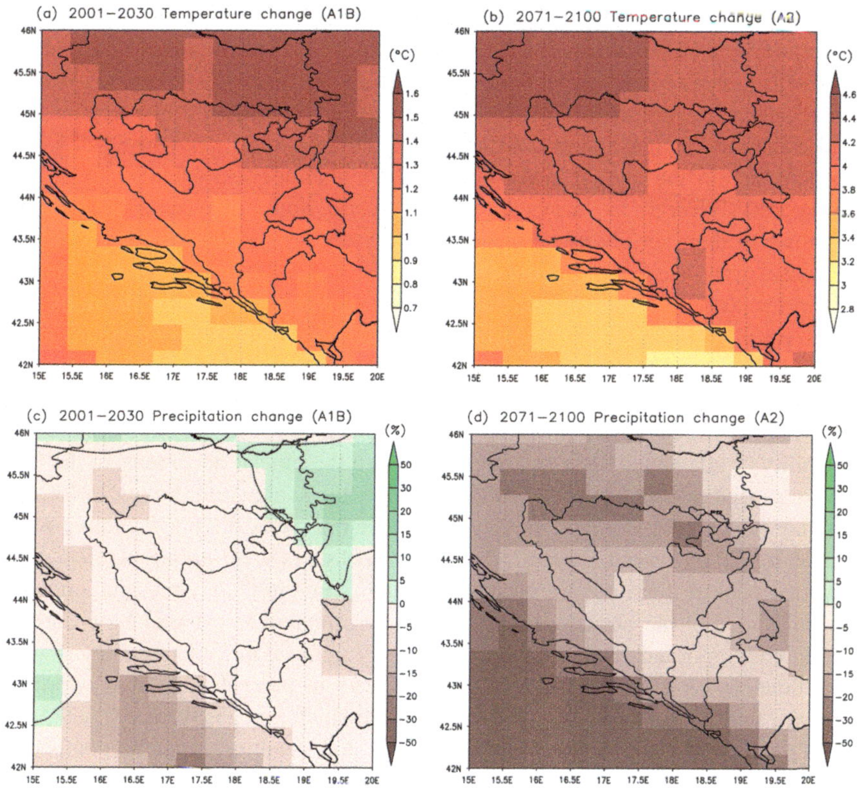

Fig. 14.2 Changes in the growing season in Bosnia and Herzegovina: average daily temperature (ΔTgs) and growing season average precipitation (ΔRgs) for 2001–2030 under the SRES A1B scenario (**a**, **c**) and for 2071–2100 under the SRES A2 scenario (**b**, **d**) relative to the 1961–1990 reference period

temperature (figures on the left) and decrease of rainfall (figures on the right) are presented in Annex 2 (Fig. 14.2).

According to the climatology forecasting models, the mean seasonal temperature changes in the period 2001–2030 are expected to be in the range of +0.8 to 1.0 °C above the average temperature. In addition to predicted warmer winters (0.5–0.8 °C), the biggest temperature changes are expected during the summer months—June, July and August, with the forecasted changes of +1.4 °C in the northern areas and +1.1 °C in the southern area. Rainfall is anticipated to be reduced by 10% in the western parts of the country, while a 5% increase is expected in the east. It is anticipated that the seasons of autumn and winter will have the greatest decrease in precipitation (Avdic et al. 2013). Results for the A1B scenario for the period 2071–2100 show the spatial structure of meteorological parameters (especially temperature) changes, similarly to the previously observed period from 2001 to 2030, but at higher scale. For the last three decades of this century, the change in temperature ranges from +2.8 to +3.6 °C. The biggest temperature increase of +3.6 °C and above is projected for the summer period. During the winter season, the highest increase is forecasted for central regions with values up to 2.4 °C. For spring season these changes range from 2.4 to 2.6 °C, while the entire territory of the Republic of Srpska autumn changes are in the range from 2.0 to 2.4 °C.

According to the annual index of global adaptation (GAIN Index) for 2014, Bosnia and Herzegovina ranks 84 in the world and penultimate in Europe, according to their sensitivity and readiness to response to climate change (http://index.gain.org/). According to the Global Index Risk, during 2014, B&H ranked third in the world in terms of vulnerability when it was hardest hit by the rains and floods since the beginning of measurements 120 years ago (https://germanwatch.org/en/download/13503.pdf).

Based on everything aforementioned, it is clear that Bosnia and Herzegovina is under a great threat and vulnerable to climate changes. The sensibility and the possibility of adaptation were firstly mentioned in the official documents in the first and the second national communications on climate change according to UNFCCC (Vukmir et al. 2009; Avdic et al. 2013). The sectors under the biggest threat were defined in the said sectors and the adaptation measures were proposed. However, only a number of measures were carried through. The problems of adaptation were also addressed in the Strategy of Adaptation to Climate Change and Low-Carbon Development (Jordan et al. 2013). This Strategy defines, in a more complex way, the influences on individual sectors and describes in more detail a framework of measures and project activities. The necessary financial means were evaluated for the implementation of the measures, however, the deadlines and the competent institutions for implementing and monitoring the implementation of measures were not precisely defined. Drafting of the National Adaptation Plan (NAP) for Bosnia and Herzegovina is currently underway. Its completion and adoption is expected by the end of 2016. NAP is planned to clearly and precisely define the current and expected influences of climate changes on the most jeopardized sectors (according to climate models and scenarios). Also, the plan is to propose short-term (1–2 years of implementation period), mid-term (2–5 years of implementation period) and long-term (5–10 years of implementation) measures. The experts who participate in drafting of NAP were given the guidelines from the Fifth IPCC Report (IPCC AR5, 2013). Special attention was paid to the results and recommendations in the report of the second working group (WGII) to Chapters "13", "14", "15" and "16" (IPCC AR5, WGII, 2013). Chapter "15"—Adaptation Opportunities, Constraints, and Limits will be specially addressed with regard to the most threatened sectors of agriculture and water resources management (Klein et al. 2014). The proposed measures and activities would include necessary finances, possible financing sources and relevant institutions for their implementation. The drafting of NAP was strongly supported by the policy designers, local communities, academic community, non-governmental sector and citizens.

The main goal of the paper is to point out at, on the basis of the current and projected climate changes, the limitations and barriers of adaptation in Bosnia and Herzegovina. One of the barriers in the paper is represented by a lack of adequate data and scientific papers treating the adaptation to climate changes in Bosnia and Herzegovina.

The methodology in the paper was based on the analysis of the existing strategic documents, legislation and regulations, scientific papers, curricula and climate data. The authors of this paper worked on designing the climate models for national communications, while the projection of air temperature and quantity of precipitation for the vegetation period (April–September) was done for the climate scenarios A1B and A2.

Results and Discussion

The Possibility to Adapt to Climate Change

The most intensive influence of climate changes and at the same time the possibility of adaptation is in the agricultural sector (Trbic et al. 2014). During the previous period the agricultural producers demonstrated a certain capability to quickly adjust to climate change and other ecological factors. Farmers may be expected to continue to behave in such a way in future too. However, a key issue for agriculture is whether the adaptation will be previously planned or spontaneous. The planned adjustment at all levels may yield better effects compared to individual adjustment of agricultural subjects, but in both cases a question is raised of providing the necessary funds to cover the expenses. Numerous studies related to this problem showed that the political and economic factors were dominant in decision-making on taking adequate measures for adjustment to climate changes. The developing countries and the poorest countries in the world have most problems because they have lower capacity to adjust, so they are more prone to the damage caused by climate changes, in the same way that they are more vulnerable to other stresses. The key elements in the approach to solving the problems of vulnerability and adjustment to climate changes faced by the agricultural sector do not include only the changes of the average climate conditions, but also the climate extremes, variability and the speed of climate changes (Spasov et al. 2007).

The studies on influences in agriculture carried out during the past decade considered certain technological options for adjustment of agriculture to climate change. The prolonging of the warm season and consequently, the increase in the length of the vegetation period will provide bigger possibilities for agriculture to adjust, especially in higher and at medium geographical latitudes and higher altitudes, where the agricultural production is currently limited by insufficient length of the frost-free period. A longer vegetation season enables a change of species and hybrids from the early ones to those that mature later and that, as a rule, give higher yield. Besides, according to the farmers, it is already now possible to have two harvests with the crops with shorter vegetation.

As for the winter conditions for the crops, an increase in temperature results in a lower risk from damaging winter crops by the frost and other adverse conditions of the cold season. Unfortunately, no detailed studies have been carried out in Bosnia and Herzegovina on adapting agriculture to expectedly warmer winters, although it is known that, as a result of greenhouse effect, the temperatures will increase during the winter more than during the summer. Warmer winters will have a negative effect on dormancy of vegetation, while in the past years early blossoming of certain fruit species has been noticed (in 2016, even in mid-February).

According to the numerous experts in the field of agriculture, a genetic base for introduction of new species of crops that can adapt to the expected climate conditions is quite large, so that significant effects may be expected in that regard. It is also noteworthy that climate change will decrease the available water resources for irrigation in many regions, because the evaporation increase will be higher than precipitation (Vukmir et al. 2009). In such conditions, the question of the application of the technologies for more rational use of water in irrigation, the introduction of 'drop by drop' system etc. will become increasingly topical. The "drop by drop" irrigation system has especially become topical with the production of vegetables in greenhouses and with the production of fruits and grapes.

In order to avoid draught, crops and plantations that reach maturity early can be used. The examples of such species may be found in all species of plant production (early potato, salad, green onion, early fruits, etc.). Some domestic (autochthonous) or old plant species and populations have specific forms of adaptation to local conditions of production, including adaptation to diseases and climate variation, i.e. the occurrence of high temperatures and draught. Such species are frequently not on sale but are kept "on farm' and in exchanges of seeds between farmers. Generally speaking, to raise a draught-resistant species does not mean that a high yield will be obtained as when another non-resistant species in the conditions of intensive agricultural production (irrigation) (Bradshaw et al. 2004). However, the draught-resistant species are better in the conditions without irrigation compared to many popular species, because they give the stable yield during the variation of external conditions, and especially in air and land draught.. In principle, the climate change is expected to have a positive influence on the yield with most winter crops, while with the spring crops the summer draughts will prevail and result in significant decrease of yield. For that reason, a decrease in the share of spring and increase in the share of winter crops is proposed as an adaptation measure.

In the areas with summer deficit or lack of water for normal development of agricultural species, and where there are no possibilities for irrigation, the choice of cultures should be adjusted to the natural schedule of precipitation, i.e. to the dynamics of land-water balance. This is the so-called *rainfed agriculture* or *crops to climate and soil suitability* (Singh et al. 2015; Tow et al. 2011).

Limits to Adaptation on Climate Change in Bosnia and Herzegovina

Although the possibilities of adaptation to climate changes in Bosnia and Herzegovina are evident, certain barriers and limitations are nevertheless present. We can generally group the limitations in four different groups:

- Technical and technological limitations,
- Legislative limitations,
- Limitations in education and research,
- Financial limitations.

Technical and Technological Constraints

We have underlined above that the biggest effect that the climate changes in Bosnia and Herzegovina have is on water resources. The dominant influence is on potable water resources but also on water for irrigation and the production of electricity (hydro-energy). Intensive precipitation also creates an increasing problem causing the occurrence of landslide.

The development and improvement of appropriate irrigation and water supply systems create a primary problem in terms of technical and technological limitations. The existing systems are quite outdated and do not keep up with the modern developments.

Significant improvements in the processes of monitoring, analysis and modeling of data are necessary for adequate assessment of climate changes and their influence on the most jeopardized sectors. In the past few years strengthening of the capacities of the Republic of Srpska and Bosnia and Herzegovina Federation has been noticeable. Strengthening of the capacities is reflected through the improvement of monitoring, climate bases, weather forecast and early notification system. However, the existing capacities are still not at such a level to be able to keep up with the present climate changes. A positive move in overcoming the said barriers can be seen through strengthening of regional cooperation and exchange of information, knowledge and experiences.

The application of contemporary solutions that correspond to specific, local conditions is necessary for efficient adaptation to changed climate conditions. The farmers will have to adopt new knowledge and keep up to date with new scientific and technological solutions, in order to adjust their production to the changed environment conditions. Irrigation will certainly present one of the key mechanisms for adaptation. However, the protection against flood and draining excess waters from the land plot, as well as the regulation of water air regime in general is a priority question in water resources management. Construction or upgrading of the existing complex hydro melioration systems as well as the watershed approach to space development is posed as a key strategic issue.

Legislative Framework as a Limiting Factor in Adaptation to Climate Change

Bosnia and Herzegovina institutions are facing significant challenges leading to the decreased capacities for the adjustment and the capacities to carry out and additionally develop the strategies, plans and programs of adjustment to climate change (Smit and Wandel 2006). A strong institutional, political and legislative framework is required to manage the risks arising as a consequence of climate change and adjustment to climate change.

The existing legislative framework in Bosnia and Herzegovina does not provide an adequate basis that would appropriately treat the influence of climate changes, which creates an additional problem to the process of adaptation to changed climate conditions. According to Bosnia and Herzegovina Constitution, the legislative competence for the field of the environment belongs to BiH entities (the Republic of Srpska and Bosnia and Herzegovina Federation) and to BiH Brcko District. Analyzing the overall legislative framework in BiH, it is noticeable that all legal acts or by-laws in BiH treat this problem as part of the term "climate" which is certainly not enough. The only document that in its contents really addresses the issue of climate changes is The Rulebook on the Method of Design, Content and Forming Spatial Planning Documents (2013) in the Republic of Srpska, which clearly defines the methodological approach to designing the spatial development documents, underlining that the climate changes are one of the conditions to be considered when defining the content of the space and spatial documents. (Official Gazette of Republic of Srpska, No. 69/13).

In accordance with that, the Republic of Srpska Spatial Development and Construction Act, as a related enactment, treats the issue of climate changes only minimally too. (Official Gazette of Republic of Srpska, Nos. 40/13, 2/15, 106/15, 3/16). In the other entity (Bosnia and Herzegovina Federation) the key law regulating this issue is the Law on Spatial Planning and Use of Land at the Level of the Federation and Bosnia and Herzegovina. However, in this Law the problem of climate change is not mentioned at all (Official Gazette of Federation of Bosnia and Herzegovina, Nos. 2/06, 72/07, 32/08, 4/10, 13/10, 45/10).

The spatial plan is clearly the most important strategic plan document of a country or a region. However, the previous analysis of the climate conditions in spatial plans, both at the entity level and at the local level, implied addressing the temperature of air, quantity of precipitation and possibly the speed and direction of the wind. In other words, climate change as a dynamic factor of influence has not been addressed at all in the existing spatial plans in BiH. Therefore, in the existing legislation and the accompanying rulebooks, there are no chapters on the issue of climate change, nor is this issue treated in any way as a dynamic factor of influence.

In reference to that, it is necessary to initiate the procedures of amendments to the valid legal solutions in BiH in the sense of integrating the terms "climate change" and "adaptation to climate change" in the most jeopardized sectors. The implementation of the project "Integration of Climate Changes in the Problem of Floods in the River Vrbas Basin" carried out by UN Development Program (UNDP), and financed by GEF, is currently underway. The realization of the project activities will, among others, offer the amendments to legal solutions which will appropriately treat and integrate the climate changes in the legislation governing the field of planning and development of space, as well as in the spatial, regulation and zoning plans.

As part of the proposed solutions, recommendations will be offered according to which spatial and regulation plans will have to incorporate and regulate, besides the analysis of climate conditions, the issues of climate extremes and climate models, i.e. projecting the climate change. By addressing the climate extremes and projected climate changes, a different strategic approach will be created toward the development of certain economic branches in BiH that are affected by the climate change influences. In such a case, the strategic development of certain economic branches (agriculture, water management, energy sector, etc.) will be followed by appropriate action measures and plans that will also include the adaptation to climate change.

In addition to that, integrating the climate changes in strategic and spatial plans at the entities' level will also result in integrating the climate change in regulation and spatial plans at the municipal level.

Limitations in the Field of Education and Research

Improvement of education, enhancing knowledge and raising awareness are some of the key segments in terms of limitations to adjustment to climate change. The education system in Bosnia and Herzegovina pays very little attention to climate changes or related issues of the environment protection. There is a need to improve the curricula and syllabi at all education levels (primary, secondary and higher education).

The analysis of the presence of the content related to the protection of the environment in educational curricula in Bosnia and Herzegovina has showed a number of interconnected basic characteristics. The first one is insufficient presence of the contents dealing with the problem of protection of the environment. Another characteristic relates to a dominant role of theory of material education during the selection of educative content and organization of teaching subjects in general, including those that deal with the problems of protection of the environment. Namely, the focus is on acquiring the factual knowledge, while the process-based, integrity and fundamental interconnection within all the parts of geosystem are being neglected. We can generally say, based on this preliminary overview, that the primary interest is on cognitive functions, primarily on acquiring the facts, while the emotional components is largely neglected, i.e. this sphere of material education includes more education on the environment rather than for the environment in the sense of formative and educational effect of ecological education, rather than the material only. Such an approach necessarily leads to the formalization of knowledge itself, as well as to something that is more dangerous, i.e. the concept of education itself, "education on the environment" rather than "education for the environment". This is present in the structure itself of learning the contents in the field of protection of the environment, which is, both in primary and secondary school education, designed on the system of "concentration on a single subject". In primary school this is biology for the eighth grade and in the secondary school ecology and protection of the environment—they are designed as supporting pillars of education in the field of protection of the environment. However, an edifice of successful education cannot rely on two pillars, while the horizon of such education is narrowed especially if we take into account the sporadic character of studying the mentioned contents through the syllabi of other subjects. A third, implicit characteristic refers to exactly to the lack of inter-curricular approach to the formation of an individual for the environment, i.e. his/her sensibilization for the problems that are posed before the future development and orientation d of mankind. Instead of focusing on one fundamental subject, a principle of interweaving of all subjects that have to do with the mentioned problematic is necessary. Along these lines, a Decade of Education for Sustainable Development 2005–2014 has been declared by UN, providing guidelines and recommendations for integration of the very same ecological contents in educational subjects at all levels, while placing an accent on intercurricular approach and interdisciplinarity.

Generally speaking, climate changes, the risks and sensitivities related thereto, have been insufficiently integrated in the educational curricula, capacity-building and awareness raising programs. Overcoming of this barrier can result in increased knowledge, better awareness of the public and better climate change management. The universities and schools, i.e. the educational system in general have a key role in overcoming these limitations.

The research in the field of climate changes in Bosnia and Herzegovina has been actualized in the past decade, when the information was systematized for the needs of the development of National Communications on Climate Changes according to UNFCCC. With the development of local climate models and the analysis of their influence on the most affected sectors, numerous scientific issues have been only opened. Future researches should be more systematic, comprehensive and better directed toward an increase in knowledge and to providing support to policy designers and decision makers. Adequate research tools, models and reliable data will be available to ensure that the knowledge of the climate change and the risks related thereto is enhanced, thus ensuring a solid basis for the development of the policies based on scientific proofs.

Financial Constraints

The actual implementation of the measures and the action plan of adaptation to climate changes require financial means, that are higher than the means currently available in Bosnia and Herzegovina. Additional finances will have to be provided to enable efficient implementation and actualization of the programs and plans of adaptation to climate changes. It is necessary to underline the possibilities of co-financing, public-private partnership and economic activity of social enterprises. Innovative partnerships will have to be developed with multilateral financial agencies, that currently review their possibilities of development assistance in the context of resistance of development to climate change. Certain adaptation measures have been defined through national communications on climate change and in the Strategy of Adaptation to Climate Change. Currently, the National Plan of Adaptation for Bosnia and Herzegovina is being developed, which is expected to provide more specific measures and possible financing sources. The envisaged measures with adjustment to climate changes offer both business opportunities and improvement of social and economic development of Bosnia and Herzegovina. Co-financing arrangements will be requested from the main donors such as: the World Bank, European Bank for Reconstruction and Development, GEF, Green Climate Fund (FCF), the European Commission, USAID, SIDA and other donors and financial institutions.

Adequate resources will be provided to ensure that the strategy of adaptation to climate change and related plans and programs are efficient and delivered on time, with appropriate monitoring, evaluation and upgrading measures, that will help achieve both general and specific goals and objectives.

Conclusions

Bosnia and Herzegovina is at an extreme risk in terms of climate change. A significant trend increase in temperature in all seasons has already been observed, as well as the changes in distribution of rainfall during the year. In addition, the more intense and more frequent occurrence of floods and droughts were noticed, which put pressure on the environment, people and their activities. Projected climate change suggests that climate change will continue in the future with different intensity, depending on the climate scenario.

The previous adaptation to climate change was more chaotic and there was no strategic plan. By creating strategic documents and action plan to adapt to climate change, and providing insufficient financial resources, the process of adjustment will be thoroughly planned. Key constraints are reflected in the lack of knowledge of the law, legislation, techniques and technology, and financial resources. Also, the limitation is presented by the lack of the data on losses and damage caused by climate change. Past experience has shown that damage can only be estimated, and that there is no uniform methodology for its calculation. The problems of adaptation were not treated with the Intended Nationally Determined Contributions (INDC) for Bosnia and Herzegovina. Chapter Loss and Damage is expected to be included in the upgraded version of INDC.

In order to achieve the previous objectives and overcome constraints, a strong institutional, legal and scientific framework for risk management and adaptation to climate change is necessary, as well as sustainable financial mechanisms for their implementation. An increase in hazard and risk, which is primarily manifested in increased intensity and frequency of climate extremes, suggests that the adaptation in the future must be clearly planned, based on scientific facts and modern technological solutions. These conclusions are consistent with the generally accepted viewpoint on climate change and the need for adaptation, provided that the adaptations are much more successful if implemented at lower regional and local level, and not globally.

Acknowledgements The authors are grateful to the Ministry of Science and Technology of the Government of Republic of Srpska, United Nations Development Programe and Faculty of Science University of Banja Luka for the assistance provided in the research.

References

Agency for Statistics of Bosnia and Herzegovina, Sarajevo. (2015). http://www.bhas.ba/?option=com_publikacija&id=1&lang=en.

Avdic, S., et al. (2013). Second National Communication Bosnia and Herzegovina under UNFCCC. http://www.ba.undp.org/content/bosnia_and_herzegovina/bs/home/library/environment_energy/-2013.html.

Baethgen, W. E., Meinke, H., & Gimene, A. (2003). *Adaptation of agricultural production systems to climate variability and climate change: Lessons learned and proposed research approach.* Paper presented at climate adaptation.net conference "insights and tools for adaptation: Learning from climate variability," November 18–20, 2003. Washington, DC. http://www.climateadaptation.net/docs/papers/Baethgen%20Meinke%85z%20NOAA%202003.pdf.

Bradshaw, B., Dolan, H., & Smit, B. (2004). Farm-level adaptation to climatic variability and change: Crop diversification in the Canadian prairies. *Climatic Change, 67,* 119–141.

Climate Change Adaptation and Low-Emission Development Strategy for Bosnia and Herzegovina. http://www.ba.undp.org/content/bosnia_and_herzegovina/en/home/library/environment_energy/climate-change-adaptation-and-low-emission-development-strategy-.html.

Cramer, W., et al. (2001). Working group II: Impacts, adaptation and vulnerability, IPCC AR3, Chapter 13-Europe. https://www.ipcc.ch/ipccreports/tar/wg2/pdf/wg2TARchap13.pdf.

Djurdjevic, V., & Rajkovic, B. (2008). Verification of a coupled atmosphere-ocean model using satellite observations over the Adriatic Sea. *Annals Geophysics*, *26*, 1935–1954, 2008 © European Geosciences Union 2008. www.ann-geophys.net/26/1935/2008/.

Documentation and materials, Hydrometeorological Institute of Republica Srpska. http://rhmzrs.com/.

Documentation and materials Federal Hydrometeorlogical Institute of Bosnia and Herzegovina. http://www.fhmzbih.gov.ba/.

European Comission (EC). (2014). Bosnia and Herzegovina recovery needs assessment, floods 14–19, May. Executive summary. http://ec.europa.eu/enlargement/pdf/press_corner/floods/rna-executive-summary.pdf.

Federation of Bosnia and Herzegovina, Institute of Statistics. (2015). http://www.fzs.ba/Eng/index.htm.

INDC, Bosnia and Herzegovina. (2015). http://www4.unfccc.int/submissions/INDC/Published%20Documents/Bosnia-Herzegovina/1/INDC%20Bosnia%20and%20Herzegovina.pdf.

Jordan et al. (2013) Climate Change Adaptation and Low Emission Development Strategy for Bosnia and Herzegovina, Technicalpapre, DOI: 10.13140/RG.2.1.1235.5443

Klein, R. J. T., Midgley, G. F., Preston, B. L., Alam, M., Berkhout, F. G. H., & Dow, K., et al. (2014). Adaptation opportunities, constraints, and limits. In C. B. Field, V. R. Barros, D. J. Dokken, K. J. Mach, M. D. Mastrandrea, T. E. Bilir, M. Chatterjee, K. L. Ebi, Y. O. Estrada, R. C. Genova, B. Girma, E. S. Kissel, A. N. Levy, S. MacCracken, P. R. Mastrandrea & L. L. White (Eds.), *Climate change 2014: Impacts, adaptation, and vulnerability. Part A: Global and sectoral aspects. Contribution of working group II to the fifth assessment report of the intergovernmental panel on climate change* (pp. 899–943). Cambridge: Cambridge University Press. http://www.ipcc.ch/pdf/assessment-report/ar5/wg2/WGIIAR5-Chap16_FINAL.pdf.

Kreft, S., & Eckstein, D. (2013). Global climate risk index 2014. Bonn: Germanwatch. https://germanwatch.org/en/download/13503.pdf, http://index.gain.org/.

Marsland, S. J., Haak, H., Jungclaus, J. H., Latif, M., & Roske, F. (2003). The Max-Planck-Institute global ocean/sea ice model with orthogonal curvilinear coordinates. *Ocean Modelling, 5,* 91–127. doi:10.1016/S1463-5003(02)00015-X.

Republic of Srpska, Institute of Statistics. (2015). http://www.rzs.rs.ba/.

Roeckner, E., Bäuml, G., Bonaventura, L., Brokopf, R., Esch M., & Giorgetta M., et al. (2003). The atmospheric general circulation model ECHAM 5. Part I: Model description, MPI-Report No 349. http://hdl.handle.net/11858/00-001M-0000-0012-0144-5.

Singh, K. A., et al. (Eds). (2015). Climate change modelling, planning and policy for agriculture. doi: 10.1007/978-81-322-2157-9_1©Springer India 2015. http://www.springer.com/us/book/9788132221562.

Smit, B., & Wandel, J. (2006). Adaptation, adaptive capacity and vulnerability. *Global Environmental Change, 16.* http://www.uio.no/studier/emner/annet/sum/SUM4015/h08/Smit.pdf.

Spasov, D., Bruci, E., Trbic, G., et al. (2008). The South East European climate change framework action plan for adaptation, SEE/VCCC—South East European virtual climate change center and regional environmental center, Belgrade. http://www.seevccc.rs/CCFAP-A/CCFAP-A.pdf, ISBN: 86-7550-058-0.

Spasov, D., Trbic, G., Trkulja, V., & Majstorovic, Z. (2007). Study for climate change impact assessment on agriculture and adaptions strategy development in Bosnia and Hercegovina, The Government of the Republic of Srpska, Ministry of Construction, Spatial Planning and Ecology, and Regional Environmental Center, Banja Luka.

The Regional Environmental Center for Central and Eartern Europe. (REC). The impacts of climate change on food production in the western Balkan region. http://documents.rec.org/topic-areas/Impacts-climage-change-food-production.pdf.

Tow, P., et al. (Eds.). (2011). *Rainfed farming system.* Berlin: Springer. doi:10.1007/978-1-4020-9132-2-1.

Trbic, G., Bajic, D., Popov, T., & Oprasic, S. (2013). Problems of drought in Bosnia and Herzegovina, Herald, 17. Banja Luka. www.gdrsbl.org.

Trbić, G., Đurđević, V., Bajić, D., Cupać, R., Vukmir, G., Popov, T. (2014). Climate change and adaptation options in Bosnia and Herzegovina—Case study in agriculture. In *The international conference "ADAPT to CLIMATE", Filoxenia conference Centre in Nicosia, Cyprus, on March 27th & 28th, 2014.*

Vukmir, G., et al. (2009). First National Communication Bosnia and Herzegovina under UNFCCC. http://www.ba.undp.org/content/bosnia_and_herzegovina/en/home/library/environment_energy/the-initial-national-communication–inc-.html.

Laws and Regulations

Law on spatial planning and construction (Official Gazette of Republic of Srpska, No: 40/13, 2/15, 106/15, 3/16).

Book of rules for the methodology of preparation, content and forming of spatial planning documents (Official Gazette of Republic of Srpska, No. 69/13).

Law on spatial planning and use of land at the level of the Federation of Bosnia and Herzegovina (Official Gazette of Federation of Bosnia and Herzegovina, No. 2/06, 72/07, 32/08, 4/10, 13/10, 45/10).

Decree on unique methodology for prepation oh planning documents (Offical Gazzeteof Federation of Bosnia and Herzegovina, No. 63/04, 50/07, 84/10)

Author Biographies

Professor Goran Trbic is a Full Professor at the Faculty of Science, University in Banja Luka. Graduate (2001) and postgraduate studies (2003) graduated from the Faculty of Science and Ph.D. (2006) at the Faculty of Geography, University of Belgrade in the field of geography and climatology. Her field of scientific interest climate change and adaptation to climate change. He participated in the development of several national reports and strategies in the field of climate change. He has published about 50 papers in national and international journals in the field of climate change and the environment. Is currently the focal point of Bosnia and Herzegovina to the IPCC.

Dr. Davorin Bajic is an Associate Professor at the Faculty of Science, University in Banja Luka. Graduate (2000) from the Faculty of Science, University of Banja Luka. Postgraduate stidies (2005) graduated from the Faculty of Geography, University of Belgrade in the field of social geography. Ph.D. (2011) graduated from the Faculty of Science, University in Banja Luka in the field social geography. Her field of scientific interest geographic information systems, spatial planning and natural resources. He participated in the development of several national reports and strategies in the field of climate change. He has published about 30 papers in national and international journals in the field of GIS, climate change and the environment.

Dr. Vladimir Djurdjevic is an assistant professor at the Department of Meteorology, Faculty of Physics, University of Belgrade. He received B.Sc., M.Sc., and Ph.D. degree at the same Department. The main area of his research is development and application of regional climate models. His scientific papers have been published in international journals in the field of climate and numerical modeling of the atmosphere and ocean. Also, as an expert he participated in preparation on several national reports on climate change for different countries.

Professor Vladan Ducić is Full Professor in the Faculty of Geography, University of Belgrade. Also, he is working in the Master Program in the Faculty of Natural Sciences and Mathematics in Banja Luka (Bosnia and Herzegovina). His current research interests include climate change, dendrochronology and animals population dynamics. He is cofounder of “Belgrade School of meteorology” edition (from 2008). He was given a Golden Charter of Serbian Biological Society in 2014. He was been a consultant in UNDP program in Bosnia and Herzegovina for drought analysis.

Ms. Raduska Cupac is a graduated economist, specialized in international finance, with a master degree in EU studies majoring in economic instruments in environmental policies (2011). Ms. Cupac is managing development projects for over a decade and her current research interests include sustainability in and economic analysis of climate resilient and low-emission development.

Mr. Markez Djordje is graduated lawyer, with Master of Science degree in Law. He works as a Secretary General of the University of Banja Luka (BiH). He participated in different international and national projects in area of higher education, RTD policy analyses and project management, funded by World Bank and different EU programmes TEMPUS, FP6 and FP7 programmes. Beside this, he was engaged in different consultancy in area of legislation analyses in water management and climate change. He is also working as a HORIZON 2020 national contact point (NCP) for Legal and Financial Issues and JRC. He is author on several scientific publications.

Goran Vukmir holds a Master degree in Environmental Economics from the University of London (2014), and a degree in Economy from the Banja Luka College (2011). He has 18 years of UN work experience, including over 10 years of experience in developing and managing energy and environment programmes and projects in Bosnia and Herzegovina, especially in the fields of climate change, national GHG reporting, low emission development, energy efficiency, renewable sources of energy and biodiversity. He is a Climate Change Policies and Measures Expert Reviewer for UNFCCC, taking part in the in-depth reviews of the information reported in the National Communications (NC) and the technical reviews of the Biennial Reports of the Annex I Parties of the United Nations Framework Convention on Climate Change (UNFCCC).

Professor Radoslav Dekic graduated in Biology in 2001 and finished his Master's degree in 2006 at the Department of Biology at the Faculty of Science, University of Banja Luka. He completed his Ph.D. in 2010 also at the University of Banja Luka. He currently holds the position of associate professor at the Faculty of science and mathematics, University of Banja Luka, at the Department of Biology and the Department of Ecology and environmental protection. His research interests involve environmental impact on animal physiology, hematology and physiological adaptations. He has coordinated and participated in numerous research projects of biodiversity, physiology, animal ecology, environmental protection and climate change.

Assistant Tatjana Popov holds a bachelor degree in Geography (2007) and a master's degree in Geospatial and Environmental Science (2010). She has been a Ph.D. student in Geosciences at the University of Belgrade (Faculty of Geography), Belgrade, Republic of Serbia since 2011 and her primary interest is the impact of climate change on plants and ecosystems. Currently, she works as a senior teaching assistant at the University of Banjaluka (Faculty of Sciences), the Republic of Srpska, Bosnia and Herzegovina.

Part IV
Limits to Climate Change Adaptation in the Pacific Region

Chapter 15
Climate Change Adaptation Limits in Small Island Developing States

Stacy-ann Robinson

Introduction

Small island developing states (SIDS) are particularly vulnerable to the impacts of climate change. SIDS are 58 countries located in three main geographic regions—the Atlantic, Indian Ocean, Mediterranean and South China Sea (AIMS), the Caribbean, and the Pacific—and comprising an unofficial list published by the United Nations Office of the High Representative for the Least Developed Countries, Landlocked Developing Countries and Small Island Developing States (UN-OHRLLS)[1] (see UN-OHRLLS 2015). As a distinct group of developing countries with interconnected vulnerabilities and development concerns (Bruckner 2013; UNGA 1994), SIDS are especially susceptible to climate impacts such as "sea-level rise, tropical and extratropical cyclones, increasing air and sea surface temperatures, and changing rainfall patterns" (Nurse et al. 2014, p. 1616). The imminent threat of climate change and a consensus that adequate adaptation is an essential component of sustainable development in SIDS, many SIDS have made it a top priority. The 1992 United Nations Framework Convention on Climate Change (UNFCCC) (also called the 'Convention') obliges its State Parties to "[f]ormulate, implement, publish and regularly update national [...] programmes [sic] containing measures to [...] facilitate adequate adaptation to climate change" [United Nations 1992, Art. 4.1(b)]. Adaptation, however, poses a "significant financial and resource

[1]There is no official list of SIDS (Bruckner 2013); likewise, there is no agreement on which countries should be classified as SIDS and on the total number. Some SIDS listed by the UN-OHRLLS are neither small, islands nor developing in the strictest of senses (Mysiak et al. 2015).

S. Robinson (✉)
Institute at Brown for Environment and Society, Brown University,
91 Waterman Street, Providence, RI 02912, USA
e-mail: stacy-ann_robinson@brown.edu

W. Leal Filho and J. Nalau (eds.), *Limits to Climate Change Adaptation*,
Climate Change Management, https://doi.org/10.1007/978-3-319-64599-5_15

challenge for islands" (Nurse et al. 2014, p. 1635). In spite of these challenges, SIDS must take action to adapt.

The Fifth Assessment Report of the Intergovernmental Panel on Climate Change (IPCC) (also referred to as 'AR5'), which "provides a clear and up-to-date view of the current state of scientific knowledge relevant to climate change" (IPCC 2016, online), defines 'climate change adaptation' as the "process of adjustment to actual or expected climate and its effects" (IPCC 2014, p. 1758). Adaptation can be incremental or transformational. Incremental adaptation aims to maintain "the essence and integrity of a system or process at a given scale"; transformational adaptation aims to change the "fundamental attributes of a system in response to climate and its effects" (IPCC 2014, p. 1758). Adaptation can take place in human or natural systems (IPCC 2014). In human systems, it "seeks to moderate or avoid harm or exploit beneficial opportunities" (IPCC 2014, p. 1758); in some natural systems, "human intervention may facilitate adjustment to expected climate and its effects" (IPCC 2014, p. 1758). This suggests that adaptation may not always be possible or sustainable—it has limits. As Klein et al. (2014, p. 907) put it, "for a particular actor, system, and planning horizon of interest, no adaptation options [may] exist, or an unacceptable measure of adaptive effort is required, [in order] to maintain societal objectives or the sustainability of a natural system". With this in mind, adaptation limits can exist in both human and natural systems and in planning horizons.

AR5's Chapter 29, which focusses on small islands, suggests that more can be learnt about adaptation drivers, barriers and limits in these countries (see Nurse et al. 2014). This study helps to fill the gap in relation to adaptation limits. It has one primary aim—to identify and discuss the nature and potential range of adaptation limits in SIDS. Using a sample of 19 SIDS across the three main geographic regions, limits are identified from these countries' National Communications (NCs) to the UNFCCC—SIDS that are non-Annex I Parties to the Convention "publish and regularly update national [...] programmes [sic] containing measures to [...] facilitate adequate adaptation to climate change" through their NCs (United Nations 1992, p. 10). To achieve its primary aim, the remainder of this chapter is divided into five sections. The first section is a review of how adaptation limits are defined and conceptualised in the grey and academic literature, and in relation to SIDS. The second overviews this chapter's methods, their limitations and the resulting constraints of the paper. The third presents its findings and the fourth discusses them, highlighting any parallels with and divergence from the literature. The fifth summarises the chapter's main lessons and outlines future prospects.

Defining and Conceptualising Adaptation Limits

There is now some consensus in the literature on the definition of 'adaptation limit'. This chapter uses AR5's definition—"[t]he point at which an actor's objectives (or system needs) cannot be secured from intolerable risks through adaptive actions"

(IPCC 2014, p. 1758). Actors include national governments, regional organisations, community organisations and individuals; their objectives, in this context, relate to climate change adaptation. In human systems, their objectives (or needs) involve changes to the governance architecture (i.e. laws, policies, regulations, standards and organisations etc.), which may "necessitate new institutions and institutional arrangements to effectively address multi-actor, multiscale risks", along with changes that preserve economic, social and cultural cohesion (Klein et al. 2014, p. 903). In natural systems, they involve ecosystem changes that preserve life and lifeforms; examples include the availability of atmospheric moisture and non-catastrophic ranges of air and sea surface temperature (Klein et al. 2014). These systems are at "risk of climate change impacts" when there is "[t]he potential for consequences where something of value is at stake and where the outcome is uncertain, recognizing the diversity of values" (IPCC 2014, p. 1772). Risk, which "results from the interaction of vulnerability, exposure, and hazard" (IPCC 2014, p. 1772), can be "routine", having limited, acceptable consequences, or "intolerable", posing "fundamental threats to actors' objectives or the sustainability of natural systems" (Klein et al. 2014, p. 902). Adaptive capacities are finite and, therefore, limits to adaptation exist (Klein et al. 2014).

The literature presents different typologies of adaptation limits. Typically, it identifies physical and ecological, technological and economic limits (NCCARF 2011); some authors, for example Inderberg and Eikeland (2009), also consider institutional limits. Physical and ecological limits are the "natural limitations to adaptation, associated largely with the natural environment, ranging from ecosystem thresholds to geographical and geological limitations" (Morgan 2011, p. 7). Technological limits occur when the "technology required for a specific approach is not available or too expensive" or where the technology simply cannot 'keep up' with the adjustments required for successful adaptation (Morgan 2011, p. 7). Technological limits interact closely with economic limits, which occur when "the costs of adaptation exceed the costs of the impacts averted" (Morgan 2011, p. 7). Adger et al. (2009), a seminal work, introduced and contextualised social limits. The authors contended that limits depend on an actor's adaptation objectives, which are values-laden, and that limits have individual and social drivers—they vary according to risk attitudes, culture, knowledge and ethics. AR5, however, primarily differentiates between 'hard' and 'soft' limits (see Klein et al. 2014). A hard limit is where "[n]o adaptive actions are possible to avoid intolerable risks"; a soft limit is where "[o]ptions are currently not available to avoid intolerable risks through adaptive action" (IPCC 2014, p. 1758). Though the literature presents clear typologies, they are not static. They are dynamic, fluid and often overlapped. This increases the complexity of identifying and typifying adaptation limits in specific geographies such as SIDS as well as the complexity of adaptation itself.

AR5 distinguishes among adaptation constraints, obstacles, barriers and limits (see Klein et al. 2014). Constraints, used interchangeably with 'obstacles' and 'barriers', are "[f]actors that make it harder to plan and implement adaptation actions" (Klein et al. 2014, p. 907). Constraints, whether singly or collectively, "can drive an actor or natural system to an adaptation limit" (Klein et al. 2014, p. 907).

AR5's Chapter 29, which focusses on small islands, briefly examines the interplay of adaptation barriers and limits in these countries but does not explicitly differentiate between the two in the small island context (see Nurse et al. 2014). Among other things, Chapter 29 identifies resource (human, financial, technical) and institutional constraints as well as constraints imposed by (1) the social and cultural acceptability of adaptation measures, (2) economic structures, (3) colonial status, (4) a preference for 'hard', shorter-term measures such as seawalls, and (5) social factors—local risk attitudes, culture, knowledge and ethics, previously explored by Adger et al. (2009) (see Nurse et al. 2014). It also suggests that adaptation can be limited by the absence of scale-appropriate modelling technologies, capacities and skills (see Nurse et al. 2014).

Though the explicit question of adaptation limits in SIDS is under-explored in the academic literature, there are several works preceding AR5 that addressed adaptation limits in various geo-contexts. Four such studies are highlighted here. First, Ford (2009), though emphasising the multiplicity of climate stressors, revealed that some populations encounter few or no limits—the adaptability of the Inuit population in Canada significantly reduced their losses from extreme events and their limits, though the author observed several adaptation barriers. The study confirmed that limits can vary within a specific population, according to socio-economic sub-groups and their livelihood activities. Second, Barnett et al. (2013) argued that limits emerge because different groups view the success of adaptation responses differently. Working broadly but briefly drawing on earlier studies that explored endogenous and exogenous perceptions of climate-related migration as an adaptation response in small islands, the authors posited that limits "are, to some extent, in the eyes of the beholder" (Barnett et al. 2013, p. 226). Third, Monnereau and Abraham (2013), studying household-level adaptation to coastal erosion in the Federated States of Micronesia, a Pacific SIDS, found householders implementing measures such as seawalls, tree planting and gabion baskets being limited by ecological, economic and technical factors; householders not implementing any adaptation measures citing, non-exclusively, a lack of resources, knowledge and know-how, and skills. The authors further noted the overwhelming impact of cultural traditions on selecting and implementing measures to adapt to erosion. Fourth, Morrison and Pickering (2013), studying the Australian Alps, found that a large proportion of current and planned adaptation actions were not implemented primarily because of ecological constraints followed by economic constraints. The authors also found that potential limits are closest associated with large-scale conservation efforts, including invasive species control and habitat rehabilitation, and that reported adaptations in the tourism sector are mostly impeded by economic and technological limits and then by the economic and social costs of diversifying the tourism product. These four pre-AR5 studies, as examples, illustrate that (1) adaptation limits are not always present as increasing an actor's (or system's) adaptability reduces its vulnerability to climate extremes and limits, (2) limits are subjective, (3) there are limits to adaptation action as well as inaction, (4) different sectors are faced with different limits, and (5) social and cultural factors are potentially stronger drivers of adaptation limits than physical and ecological factors.

The post-AR5 limits literature is in its infancy; there may be increasing academic interest in the subject, though not specifically in a SIDS context. Two related studies published in 2015 are noteworthy. First, Barnett and Palutikof (2015, p. 233), comparing limits across six Australian study areas, including the low-lying Torres Strait Islands, arrived at five key findings, three of which are that (1) limits emerge because of resource allocation and management trade-offs, (2) social and economic limits, in particular, can occur as a result of "processes that are distant in space and time", and (3) some adaptation barriers are in fact limits. Second, Barnett et al. (2015, p. 1), using the same case study areas, found that "the path-dependent nature of the institutions that govern natural resources and public goods is a deep driver of barriers and limits to adaptation". The authors conceptualised 'path-dependency' as a resistance to transformational change or adjustment. It can, therefore, contravene adaptation aims; it can create scenarios where climatic changes are faster than adaptations, which drive human and natural systems to their limits; it can also promote maladaptation, understood as today's adaptations that may "positively increase the vulnerability of other groups and sectors in the future" (Barnett and O'Neill 2010, p. 211). These two studies are important for shaping an understanding of (1) adaptation limits going forward, and (2) the trajectory and scope of future adaptation limits research.

The fact that adaptation limits in SIDS are under-explored in the academic literature should be addressed—SIDS have been recognised as a distinct group of developing countries, requiring special consideration, since the adoption of the UNFCCC (UNGA 1994; UN-OHRLLS 2011). Additionally, authors such as Biesbroek et al. (2013) and Eisenack et al. (2014) have called for more comparative studies in understanding the drivers of limits. These considerations create an opportunity for a comparative study that aims to understand the nature and potential range of climate change adaptation limits across SIDS, a suite of developing countries determined to be particularly vulnerable to climate and its effects (see Nurse et al. 2014).

Methods

Identifying Adaptation Limits in SIDS

Summative content analysis, which "involves counting and comparisons, usually of key words or content, followed by the interpretation of the underlying context" (Hsieh and Shannon 2005, p. 1277) and which is "an unobtrusive and nonreactive way to study the phenomenon of interest" (cited in Hsieh and Shannon 2005, p. 1285), was used to count and compare content across case study countries' NCs to the UNFCCC. NCs (1) are produced by governments and are considered official,

(2) contain the positions of governments, (3) are all publicly available on the UNFCCC website, and (4) are key for understanding (a) the interaction of vulnerability, exposure and hazards at the national level and (b) country-specific adaptation barriers, constraints, obstacles, risks and/or threats (following Lesnikowski et al. 2015; Robinson 2015). In this study, one primary key word, 'limit*', when appearing in relation to one or more of six other key words —'adapt*', 'barrier*', 'constrain*', 'obstacle*', 'risk*' and 'threat*' (per Klein et al. 2014, p. 907)—is identified, counted and compared across the NCs of sample countries. The asterisks served as stem/root word/truncation operators, retrieving matches where words began with the words that preceded them. 'Appearing in relation to' means (1) being contextually connected to the country's adaptation objectives and/or its system's sustainability, and (2) appearing in the same sentence or paragraph, or under a contextually-connected sub-heading. Only countries' Second or Third NCs were analysed, where at least two NCs had been submitted by that country up to December 31, 2015. A focus was placed on the adaptation-related chapters and sections contained within the NCs. The findings are reported in Section "Findings" of this chapter as summative cross-regional counts and comparisons of adaptation limits.

Selecting Case Study Countries

Nineteen of the 58 SIDS listed by the UN-OHRLLS are used in this chapter as case study countries: four AIMS SIDS (Bahrain, Mauritius, Seychelles and Singapore), eight Caribbean SIDS (Antigua and Barbuda, Bahamas, Belize, Dominica, Guyana, Jamaica, St. Lucia and Trinidad and Tobago) and seven Pacific SIDS (Federated States of Micronesia, Fiji, Kiribati, Marshall Islands, Nauru, Samoa and Tonga). These countries were selected to support cross-regional counts and comparisons of limits, as reported in their most recent NC. The selection criteria followed the steps outlined and used by Robinson (2015)—non-UN Member State SIDS ($N = 20$) were excluded as they have no obligation to produce and submit NCs. UN Member State SIDS that had not submitted at least two NCs up to December 31, 2015 ($N = 12$) in English ($N = 7$) were also excluded. Only countries' Second or Third NCs were analysed (Third NC in the case of Singapore only). Second and Third NCs provide insights into any new, recent and/or emerging (1) interactions of vulnerability, exposure and hazards at the national level and (2) country-specific adaptation barriers, constraints, obstacles, risks, threats and/or limits. While acknowledging that "[v]ulnerabilities and adaptation needs are as diverse as the variety of [small] islands between regions and even within nation states" (Nurse et al. 2014, p. 1635), the case study countries used in this chapter represent the diverse physical and ecological, technological, economic and social characteristics of SIDS. Representing this diversity is a critical part of illustrating the nature and potential range of adaptation limits across SIDS.

Limitations of Methods and Constraints of Paper

The methods used in this paper have limitations. First, the paper is limited to desk-based, qualitative analysis only—rich perspectives on adaptation limits in SIDS could be derived from other methods e.g. interviews with policy-makers and on-the-ground implementers of adaptation projects. This could form part of a future research agenda, which is explored in Section "Future Prospects" of this chapter. Second, the documents used in the desk-based analysis are limited to Second and Third NCs to the UNFCCC and to the adaptation-related chapters and sections contained within them. This could exclude other critical documents that are also produced by governments and contain their official positions on adaptation, which may not be publicly available online. The fact that NCs are open access via the UNFCCC website and countries prepare them according to guidelines approved by the Conference of the Parties (COP), the "supreme decision-making body of the Convention" are advantages—data is equally available for sample countries (UNFCCC 2014a, online; see UNFCCC 2014b). Third, many SIDS are in the process of preparing Third NCs and so those currently available may not contain the most up-to-date information, though 47% of the 19 SIDS last submitted their NCs (inclusive of any addenda or updates) between 2013 and 2015. Further, the findings of this chapter are limited to what SIDS report in their NCs. It, therefore, does not cover what countries have reported elsewhere and through other channels. Fourth, the precise use of English in NCs submitted by countries where English is either a second language or one of multiple official languages may not be universal. Fifth, while summative content analysis can provide "insights into how words are actually used", its findings may be "limited by their inattention to the broader meanings present in the data" (Hsieh and Shannon 2005, p. 1285), making this a constraint of the paper. Despite these limitations and the consequent constraints of the paper, this chapter substantially illustrates the nature and potential range of adaptation limits in SIDS, as demonstrated in Section "Findings".

Findings

Forty-eight unique key words and content relating to the 'limit*' to SIDS' adaptation objectives and/or the systems' sustainability were identified from the sample countries' NCs. There were 600 counts of 'limit*'. The 19 SIDS sampled mostly reported being limited by (1) finance/budgetary restrictions/income (64 counts), (2) natural resources/features (60 counts), (3) data/records (49 counts), (4) technical capacity/resources (40 counts), (5) human resources/manpower/turnover (34 counts), (6) land area/availability/use (34 counts), (7) knowledge/under standing/expertise (25 counts), (8) level of public education/awareness (25 counts), (9) their capacity to adapt (not otherwise specified) (20 counts), and (10) existing infrastructure (19 counts). Few SIDS reported being limited by their own inaction, culture, perceptions, the social acceptability and limited scaling up potential of proposed adaptation strategies and measures. Of the four AIMS SIDS sampled, Seychelles reported the most number of limits (45 counts). Guyana reported the most number of limits (51 counts) among the eight Caribbean SIDS. Of the seven Pacific SIDS and all 19 SIDS sampled, the Marshall Islands reported the most number of limits (60 counts). Table 15.1 shows the nature and potential range of adaptation limits across the 19 SIDS sampled.

There were 63 counts of 'limit*' in relation to AIMS SIDS' adaptation objectives, as outlined in their NCs. Sampled countries mostly reported being limited by (1) finances/budgetary restrictions/income (8 counts), (2) land area/availability/use (7 counts), (3) data/records (6 counts), (4) human resources/manpower/turnover (4 counts), (5) their capacity to predict climate impacts (4 counts), (6) modelling software/models (4 counts), (7) their size (4 counts), (8) natural resources/features (4 counts), (9) public awareness, knowledge and ownership (3 counts), and (10) their capacity to adapt (not otherwise specified) (3 counts). Seychelles, for example, mostly reported being limited by finances/budgetary restrictions/income (5 counts). Bahrain mostly reported being limited by its capacity to adapt (2 counts). There was one count each of population density, natural hazards, external shocks, international law and insufficient mainstreaming progress. In the context used, mainstreaming refers to the integration of either the adaptation and disaster risk reduction agendas or the adaptation and (sustainable) development agendas at the national level. None of the sampled AIMS SIDS reported being limited by factors such as the focus or scope of national institutions/policies/legislation/regulations, technical capacity/resources or poor inter-agency communication/coordination.

There were 202 counts of 'limit*' in relation to Caribbean SIDS' adaptation objectives, as outlined in their NCs. Sampled countries mostly reported being limited by (1) finances/budgetary restrictions/income (31 counts), (2) technical capacity/resources (21 counts), (3) data/records (16 counts), (4) natural resources/features (13 counts), (5) human resources/manpower/turnover (12 counts), (6) knowledge/expertise (10 counts), (7) focus or scope of national policies/legislation/regulations (8 counts), (8) monitoring and evaluation/enforcement capacity (7 counts), (9) their size (6 counts), and (10) public education/awareness (6 counts). Antigua and Barbuda, for example, mostly reported being limited by technical capacity/resources (11 counts). Guyana mostly reported being limited by finances/budgetary restrictions/income (10 counts). There was one count each of the limiting nature of culture, individual limits/preferences and the social acceptability of adaptation strategies and measures. None of the sampled Caribbean SIDS reported being limited by factors such as population density, the impacts of external (economic) shocks or the limited scaling up potential of adaptation strategies and measures.

There were 335 counts of 'limit*' in relation to Pacific SIDS' adaptation objectives, as outlined in their NCs. Sampled countries mostly reported being limited by (1) natural resources/features (43 counts), (2) data/records (27 counts), (3) finance/budgetary restrictions/income (25 counts), (4) land area/availability/use

Table 15.1 Nature and potential range of adaptation limits in SIDS (Based on NCs analysed in this chapter)

NO.	ADAPTATION LIMIT (AS REPORTED IN NCs)	NO. OF COUNTS
1	Finance/budgetary restrictions/income (including government expenditure, capital)	64
2	Natural resources (availability of/dependence on)/features (excluding topography)	60
3	Data/records	49
4	Technical capacity/resources	40
5	Human resources/manpower/turnover	34
6	Land area/availability/use	34
7	Knowledge/understanding (of climate effects; gained through research) and expertise	25
8	Level of public education (including formal education), awareness, knowledge, 'ownership'	25
9	Limited capacity to adapt (not otherwise specified)	20
10	Existing infrastructure	19
11	Nature of markets/economic structures	18
12	Focus or scope of national institutions/policies/legislation/regulations	17
13	Land arability/soil chemistry	16
14	Size (including population)	15
15	Individual limits/preferences	13
16	Nature and/or diversity of climate impacts	11
17	Environmental conditions/management // natural thresholds	11
18	Modelling software/models (including geographical information systems)	9
19	Monitoring and evaluation/enforcement capacity	9
20	Poor inter-agency communication/coordination	8
21	Natural hazards	8
22	Opportunities/options/choice	7
23	Insufficient mainstreaming progress (disaster risk reduction and/or adaptation into development)	6
24	Limited capacity to predict climate impacts	6
25	Limited response capability	6
26	Socio-economic conditions	6
27	Quality of stakeholder participation	6
28	Facilities/equipment e.g. medicine storage	5
29	Limits of current adaptation strategies (e.g. marine protected areas)	5
30	Frequency/intensity of climate impacts	5
31	Private sector development	5
32	Geography - sparse and scattered nature of islands	5
33	Technological (not otherwise specified)	4
34	Unavailability of insurance/under-development of reinsurance market	4
35	External (economic) shocks	4
36	Limited capacity to manage	3
37	Sector productivity	3
38	Jurisdiction of national institutions/bodies	2
39	Cost- and other effectiveness	2
40	International law e.g. limits of exclusive economic zones	2
41	Topography	2
42	Absence of champion/strong leader	1
43	Own inaction	1
44	Culture	1
45	Perceptions	1
46	Social acceptability of proposed adaptation strategies and measures	1
47	Limited scaling up potential of adaptation strategies and measures	1
48	Population density	1
	TOTAL COUNTS	600

(23 counts), (5) technical capacity/resources (19 counts), (6) human resources/manpower/turnover (18 counts), (7) existing infrastructure (16 counts), (8) level of public education/awareness (16 counts), (9) nature of markets/economic structures (15 counts), and (10) land arability/soil chemistry (15 counts). The Marshall Islands and Nauru, for example, mostly reported being limited by natural resources/features (16 and 12 counts, respectively). Tonga mostly reported being limited by technical capacity resources, and public education/awareness (9 counts each), and by the focus or scope of national institutions/policies/legislation/regulations and data/records (8 counts each). There was one count each of the limiting nature of monitoring and evaluation/enforcement capacity,

knowledge/understanding/expertise and the absence of a champion/strong leader. None of the sampled Pacific SIDS reported being limited by factors such as the underdevelopment of the insurance/reinsurance market, international law or sector productivity; and similar to the Caribbean SIDS, no Pacific SIDS reported being limited by population density or the limited scaling up potential of adaptation strategies and measures.

The above 48 limits were categorised according to the typologies proposed by authors such as Adger et al. (2009), Inderberg and Eikeland (2009) and NCCARF (2011). These categorisations are showing in Table 15.2.

Based on Table 15.2 above, SIDS mostly reported limits that may be categorised as institutional e.g. poor inter-agency communication/coordination and (the limited) jurisdiction of national institutions/bodies. Institutional limits had 232 counts and accounted for 39% of total counts. Limits such as land arability/soil chemistry and topography, categorised as physical and ecological, had 168 counts and accounted for 28% of total counts. Economic limits e.g. finance/budgetary restrictions/income and external economic shocks had the third highest count—96. This typology accounted for 16% of total counts. Limits such as the level of public education/awareness and culture, categorised as social, had 85 counts or 14% of total counts. Technological limits e.g. modelling software/models accounted for 3% of total counts. This typology had the lowest count—19. An illustration of adaptation limits by typology, in percentages, is shown in Fig. 15.1.

The 48 limits reported were also categorised according to whether they are occurring in human or natural systems. The limits occurring in human systems (technological, economic, institutional and social typologies) accounted for 432 of the total 600 counts or 72% of the limits reported. The limits occurring in natural systems (physical and ecological typology) accounted for the remaining 168 counts or 28% of the limits reported.

Discussion

The findings above are likely to raise a number of questions. Chief among them are:

1. In SIDS, are reported limits really limits?
2. What factors are potentially driving SIDS to their limits?
3. Can SIDS' adaptation objectives be secured against intolerable risks?

This discussion section aims to answer these three questions.

In SIDS, are reported limits really limits? Yes, but it depends on how one conceptualises 'limits'. This study identifies 48 unique key words and content relating to the 'limit*' to SIDS' adaptation objectives and/or the systems' sustainability. This potential range of limits highlights the diversity across SIDS and

Table 15.2 Adaptation limits in SIDS by typology. (Based on NCs analysed in this chapter)

NO.	LIMITS TYPOLOGY	NO. OF COUNTS
	Physical and Ecological	
1	Natural resources (availability of/dependence on)/features (excluding topography)	60
2	Land area/availability/use	34
3	Land arability/soil chemistry	16
4	Size (including population)	15
5	Nature and/or diversity of climate impacts	11
6	Environmental conditions/management // natural thresholds	11
7	Natural hazards	8
8	Frequency/intensity of climate impacts	5
9	Geography - sparse and scattered nature of islands	5
10	Topography	2
11	Population density	1
	Sub-Total	**168**
	Technological	
1	Modelling software/models (including geographical information systems)	9
2	Limited capacity to predict climate impacts	6
3	Technological (not otherwise specified)	4
	Sub-Total	**19**
	Economic	
1	Finance/budgetary restrictions/income (including government expenditure, capital)	64
2	Nature of markets/economic structures	18
3	Private sector development	5
4	External (economic) shocks	4
5	Sector productivity	3
6	Cost- and other effectiveness	2
	Sub-Total	**96**
	Institutional	
1	Data/records	49
2	Technical capacity/resources	40
3	Human resources/manpower/turnover	34
4	Limited capacity to adapt (not otherwise specified)	20
5	Existing infrastructure	19
6	Focus or scope of national institutions/policies/legislation/regulations	17
7	Monitoring and evaluation/enforcement capacity	9
8	Poor inter-agency communication/coordination	8
9	Insufficient mainstreaming progress (disaster risk reduction and/or adaptation into development)	6
10	Limited response capability	6
11	Limits of current adaptation strategies (e.g. marine protected areas)	5
12	Facilities/equipment e.g. medicine storage	5
13	Unavailability of insurance/under-development of reinsurance market	4
14	Limited capacity to manage	3
15	Jurisdiction of national institutions/bodies	2
16	International law e.g. limits of exclusive economic zones	2
17	Absence of champion/strong leader	1
18	Limited scaling up potential of adaptation strategies and measures	1
19	Own inaction	1
	Sub-Total	**232**
	Social	
1	Knowledge/understanding (of climate effects; gained through research) and expertise	25
2	Level of public education (including formal education), awareness, knowledge, 'ownership'	25
3	Individual limits/preferences	13
4	Opportunities/options/choice	7
5	Socio-economic conditions	6
6	Quality of stakeholder participation	6
7	Culture	1
8	Perceptions	1
9	Social acceptability of proposed adaptation strategies and measures	1
	Sub-Total	**85**

confirms that one size does not fit all. If strictly applying AR5's definition of limits, i.e. "[t]he point[s] at which an actor's objectives (or system needs) cannot be secured from intolerable risks through adaptive actions" (IPCC 2014, p. 1758), the 'real' limits in SIDS may be those categorised as physical and ecological. These

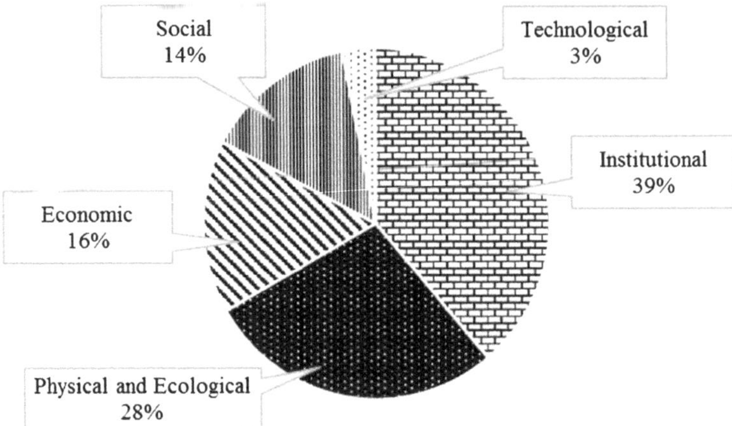

Fig. 15.1 Adaptation limits in SIDS by typology (%) (based on NCs analysed in this chapter)

limits, occurring in natural systems and relating to, for example, geography (the sparse and scattered nature of islands), cannot be secured against intolerable risks, though it may not be impossible to reduce and/or manage them. As Barnett et al. (2013, p. 226) argued, however, "to some extent, [limits are] in the eyes of the beholder". This holds true in SIDS. The Federated States of Micronesia, Marshall Islands and Nauru, for example, were the only SIDS reporting the absence of or slow private sector development as a 'limit'. Whether this should be identified and/or classified as such, is a matter of conceptualisation, which may be influenced by a country's circumstances, experiences and adaptation objectives, which are values-laden (following Adger et al. 2009; Barnett et al. 2013). Additionally, the difference between and among adaptation constraints, obstacles, barriers and limits in SIDS is negligible—the line is not distinct and often the word 'limit' is used to mean much the same as 'constraint', 'obstacle' and/or 'barrier'. On the one hand, this could be a function of language—the precise use of English in NCs submitted by countries where English is either a second language or one of multiple official languages may not be universal. On the other hand, it could also be a general consistency with AR5's small islands chapter, which briefly examines the interplay of adaptation constraints and barriers with adaptation limits in these countries but which does not explicitly differentiate between constraints or barriers and limits in the small island context (see Nurse et al. 2014). It was not uncommon for SIDS to report, in the same paragraph or under the same sub-heading, a constraint as a limit as well as being 'constrained by' and/or 'limited by' the same factors, making it possible to use the words interchangeably in a SIDS context. This illustrates that limits are conceptualised differently in and across SIDS.

What factors are potentially driving SIDS to their limits? The top two single factors potentially driving SIDS to their limits are proposed here based on the number of counts across the NCs analysed as part of this study: (1) finance/budgetary restrictions/income (including government expenditure, capital) (economic limits typology) and (2) natural resources (availability of/depended on)/features (excluding topography) (physical and ecological limits typology). This is in line with the findings of Morrison and Pickering (2013), regarding the top two reasons for a large proportion of current and planned adaptation actions not being implemented, but in reverse order—in SIDS, unlike the Australian Alps, current and planned adaptation actions appear not to be implemented primarily because of economic limits followed by physical and ecological limits.

This study identifies 64 counts of finance/budgetary restrictions/income constraints, accounting for 11% of total 'limit*' counts. Seventeen of the 19 SIDS sampled reported it as a limit. Singapore and Trinidad and Tobago were the exceptions—Guyana alone had 10 counts. This suggests that economic constraints could be the single greatest and/or single strongest driver of adaptation limits in SIDS across all three regions. It also signals a possible correlation between the availability of financing and the ability of SIDS to identify, design and implement appropriate interventions, especially in the case of soft limits. Nunn et al. (2014) and Storey and Hunter (2010) acknowledged a reliance of Pacific island states on foreign aid to support national adaptation actions. Klint et al. (2012) provided a salient example that illustrates the extent of this reliance. In conducting interviews with tourism officials in Vanuatu (a Pacific SIDS not covered in this study), an interviewee noted there is no back-up funding or funding from the government to deal with climate change and so the rest of the world would need to "pick up the Pacific" (Klint et al. 2012, p. 268). In Niue (another Pacific SIDS not covered in this study), aid has proven to be the "most critical factor in the availability of finance for adaptation to climate change, even if it constrains the way the money can be spent" (Barnett 2008, p. 45). The implications of relying on foreign aid or external finance in this way are many, including constraints on state autonomy and on adaptation spending, and the undermining of governance (Barnett 2008). This should be addressed in addition to identifying more flexible, innovative and sustainable streams of national and international adaptation financing.

This study finds 60 counts of natural resource/feature constraints, accounting for 10% of total 'limit*' counts. Fifty per cent of the AIMS and Caribbean SIDS sampled reported it as a limit. Exceptions include Bahrain and Trinidad and Tobago, respectively. Eighty-six per cent of the Pacific SIDS sampled reported it as a limit—the Marshall Islands and Nauru had the most counts, accounting for 65% of the Pacific total. This suggests that physical and ecological constraints could be the second single greatest and/or second single strongest driver of adaptation limits in SIDS across all three regions. It also suggests that physical and ecological constraints could be a greater and/or stronger driver of adaptation limits in SIDS in the Pacific, over SIDS in the other two regions. Institutional constraints, however, appear to collectively be the greatest and/or strongest driver of adaptation limits in SIDS, over physical, ecological and economic constraints, having accounted for 39% of limits reported by SIDS. This has implications for adaptation planning and decision-making at different scales. An understanding of these single and collective dynamics is important as actors could be misplacing their efforts if not enough collective attention is being paid to reducing SIDS' vulnerability to institutional limits.

Can SIDS' adaptation objectives be secured against intolerable risks? To some extent, but not completely—action in human systems (over natural systems) may produce better adaptation results, given the nature of the physical and ecological limits reported by SIDS. These countries are currently faced with a number of hard limits in the natural system. Examples include natural resources (availability of)/features, land area/availability, and land arability/soil chemistry. Securing SIDS' adaptation objectives against intolerable risks will require purposive, and possibly even culturally-confronting action (not inaction) in areas in which the negative human impacts on natural and human systems can be mitigated. And "[b]ecause much cannot be done to combat the physical aspects of climate change, there is the need to manage properly [sic] and to limit negative human impact on these systems [...]" (Government of Belize 2011, p. 62). As such, two recommendations for purposive action in human systems are made in the "Recommendations" section below.

Recommendations

The first recommendation is that SIDS should prioritise reducing their vulnerability to institutional limits. This will involve the establishment and operationalisation of appropriate, national-level institutions that are likely to "facilitate adequate adaptation to climate change" (United Nations 1992, p. 10). Belize, for example, had anticipated that the operationalisation of its National Climate Change Committee would help to reduce constraints to its adaptation objectives through the "formulation and implementation of a national climate change adaptation strategy; with complementary execution of programmes/projects [sic]" (Government of Belize 2011, p. 138). Though the extent to which these objectives in Belize and other SIDS is being achieved falls outside the scope of this paper, Robinson (2015) had found 28% of adaptation actions in SIDS being focused on the observation and assessment of climate impacts with only 19% of actions being focused on planning, institutions and policies. This suggests that SIDS are more heavily involved in observation and assessment activities and that opportunities for reducing their vulnerability to institutional limits could exist and should be explored and acted on.

The second recommendation is that SIDS should (continue to) actively pursue strategic alliances between and among national governments and other stakeholders at the local, national, regional and international levels in order to mobilise the resources necessary for adequate adaptation. At the local and national levels, strategic alliances have helped countries such as Belize to fill data gaps through research in, for example, mangrove rehabilitation (Government of Belize 2011). Antigua and Barbuda's NC further noted that "[r]egional and international cooperation and implementation will be essential given the limited capabilities that exist [in SIDS]" (Government of Antigua and Barbuda 2009, p. 235). Pursuing greater regional and international cooperation could increase the importance and relevance of regional organisations such as the Caribbean Community Climate Change Centre and the Secretariat of the Pacific Regional Environment Programme. Robinson and Gilfillan (2016), however, found regional organisations to be more effective with respect to their adaptation-related inputs and outputs but less effective with respect to designing, implementing, monitoring and evaluating national-level adaptation programs and projects. This could be a key consideration when defining the future role of regional organisations in this regard, and for national governments deciding on the extent and nature of the strategic adaptation alliances they will pursue at and through this scale.

Conclusion

Small island developing states (SIDS) are 58 countries located in three main geographic regions—the AIMS, Caribbean and Pacific—that are among the most vulnerable to climate change and required to adapt to its impacts. The small islands chapter in the Fifth Assessment Report of the Intergovernmental Panel on Climate Change (AR5) suggests that more can be learnt about adaptation drivers, barriers and limits in these countries. This study helps to fill the gap in relation to adaptation limits. It has one primary aim—to identify and discuss the nature and potential range of adaptation limits in SIDS. Using summative content analysis, it analyses limits, as reported in the National Communications (NCs) of 19 SIDS. It finds 48 unique key words and content relating to the 'limit*' to SIDS' adaptation objectives and/or the systems' sustainability. Countries most commonly reported being limited by finance/budgetary restrictions/income, natural resources/features and data/records; their own inaction, culture, perceptions and the social acceptability of adaptation strategies and measures were among the least commonly reported limits. Of the 600 counts of 'limit*', 39% were categorised as institutional limits, 28% as physical and ecological, 16% as economic, 14% as social and 3% as technological limits. Further, limits occurring in human systems (technological, economic, institutional and social typologies) accounted for 72% of the limits reported; limits occurring in natural systems (physical and ecological typology) accounted for 28%.

There are four main lessons emanating from this chapter's findings. The first is that, if strictly applying AR5's definition of limits, the 'real' limits in SIDS may be those categorised as physical and ecological—those occurring in natural systems and relating to, for example, geography, which cannot be secured against intolerable risks. The second is that limits are conceptualised differently in and across SIDS. These conceptualisations are influenced by countries' circumstances, experiences and adaptation objectives, which are values-laden (following Adger et al. 2009; Barnett et al. 2013). The third is that the difference between and among adaptation constraints, obstacles, barriers and limits in SIDS is negligible—the line is not distinct and often the word 'limit' is used to mean much the same as 'constraint', 'obstacle' and/or 'barrier'. This could be a function of language—the precise use of English in NCs submitted by countries where English is either a second language or one of multiple official languages may not be universal. The fourth is that institutional constraints are likely to be collectively the greatest and/or strongest driver of adaptation limits in SIDS. This suggests that national SIDS governments could be misplacing their efforts if not enough collective attention is being paid to reducing their vulnerability to institutional limits. These four lessons are important for adaptation planning and decision-making at different scales. An understanding of the nature and potential range of adaptation limits in SIDS can help local, national, regional and international actors to (1) determine which adaptation actions are likely to be feasible, cost-effective and/or sustainable, (2) prioritise adaptation actions in the interest of sustainable development, (3) better design adaptation interventions, making the best possible use of scarce resources, and (4) identify and seize opportunities to increase the adaptive capacities of SIDS.

Future Prospects

This study presents opportunities for future research in adaptation limits in SIDS. Three such opportunities are cited here. First, conducting interviews with policy-makers and on-the-ground implementers of adaptation projects in SIDS in order to better understand how limits are conceptualised. These interviews could also aim to test whether there is an institutional path-dependency (i.e. a resistance to transformational change or adjustment) in SIDS and, if so, whether it has the potential for maladaptation in these countries (see Barnett and O'Neill 2010; Barnett et al. 2015). Second, linking limits to specific adaptation measures and sectors (see Morrison and Pickering 2013). Third, quantitatively testing whether (1) institutional constraints are the greatest and/or strongest driver of adaptation limits in SIDS and/or (2) a correlation exists between the availability of financing and the ability of SIDS to identify, design and implement appropriate adaptation actions. Exploring these and other opportunities would help to deepen our understanding of the nature and range of adaptation limits in SIDS.

Acknowledgements The author would like to acknowledge: funding support from the Australian Department of Foreign Affairs and Trade and the Fenner School of Environment and Society at The ANU; A/Prof Jamie Pittock and Dr. Tony Weir at The ANU along with the anonymous reviewers for feedback on earlier versions of the paper; and Daniel Ferris for proofreading. The author takes full responsibility for any and all errors.

References

Adger, W. N., Dessai, S., Goulden, M., Hulme, M., Lorenzoni, I., Nelson, D., et al. (2009). Are there social limits to adaptation to climate change? *Climatic Change, 93*(3–4), 335–354. Available at: http://dx.doi.org/10.1007/s10584-008-9520-z. Accessed 19 February 2014.

Barnett, J. (2008). The effect of aid on capacity to adapt to climate change: Insights from niue. *Political Science, 60*(1), 31–45. Available at: http://dx.doi.org/10.1177/003231870806000104. Accessed 7 March 2015.

Barnett, J., Evans, L. S., Gross, C., Kiem, A. S., Kingsford, R. T., Palutikof, J. P., et al. (2015). From barriers to limits to climate change adaptation: Path dependency and the speed of change. *Ecology and Society, 20*(3), 1–11. Available at: http://dx.doi.org/10.5751/ES-07698-200305. Accessed 20 July 2015.

Barnett, J., Mortreux, C., & Adger, W. N. (2013). Barriers and limits to adaptation: Cautionary notes. In S. Boulter, J. Palutikof, D. J. Karoly, & D. Guitart (Eds.), *Natural disasters and adaptation to climate change* (pp. 223–235). Cambridge: Cambridge University Press.

Barnett, J., & O'Neill, S. (2010). Maladaptation. *Global Environmental Change, 20*(2), 211–213. Available at: http://dx.doi.org/10.1016/j.gloenvcha.2009.11.004. Accessed 29 April 2014.

Barnett, J., & Palutikof, J. P. (2015). The limits to adaptation: A comparative analysis. In J. P. Palutikof, S. L. Boulter, J. Barnett, & D. Rissik (Eds.), *Applied studies in climate adaptation* (pp. 231–240). Somerset, New Jersey: Wiley.

Biesbroek, G. R., Klostermann, J. E. M., Termeer, C. J. A. M., & Kabat, P. (2013). On the nature of barriers to climate change adaptation. *Regional Environmental Change, 13*(5), 1119–1129. Available at: http://dx.doi.org/10.1007/s10113-013-0421-y. Accessed 22 March 2014.

Bruckner, M. (2013). Effectively addressing the vulnerabilities and development needs of small island developing states. In *CDP Background Paper No. 17*, United Nations, New York. http://www.un.org/en/development/desa/policy/cdp/cdp_background_papers/bp2013_SIDS2.pdf. Accessed 24 April 2014.

Eisenack, K., Moser, S. C., Hoffmann, E., Klein, R. J. T., Oberlack, C., Pechan, A., et al. (2014). Explaining and overcoming barriers to climate change adaptation. *Nature Climate Change, 4*(10), 867–872. Available at: http://dx.doi.org/10.1038/nclimate2350. Accessed 26 January 2016.

Ford, J. D. (2009). Sea ice change in Artic Canada: Are there limits to Inuit adaptation? In W. N. Adger, I. Lorenzoni, & K. L. O'Brien (Eds.), *Adapting to climate change: Thresholds, values, governance* (pp. 114–128). Cambridge, New York: Cambridge University Press.

Government of Antigua and Barbuda. (2009). *Antigua and Barbuda's second national communication on climate change*. St. John's: Government of Antigua and Barbuda. Available at: http://unfccc.int/resource/docs/natc/antnc2.pdf. Accessed 15 October 2015.

Government of Belize. (2011). *Second National Communication to the Conference of the Parties of the United Nations Framework Convention on Climate Change*. Belmopan: Ministry of Natural Resources and the Environment. Available at: http://unfccc.int/resource/docs/natc/blznc2.pdf. Accessed 29 March 2014.

Hsieh, H.-F., & Shannon, S. E. (2005). Three approaches to qualitative content analysis. *Qualitative Health Research, 15*(9), 1277–1288. Available at: http://dx.doi.org/10.1177/1049732305276687. Accessed 29 April 2016.

Inderberg, T. H., & Eikeland, P. O. (2009). Limits to adaptation: Analysing institutional constraints. In W. N. Adger, I. Lorenzoni, & K. O'Brien (Eds.), *Adapting to climate change: Thresholds, values, governance* (pp. 433–447). Cambridge: Cambridge University Press.

IPCC. (2014). Annex II: Glossary. In V. R. Barros, C. B. Field, D. J. Dokken, M. D. Mastrandrea, K. J. Mach, T. E. Bilir, M. Chatterjee, K. L. Ebi, Y. O. Estrada, R.C. Genova, B. Girma, E.S. Kissel, A.N. Levy, S. MacCracken, P.R. & Mastrandrea, L.L. White (Eds.), *Climate change 2014: Impacts, adaptation, and vulnerability. Part B: Regional aspects. Contribution of Working Group II to the Fifth Assessment Report of the Intergovernmental Panel on Climate Change* (pp. 1757–1776). Cambridge, New York: Cambridge University Press.

IPCC. (2016). *About the fifth assessment report (AR5)*. Geneva: Intergovernmental Panel on Climate Change. Available at: http://www.ipcc.ch/. Accessed 18 January 2016.

Klein, R.J.T., Midgley, G.F., Preston, B.L., Alam, M., Berkhout, F.G.H., & Dow, K., et al. (2014). Adaptation opportunities, constraints, and limits. In C. B. Field, V. R. Barros, D. J. Dokken, K. J. Mach, M. D. Mastrandrea, T. E. Bilir, M. Chatterjee, K. L. Ebi, Y. O. Estrada, R. C. Genova, B. Girma, E. S. Kissel, A. N. Levy, S. MacCracken, P. R. Mastrandrea, & L.L. White (Eds.), *Climate change 2014: Impacts, adaptation, and vulnerability. Part A: Global and sectoral*

aspects. Contribution of Working Group II to the Fifth Assessment Report of the Intergovernmental Panel of Climate Change (pp. 899–943). Cambridge, New York: Cambridge University Press.
Klint, L. M., Wong, E., Jiang, M., Delacy, T., Harrison, D., & Dominey-Howes, D. (2012). Climate change adaptation in the Pacific Island tourism sector: Analysing the policy environment in Vanuatu. *Current Issues in Tourism, 15*(3), 247–274. Available at: http://dx.doi.org/10.1080/13683500.2011.608841. Accessed 2 February 2015.
Lesnikowski, A. C., Ford, J. D., Berrang-Ford, L., Barrera, M., & Heymann, J. (2015). How are we adapting to climate change? A global assessment. *Mitigation and Adaptation Strategies for Global Change, 20*(2), 277–293. Available at: http://dx.doi.org/10.1007/s11027-013-9491-x. Accessed 19 July 2014.
Monnereau, I., & Abraham, S. (2013). Limits to autonomous adaptation in response to coastal erosion in Kosrae, Micronesia. *International Journal of Global Warming, 5*(4), 416–432. Available at: http://dx.doi.org/10.1504/IJGW.2013.057283. Accessed 27 August 2015.
Morgan, C. (2011). *Limits to adaptation: A review of limitations relevant to the project "Building resilience to climate change—Coastal Southeast Asia"*. Gland: International Union for Conservation of Nature and Natural Resources.
Morrison, C., & Pickering, C. (2013). Limits to climate change adaptation: Case study of the Australian Alps. *Geographical Research, 51*(1), 11–25. Available at: http://dx.doi.org/10.1111/j.1745-5871.2012.00758.x. Accessed 17 January 2016.
Mysiak, J., Calliari, E., & Amadio, M. (2015). How small is small and how much island is a coast: Which countries from among the Small Island Developing States warrant special attention? Conference presentation at the *16th Annual Sir Arthur Lewis Institute for Social and Economic Studies (SALISES) Conference,* Castries/Rodney Bay, St. Lucia, 14–16 January 2015.
NCCARF. (2011). *Limits to climate change adaptation: Key findings, synthesis report*. Gold Coast, Australia: National Climate Change Adaptation Research Facility. Available at: https://www.nccarf.edu.au/sites/default/files/attached_files_publications/Limits%20-%20synthesisWEB.pdf. Accessed 23 January 2016.
Nunn, P. D., Aalbersberg, W., Lata, S., & Gwilliam, M. (2014). Beyond the core: Community governance for climate-change adaptation in peripheral parts of Pacific Island Countries. *Regional Environmental Change, 14*(1), 221–235. Available at: http://dx.doi.org/10.1007/s10113-013-0486-7. Accessed 20 February 2014.
Nurse, L. A., McLean, R. F., Agard, J., Briguglio, L. P., Duvat-Magnan, V., & Pelesikoti, N., et al. (2014). Small islands. In V. R. Barros, C. B. Field, D. J. Dokken, M. D. Mastrandrea, K. J. Mach, T. E. Bilir, M. Chatterjee, K. L. Ebi, Y. O. Estrada, R. C. Genova, B. Girma, E. S. Kissel, A. N. Levy, S. MacCracken, P. R. Mastrandrea, & L. L. White (Eds.), *Climate change 2014: Impacts, adaptation, and vulnerability. Part B: Regional aspects. Contribution of Working Group II to the Fifth Assessment Report of the Intergovernmental Panel of Climate Change* (pp. 1613–1654). Cambridge, New York: Cambridge University Press.
Robinson, S.-A. (2015). Climate change adaptation trends in small island developing states. *Mitigation and Adaptation Strategies for Global Change*. Available at: http://dx.doi.org/10.1007/s11027-015-9693-5. Accessed 20 December 2015.
Robinson, S.-A., & Gilfillan, D. (2016). Regional organisations and climate change adaptation in small island developing states. *Regional Environmental Change*. Available at: http://dx.doi.org/10.1007/s10113-016-0991-6. Accessed 13 June 2016.
Storey, D., & Hunter, S. (2010). Kiribati: An environmental 'perfect storm'. *Australian Geographer, 41*(2), 167–181. Available at: http://dx.doi.org/10.1080/00049181003742294. Accessed 24 April 2014.
UN-OHRLLS. (2011). *Small Island developing states: Small Island, Big(ger) Stakes*. NewYork: United Nations Office of the High Representative for the Least Developed Countries, Landlocked Developing Countries and Small Island Developing States. Available at: http://unohrlls.org/custom-content/uploads/2013/08/SIDS-Small-Islands-Bigger-Stakes.pdf. Accessed 17 August 2014.

UN-OHRLLS. (2015). *About SIDS: Country Profiles*. Geneva: United Nations. Available at. http://unohrlls.org/about-sids/country-profiles/. Accessed 18 June 2015.
UNFCCC. (2014a). *About the Conference of the Parties (COP)*. Bonn: United Nations Framework Convention on Climate Change Secretariat. Available at: http://unfccc.int/bodies/body/6383.php. Accessed 1 May 2016.
UNFCCC. (2014b). *Guidelines and manuals for the preparation of Non-Annex I national reports and international consultation and analysis,* Bonn: United Nations Framework Convention on Climate Change Secretariat. Available at: http://unfccc.int/national_reports/non-annex_i_natcom/guidelines_and_user_manual/items/2607.php. Accessed 1 May 2016.
UNGA. (1994). *Report of the Global Conference on the Sustainable Development of Small Island Developing States*. Bridgetown: United Nations. Available at: http://www.un.org/esa/dsd/dsd_aofw_sids/sids_pdfs/BPOA.pdf. Accessed 2 January 2015.
United Nations. (1992). *United Nations Framework Convention on Climate Change: Text of the Convention*. Rio de Janiero: United Nations. Available at: http://unfccc.int/files/essential_background/background_publications_htmlpdf/application/pdf/conveng.pdf. Accessed 16 March 2014.

Author Biography

Stacy-ann Robinson is a Voss Postdoctoral Research Associate in Environment and Society at Brown University in the United States. Prior to her appointment, she was the 2016/2017 Fox-Zucker Fellow at Yale University in the United States where she researched climate resilience and transformation in Caribbean small island developing states. Stacy-ann is Jamaican and holds a PhD from The Australian National University (ANU) in Canberra, Australia. This paper was written while she was a PhD Scholar in the Fenner School of Environment and Society at The ANU.

Chapter 16
Limits to Coastal Adaptation in Samoa: Insights and Experiences

Richard Crichton and Miguel Esteban

Introduction

The Independent State of Samoa is a Pacific nation located in the central Pacific, otherwise referred to as Samoa. The Samoan islands towards the east are a territory of the United States of America, officially known as American Samoa. Samoa received special assistance for national climate change adaptation projects under the Least Developed Countries (LDC) Funds to combat the negative effects of climate change.[1]

The definitions of coastal area varies between developing and developed countries or regions, and involves particular set of issues and geographic facts (Kay and Alder 2005). For the purpose of this paper, the term would adhere to the description by the Food and Agriculture Organization of the United Nations (FAO), whereas "Coastal areas are commonly defined as the interface or transition areas between land and sea, including large inland lakes. Coastal areas are diverse in function and form, dynamic and do not lend themselves well to definition by strict spatial boundaries. Unlike watersheds, there are no exact natural boundaries that unambiguously delineate coastal areas (FAO 1998)".

Coastal areas play a significant role in ensuring food security for island nations. They provide livelihoods, enable people located next to them to increase their purchasing power capacity and diversify their food sources, and reduce the potential for malnutrition. The Pacific region depends heavily on the benefits derived from

[1]Though it should be noted that in January 2014 Samoa graduated out of LDC status to join the ranks of lower middle-income countries.

R. Crichton (✉) · M. Esteban
University of Tokyo, 5-1-5 Kashiwanoha, Kashiwa, Chiba Prefecture 277-8563, Japan
e-mail: richard.crichton@s.k.u-tokyo.ac.jp

M. Esteban
e-mail: esteban.fagan@gmail.com

W. Leal Filho and J. Nalau (eds.), *Limits to Climate Change Adaptation*,
Climate Change Management, https://doi.org/10.1007/978-3-319-64599-5_16

catching or growing fish and shellfish, with an average of 47% of households in coastal communities earning their first or second sources of income from fisheries. (Johnson et al. 2012). The fisheries sector is an important source of job opportunities, and more than 12,000 people are employed in tuna canneries and processing facilities in the region, with a significant proportion of the gross domestic product (GDP) derived from licence fees and fishing activities from distant-water fishing nations (Johnson et al. 2012). Coral reefs, present along the coastal areas of many island nations, are a key component of tropical coastal ecosystems, acting as a submerged breakwater that can offer protection for the coastline, residential homes and tourism infrastructure that are common along the Samoan coast line.

Adaptation to climate change has become a national priority and lead to many policies and regulations development to meet funding criteria. Adaptation programs in Samoa are concentrated along the coastal area, given that more than 70% of the population lives within 1 km coastline, including most government assets (UNDP 2012), with various international projects being channelled towards building resilience. This has included pilot projects funded by the United Nations Development Program (UNDP) and Global Environment Facility (GEF), such as the Community-based Adaptation (CBA) program to combat the effects of climate change on local communities. Another project of the UNDP, in partnership with the Secretariat of Pacific Regional Environmental Program (SPREP) as implementing partner, established a Pacific-wide regional adaptation project called the Pacific Adaptation to Climate Change (PACC) project. PACC adaptation priorities were national driven; whereas CBA had a bottom-up approach.

However, through such programs it has become evident that there are limits to the effectiveness of coastal adaptation strategies in Samoa, something which has not been the subject of academic discussion in the past. Thus, in this present chapter the authors set out to highlight these challenges by drawing on personal observations and experience from both CBA and PACC projects. The overall purpose is to highlight past experiences and lessons learnt from adaptation and coastal zone management projects undertaken in Samoa, and to share such knowledge with the wider academic community. By doing so the authors hope to foster discussion on adaptation projects and their limitations, including how to improve current adaptation mechanisms by making use of all available technological advancements to address current and future limits to adaptation. The methodology undertaken for this review stems from a comprehensive literature review of available resources, including a critical analysis of multi-level development projects in Samoa, including regional, national and community-based projects. It is noted that a major constraint in this review are the limited number of publication and literature available within the Pacific region. This fact contributes to a serious gap in literature, and the authors hope that the present chapter will help to disseminate some knowledge regarding the subject.

The structure of this chapter will be as follows: Section "Climate Change and Its Effect on the Coastal Zones of Small Island Developing States" will discuss climate change and its effects on SIDS and especially coastal zones. Then Section "Case Study of Samoa" will include a description of Samoa and its challenges, followed by the adaptation strategies undertaken in Section "Adaptation Strategies". In Section "Discussion on Future Adaptation Strategies", the authors will discuss the implications of the issues outlined and lastly present some conclusions.

Climate Change and Its Effect on the Coastal Zones of Small Island Developing States

Small Islands Developing States (SIDS) contribute the least to global carbon and greenhouse gas emissions, but are at the frontline of impacts brought by anthropogenic climate change. These impacts vary across each island depending on their geographical area, location, island type and available resources (CSIRO et al. 2014). Atoll islands states are some of the most vulnerable of islands nations. Often being situated only a few metres above sea level, and threats to them include salt water intrusion and inundation (that can cause the failure of crops due to increases in salinity), coastal erosion, natural disasters or sea level rise (Yamamoto and Esteban 2014). Larger islands may have a mountainous area and tend to have larger populations, though they can face major issues regarding food insecurity and coastal management (Morrell 2010).

In all of these islands coastal and marine systems offer both provisioning and regulating services—provisioning services largely through habitats, and regulation services by acting as ecological buffers from extreme weather events. Ecosystems fulfil important functions that underpin food security in a number of ways. Marine and coastal ecosystems provide provisioning services, meaning providing a supply of food that is sufficient to meet nutritional and dietary requirements. Reefs or coastal areas provide a first point of contact to disperse energy from extreme events which is a regulating service (Daly et al. 2010).

The PACC project, which involved 14 countries in the Pacific, showed the diversity of the challenges facing each Pacific island nation. As explained earlier, the PACC was a multimillion dollar project funded from the United Nation Development Program to address the Pacific nations' primary vulnerability to climate change. Many of the participating countries—including the Cook Islands, Federal States of Micronesia, Samoa and Vanuatu—identified Coastal Zone Management as a key adaptation priority. The majority of the larger countries, such as Fiji, Papua New Guinea and Solomon Islands, with the exception of Palau, desired projects on food production and food security. The remaining countries, mostly smaller islands which included the Marshall Islands, Nauru, Niue, Tokelau, Tonga and Tuvalu, opted to utilize their national allocations to address the management of water resources (Buncle 2013). The diverse priorities of national projects amongst the variety of participating island nations is a clear indication that each country is impacted differently by anthropogenic climate change. The present chapter will not be covering the entire range of expected impacts, but instead will discuss the impacts that climate change is expected to have on coastal zones, using Samoa as a case study.

The Samoan islands are amongst the larger volcanic islands. Global climate modelling for Samoa suggests that annual mean temperature increases, extreme rain events, ocean acidification, coral bleaching and sea level are all projected to increase (CSIRO et al. 2014). There is little consensus on whether La Niña and El Niño events, which could lead to a higher exposure to tropical cyclones, amongst other effects, are set to increase and intensify for Samoa (CSIRO et al. 2014). It is likely that such impacts, together with increasing ocean temperatures and acidification, will result in increased coral mortality and affect fisheries resources, impacting food security, livelihoods, and dietary intake (Yamamoto and Esteban 2014). However, it is clear that mutual benefits can be derived from the relationship between humans and a healthy marine environment. Wells describes this benefit and highlights the buffering capacity of coastal areas to reduce the intensity of climate-related events such as storm surges before reaching human settlements, even though the ecosystems themselves are usually damaged in such events (Wells et al. 2006). Hence, these ecosystems themselves are vulnerable to disasters, and their buffering capacity is dependent on a balance of both their resilience and vulnerability (Wells et al. 2006).

In recent times, the frequency with which these natural disasters are taking place has resulted in a constant reliance on foreign aid assistance. Category 4 Tropical Cyclone (TC) Evan affected Samoa, Tonga and Fiji in 2012, and since then a further three high category tropical cyclones have occurred in the central Pacific within a two year period. These include Tropical Cyclone Ian, a category 4 that devastated the Kingdom of Tonga and Fiji in 2014. and TC Pam, a category 5 that destroyed much of the southern Vanuatu islands as well as other surrounding island countries in early 2015. More recently, in February 2016 TC Winston—the strongest cyclone to hit Fiji to date, devastated many of the eastern and northern islands in the country. Following such large scale disasters a shift in reliance towards government and international assistance is not uncommon. These natural disasters have often resulted in the government adopting new building codes, though this has imposed the additional costs to rebuild on the individual families affected.

Climate Change, IPCC and the Paris Agreement

The Earth experiences natural cycles of warming and cooling, though there is now abundant proof that the atmosphere and oceans are gradually warming due to human activities, as last summarised in the Intergovernmental Panel on Climate Change 5th Assessment Report (IPCC 5AR). This process by which the global environment is being changed due to human activities is often referred to as anthropogenic climate change, with the warming being caused by increased amount of greenhouse gasses which are trapping the sun's energy around the planet. This process is happening so rapidly that it is feared that most ecosystems are in serious danger of collapsing, as they simply cannot adapt quickly enough (such as for example in the case of atoll islands and coral reef systems, see Yamamoto and Esteban 2014).

Anthropogenic climate change has been the subject of international discussions for the last few decades, eventually resulting in the Paris agreement in 2015, essentially the latest achievement in global agreements on the subject. The first significant milestone was achieved when the Montreal protocol entered into force in 1987, essentially a call for the international community to prevent chlorofluorocarbon (CFC) gas emissions, initially agreed upon by 29 industrialized countries and the European Commission. Many significant events occurred between the Montreal protocol and the Paris Agreement, which do not warrant further discussion here, but generally progress was gradually amplified, with the establishment of the Intergovernmental Panel on Climate Change (IPCC) in 1988. The IPCC issued its First Assessment Report in 1990, and have released a further four updates after that, clearly highlighting the scientific community's view that green-house gasses (GHG) are significantly responsible for the warming the Earth is experiencing.

In addition to this the creation of the Small Island Developing States (SIDS) group has played a key part in climate change negotiations. The creation of the SIDS in 1998 was a key attainment by the Alliance of Small Islands States (AOSIS), together with the United Nations Environmental Program (UNDP). SIDS empowered the voices of the island nations and according to UNFCCC contains 51 member countries and territories as of 2011. AOSIS and members of SIDS have become a strong force to support their political agenda at the international arena. Recently in the Paris Conference they pushed for a reduction from 2 to 1.5 °C of the cap target for the allowed atmospheric temperature increase, with many nations such as Kiribati and Tuvalu being extremely vocal on this point (Clarke 2015). This motion was welcome and received great support from many world leaders, even though current combined national pledges and efforts fall short of staying within this limit (UNFCCC 2015).

Rising Sea Surface Temperatures and Rising Seas

Warmer atmospheric temperatures will directly lead to increase in sea surface temperatures, causing sea level rise. Essentially, warmer seawater expands and increases the volume of oceans globally, which is also known as thermal expansion (or steric) sea level rise. The warming of atmosphere also causes polar ice and glaciers to melt, which further contributes to global sea level rise (IPCC 5AR). These conditions stress corals and their ability to adapt to a drastic changing environment, and may result in mass extinction of coral species (Yamamoto and Esteban 2014). These changes are already being felt in Samoa, and in a documentary produced by the community members involved in the CBA project local residents reported how they are seeing changes to the coastal area due to sea level rise (Community-based Adaptation 2009). Scientific evidence corroborates this, and Samoa's atmospheric temperature has increased at a rate of 0.22 °C per decade in the vicinity of Apia (CSIRO et al. 2011), and satellite data also indicates that sea level has risen by about 4 mm since 1993 (CSIRO et al. 2011) and further rise is expected to be in the range of approximately 7–17 cm by 2030 and 40–87 cm by 2090 (CSIRO et al. 2014). Temperature increase is projected to be between a minimum of 0.3 and 4.0 °C in the year 2090 (CSIRO et al. 2014), which will clearly have a large impact on ocean and coastal areas. According to Morrell's Pacific Food Security Toolkit (2010) increases in ocean surface temperature can reduce vertical mixing of nutrient-rich cooler waters with upper levels of the ocean, leading to disturbance in food chains and migratory patterns of fish.

Effects of Climate Change on Coral Reefs

A temperature rise of 1–2 °C above normal maximum for a few weeks can lead to devastating coral bleaching events, resulting in mass coral mortality if such stress conditions continue for longer period, leaving only behind the coral skeleton (Wilkinson and Brodie 2011). Essentially, when stressed corals expel the symbiotic algae on which they depend for survival, and thus the long-term survival of various coral species appears to depend on their ability to survive the bleaching events and cope with climate variability (Hoegh-Guldberg et al. 2011). Episodes such as El Niño and La Niña, which induce extremes in sea surface temperatures in the Pacific region, appear to be related to mass bleaching events, and further global warming could affect and exacerbate these natural oscillations of the weather system (IPCC 5AR, Belva 2015). The cyclical oscillation associate of sea temperatures in the Pacific is referred to as the El Niño-Southern Oscillation (ENSO), which is responsible for the major source of inter-annual climate variability in the Pacific region (Climate Prediction Centre 2005; Morrell 2010). Impacts include the production and distribution of phytoplankton and zooplankton in the tropics and subtropics water. This can have an impact on the various varieties of tuna, and could for example shift the areas of the ocean where skipjack tuna is present, which follow warmer waters to the central Pacific (Morrell 2010).

In 2015, Samoa, together with many other islands in the Pacific, experienced massive coral bleaching due to a strong El Niño event. The National Oceanic and Atmospheric Administration (NOAA) declared 2015 the third major bleaching event following 2010 and 1998, which was also influenced by a strong El Niño and an equally strong La Niña (Belva 2015). Coral bleaching occurred and is likely to have been severe (see Fig. 16.1), as anecdotal evidence suggests that as much as 70% of corals died in some shallow-water areas during the 1998 event (SPREP 1999). Coral bleaching events significantly alter the health of the coral reefs as a whole, affecting fisheries and potentially leading to food insecurity.

Ocean acidification also negatively affects coral polyps. As the ocean absorbs the extra carbon dioxide in the atmosphere that has been brought about by human

Fig. 16.1 Partly bleached coral in Samoan waters (*photo* taken in October 2015)

activities, it effectively acts as a carbon sink. However, this increases its overall acidity, which impacts coral reefs by decreasing the rate at which corals form their calcium carbonate skeletons and retards the capacity of coral reefs to keep up with sea level rise (Kleypas et al. 1999; Kleypas and Gattuso 2010). Essentially, acidic water has less aragonite, a necessary compound corals need to build their skeletons (Burke et al. 2011).

Case Study of Samoa

Vulnerability of Samoa to Climate Change

Samoa became the first independent Pacific nation in 1962. It consists of 10 islands, of which four are inhabited, with a total population of 187,820 (see Fig. 16.2). Its land area is 2830 km^2, with the highest elevation point at 1859 ms, an average annual rainfall of 2880 mm, average humidity of 80% and an annual temperature range between 22 and 30 °C (Samoa Statistics Bureau 2011). It has a steadily growing population, projected to surpass 200,000 by around 2020, according to Samoa Statistics Bureau (2011) statistics. This continuous growth will add to the ever increasing pressure on natural resources and the environment and require more effective management and planning on resources.

Samoa is a country that it highly vulnerable to natural disasters (Daly et al. 2010), which affect fishing, subsistence agriculture and residential areas near the coast and low-lying areas. Further vulnerabilities are brought about by isolation, including the distribution of disaster warning information and supplies following a

Fig. 16.2 Map of Independent State of Samoa (Map data © 2015 Google)

disaster, with climate change potentially further increasing the vulnerability of isolated settlements. The vulnerability of Samoa to these natural disasters and climate change impact are highlighted in the Vulnerability Profile of Samoa (UNCTAD 2006). However, it is worth noting that current trends in natural disasters are not necessarily easily to perceive. Some contradictions regarding vulnerability exist between UNCTAD report (2006) and CSIRO report (2014)- produced by the Samoa and Australian Bureau of Meteorology- with the CSIRO report actually suggesting that a decline in drought events could take place in the future, as opposed to the UNCTAD (2006) report.

Disaster events can have very large impacts on the small island nation. Events such as the 2009 earthquake and tsunami caused a national panic and perpetuated the idea that communities should start relocating inland (Mikami and Shibayama 2015). The tsunami affected areas in the southeast parts of the island of Upolu, and in this area there has been some success in relocating some settlements, with the assistance from governments, and humanitarian aid towards rebuilding homes and establishment of communities further inland for survivors (see Fig. 16.3).

However, in other parts of the country that were not affected the need to relocate inland did not receive the same financial support. Therefore, any relocation in such areas was carried out on a voluntary basis, with financing depending on the individual capacity of each family to do so. In most cases, remittances from family members living aboard were the main source for financing such housing activities (Connell 2015). At this point, it is worth noting that in general the Samoan economy is heavily subsidised by remittances (Daly et al. 2010). Nevertheless, some projects were available to assist other communities with relocation, such as how the CBA project funded a request from Lelepa village on the northern side of Savaii Island for an inland road that would assist the community in relocating inland, encouraging members to move following the 2009 tsunami. The secondary phase of the relocation was voluntary and self-funded by the families themselves.

Fig. 16.3 Relocated church in southern Upolu island following the 2009 tsunami. *Photo credit* Miguel Esteban, 2015

Otherwise, coastal erosion is becoming a common problem for many of Samoa's coastal communities, leading them to request the national government to provide assistance for the construction of coastal protection in the form of revetments (usually constructed using large rock boulders). To address coastal vulnerabilities, the government developed district Coastal Infrastructure Management (CIM) Plans in 2005 to address community needs under the National Adaptation Program of Action (NAPA), funded under the LDC Adaptation funds. Coastal erosion is exacerbated by natural disasters and the development of CIM plans were supposed to address some of these challenges, including support for building some large boulder-stone seawall defences. However, according to the Pacific Food Security Toolkit 2010, the seawalls proposed are not an environmentally sustainable solution to protect sea-level rise and inundation. Constructions of seawall are usually costly and ineffective in the long-term, as well as harmful to the environment and likely to transfer the erosion or flooding problems to a neighbouring area (see for example Takagi et al. 2015, coastal seawalls will be discussed further in the adaptation section).

Problems Faced by Fisheries

Many Samoan's urban areas are showing slow but steady growth, with the highest growth of 2.2% increase in the north-west Upolu districts. Despite the capital city of

Apia showing a negative population growth of −0.5%, the national average growth rate is 0.8% (Samoa Bureau of Statistics 2011). Population growth increases pressure on natural resources, such as fisheries for consumption, adding to the problems being brought about by climate change. Members of different communities around Samoa have identified a change in the quantity of the fish they are catching and the time required to catch fish. This is supported by the SBEC fisheries report (2010), which suggests overfishing due to poor education and unsustainable fishing practices.

Traditional fishing techniques are usually sustainable and only take what is needful for a day, typically sharing excess catch among community members or exchanging products; for example, fish trade with farmers. These traditional practices have become increasingly rare in recent times and have been replaced with commercial monetary exchanges, with foreign modern fishing practices gaining popularity. Traditional fishing practice has a clear division of roles according to gender, and many such aspects can still be seen at present: women typically fish within the vicinity of the lagoon, the area between the beach and the reef. Their target catches are usually small reef fish, shellfish and molluscs, while men mostly fish near the end of the reef and in open waters.

Essentially, a healthy reef was traditionally able to support traditional fishing, though the reef is increasingly degraded and unable to provide such ecosystem services. For example, in the case of the village of Ululaelae fishermen have indicated how the reef area has degraded over time, which may explain the number of fish caught. Also, the white sandy beach of Utulaelae, which decades ago was a popular tourist destination, has eroded leaving only a rocky surface (Puipuiga o le Siosiomaga o Utulaelae 2011). Although a direct links between climate change and reef degradation in Utulaelae is yet to be established, current scientific knowledge on the sensitivity of coral reefs to a changing climate might support such a hypothesis (Guinotte et al. 2003). However, it is still not clear whether overfishing, unsustainable fishing practices (as highlighted in the SBEC report), tourism activities, natural disaster or a combination of such activities may be responsible for the current situation in Utulaelae.

As a consequence of these problems, many people are nowadays relying on imported goods, such as tin fish or imported chicken, that can be easily bought from small family stores. This further contributes to the problem of the traditional fishing sector, as fishermen have limited access to valuable markets, resulting in catches being sold at the government-operated fish markets (which may be some distance away for fishermen to travel), through street vending, or nearby tourism operations such as resorts and hotels (SBEC 2010). Furthermore, the establishment of tuna canneries factories in American Samoa has resulted in exports from Samoa to the USA territory, given the short distance between them, which ultimately become export commodities. As a result there has been a noticeable increase in the number of fishing fleets and domestic fishing business establishments, replacing traditional canoes and small scale community fishing. SBEC originally supported many of these establishments and fishing businesses through small loan schemes, though this development has not been without commercial problems. The federal minimal salary increase mandated by the US government on American Samoa caused the shutdown of two of the three tuna canneries, which had a huge impact on tuna exports from the islands of Samoa. Thus, it is clear that commercial fishing can be subjected to important external shocks that can cause livelihood instability for small scale fishing businesses.

Adaptation Strategies

Examples of Adaptations Strategies Attempted so Far

A number of official reports, such as the IPCC 5AR, outline a series of adaptation measures that might be attempted by coastal communities throughout the planet to adapt against climate change. However, it is worth noting that a number of such strategies have already started to be implemented in Samoa to solve some of the problems that are expected to become worse as a consequence of climate change. In order to understand the potential limitations of adaptation strategies it is worth to briefly review some of the projects and initiatives that have been carried out so far.

In 1990s, tropical cyclone Ofa and Val caused large damage to the economy of the country. These two natural disasters called for better coastal protection, as the coastal walls at the time were deemed insufficient. Seawalls were erected to minimize damage to large building and infrastructure in urban areas, and it did not take

Fig. 16.4 Coastal seawall of Poutasi village on south coast of Upolu island. *Photo credit* Miguel Exteban, 2015

long before this form of coastal defence became popular and desirable by many coastal communities (see Fig. 16.4).

Fish and fishing are important culturally and economically for Samoa. Fish can be obtained from two sources: the traditional ocean seafood catches and new initiatives for fresh fish production of tilapia fish. With little national capacity to enforce national laws and coastal marine resources being mainly managed by the community village council, the government relies on local governing systems to enforce sustainable practices and other coastal marine activities. In this sense, local by-laws have provided protection and regulation of Marine Protected Areas (MPAs) to sustainably use fisheries resources. It is projected that fish supply will be insufficient to meet the population demands by 2030, as Samoa has one of the highest fish consumption per capita in the region according to Bell et al (2009). Aquaculture as an alternative adaptation measure to meet fish consumption needs is one of the fastest-growing primary industry sectors in the country, tripling overall production in the last 20 years, and currently producing about 78 million tonnes of fish (FAO 2016).

Limit of Current Adaptation Strategies

There are limits to the effectiveness of the adaptation strategies outlined in the previous section. For example, the village of Vaiala, which is located in the vicinity of the city of Apia, erected a coastal wall and also conducted tree replanting to minimize coastal erosion. The wall suffered much damage during its 30 year lifespan (see Fig. 16.5), until eventually high swells and the impact of Tropical Cyclone Evan in 2012 and high seas and pounding waves in January 2015 destroyed the already decayed wall, eroding much of the soil and threatening the main road (Keresoma 2015). A new replacement wall was erected with funding and support from both the United Nations Development Program and the Government of Samoa, which was completed by end of 2015 (see Fig. 16.6). The new wall, which is larger than the previous, suggests that coastal defences will require constant upgrades and enlargement to deal with future impacts (as explained for example in Yamamoto and Esteban 2014). Newer defences, such as these hard engineering solutions, become progressively more expensive and at some point the cost of adaptation would exceed the benefits received from them, making them unsustainable. This would present a clear economic limit to this form of adaptation.

However, it is worth noting how even at present the construction of coastal revetments has not proceeded without problems. For example, in the village of Saleia in Savaii Island, a coastal protection wall was constructed to protect homes and the main road from waves. The construction of a raised stone seawall offered protection from the ocean, but limited direct outlet for a river which runs parallel to the coastal seawall. Flooding then became problematic for local residents, especially in rainy season. Eventually, Saleia village became one of the benefactors of CBA, who constructed a riverbank wall to minimize the direct overflow of the river

Fig. 16.5 Damaged coastal seawall in Vaiala. *Photo credit* Tomoya Shibayama

Fig. 16.6 Vaiala seawall was re-constructed in 2015, replacing the old Vaiala seawall (see Fig. 16.5). *Photo credit* Motisha Aiono-Solo

into residential area. Basically, in this case an adaptation measure required additional projects to rectify the problems that were created due to poor design.

In other areas, tourism operators are finding a decrease in numbers of visitors as the beaches are replaces with seawalls, which are not attractive and limit beach access. For example, in the village of Lalomalava a prominent seaside hotel resort received a coastal protection seawall as part of the national project under the PACC program, even though tourists do not necessarily find it attractive. Furthermore, the wall was not able to cover the entire village, and as a result many homes beyond the extent of the wall were constantly under threat from inundation and erosion caused by strong waves. At the same time, the tourism operators are facing the challenge of attracting visitors to their establishment as sandy beaches are replaced with rock walls, affecting the livelihoods of those involved in tourism businesses. Such problems highlight the clear limitations of adaptation strategies. Essentially, money is required to build sea defences and other such adaptation strategies, yet these can lead to a decrease in the natural beauty of the area (limiting tourism income) or damage the local environment (affecting local livelihoods). Eventually, a limit point will be reached where the country will be unable to pay for the ever-increasing cost of protecting coastal areas, given the loss of revenue from them. Again, this represents an economic limit to adaptation.

As an alternative to hard structures, soft forms of protection can be attempted along a coastline, by using vegetation for example. Replanting of trees for coastal protection can be done using native coastal trees such as coconuts, tropical almonds and other salt-tolerant species. However, coastal replanting sites are often damaged by free-range animals such as pigs and chicken, or by big swells (see Fig. 16.7). Failure rates in coastal replanting can be quite high, often limiting the value of such strategies as an adaptive solution, despite their cheapness. In addition, rising sea level brings about the question of how much salt-water concentration these species

Fig. 16.7 Coastal erosion and vegetation damages due to ocean swells. *Photo credit* Richard Crichton, 2014

are able to tolerate, which may affect the chances of success as an effective adaptation option. This effectively presents an ecological and physical limitation.

Other types of adaptation projects, such as the establishment of Marine Protected Areas (MPA), are common in village communities, and while there has been some success stories, others have failed to achieve their purpose. Successful MPA usually includes some form of re-introduction of marine species. Savaia village MPA on the island of Upolu, which received national attention, is a successful MPA because of the reintroduction of marine species, particularly giant clams brought in from neighbouring country of Tonga. Meanwhile, another similar project in Faga district, which did not introduce any new species, saw little to no improvement in coral reef and fisheries, which were unable to return in numbers or in a healthy state.

Other projects attempted to replant coral species off the island of Manono, though most did not survive. While warm ocean temperatures, acidification and even human activities such as pollution could be factors to consider, the linkage between these issues and the outcome of the projects is yet to be scientifically established. However, it is clear that in the future there will be limitation to adaptation due to the complete lack of ability of these communities to address the immediate root causes of rising ocean temperatures and acidification (Veron et al. 2009). Even at the national and regional level reducing gas emissions (a mitigation strategy) is unlikely to have a significant effect, as the Pacific region accounts for only 0.03% of total emissions. All of these represent ecological limitations.

In addition, regarding small island states such as Samoa, the challenge to successfully implement adaptation strategies is not only related to cost, but also the availability of scientific research to support best practice as well as the human capacity and technical know-how. Much research is available worldwide, but often that information is not shared or is not specific to small island needs. There is also a need for research to be localized and be specific to local circumstances. The PACC project report highlights this same challenge of having the right capacity to address such issues. Finally, it is clear that further research is needed in general, for example to address the case of what is happening to the coral reef in Manono island, as stated above. Basically, financial capacity and knowledge capacity constitute limits to adaptation.

Discussion on Future Adaptation Strategies

Adaptation is one way for humanity to face changing climate change conditions, and is crucial for small island nations such as Samoa. It is important that any solutions proposed come with proper assessments and consider current systems that may impact or alter the results of these adaptation approaches. Learning from past mistakes and recognizing the limitations of current adaptation is important when moving forward and addressing the next stage of development of adaptation strategies.

The adaptation activity space, as prescribed by Pelling et al. (2014), should include: Institutions (regulatory and cultural); Individuals (values and identity) Environment (biotic and abiotic); Behaviour (practices and routines); Discourse (popular and policy); Livelihoods (production and labour processes); and Technology (material and organizational). Transformational adaptation, as suggested in the IPCC 5AR report, as a form to address the limits to adaptation is necessary because of the significant vulnerability of these small islands to climate change and the severe impacts that can be expected as a result, which could overwhelm the capacity of human–environment systems to cope.

Transformational adaptation is important to address adaptation needs for countries like Samoa where the significant influence of culture, livelihoods and popular behaviour need to be incorporated into adaptation strategies. It requires a careful examination of the interactions between different activity spheres, such as politics and power, knowledge and science to be applied to adaptation. Research (scientific knowledge) to address local climate change would provide knowledge for the best adaptation strategies to be bridged with cultural (including local decision-making institutions) and political entities.

Current adaptation mechanisms are not always effective, and in some cases have caused an additional need for new adaptation projects that address the shortcoming or problems created by poorly formulated projects. Low cost and soft solutions to adaptation measures can sometimes be less effective than hard solutions, though these can often prove too costly and unattainable for the economies of small island nations.

Conclusion

It is clear that anthropogenic climate change will inevitably bring about impacts and stresses on the environment, from which Samoa cannot escape. To overcome these, it is imperative that research and innovation carried out in foreign countries should be accessible to Small Islands States. These technologies should be attainable and low-cost to meet the needs of small island nations with small economies. Furthermore, all adaptation should be localized, to consider the various dynamics in transformational adaption of the communities concerned. When it comes to foreign assistance and funding it is clear that there needs to be further clarity on the sources and predictability of this support, and that procedures for direct access must be simplified. Finance must be scaled up dramatically to meet the adaptation needs of these small islands, particularly given the ecological, physical, economic and technological limits to adaptation summarised in this chapter. While adaptation is essential for Small Islands States to deal with current environmental pressures, further commitment to climate change mitigation is essential in the long term. Nevertheless, it is important to recognise that there will probably eventually be a limit to adaptation, and that planned adaptation will require carefully designed strategies and a need to manage the consequences of exceeding the adaptation limits.

References

Australian Bureau of Meteorology, & Commonwealth Scientific and Industrial Research Organisation (CSIRO). (2011). *International Climate Change Adaptation Initiative: Pacific Climate Change Science Program.*

Australian Bureau of Meteorology, & Commonwealth Scientific and Industrial Research Organisation (CSIRO). (2014). *Climate variability, extremes and change in the western tropical pacific: New science and updated country reports. Pacific-Australia Climate Change Science and Adaptation Planning Program.* Melbourne, Australia: Australian Bureau of Meteorology and Commonwealth.

Bell, J. D., Kronen, M., Vunisea, A., Nash, W., Keeble, G., Denmke, A., et al. (2009). Planning the use of fish for food security in the Pacific. *Marine Policy, 22,* 64–69. doi:10.1016/j.marpol.2008.04.002.

Belva, K. (2015, October 8). *NOAA declares third ever global coral bleaching event.* Retrieved from http://www.noaanews.noaa.gov/stories2015/100815-noaa-declares-third-ever-global-coral-bleaching-event.html.

Buncle, A. (2013). *Informing climate-resilient development: The application of cost–benefit analysis (CBA) in the PACC program* (Technical Report No. 2). Retrieved from Secretariat of the Pacific Regional Environmental Program Website. https://www.sprep.org/pacc/publications/technical-reports.

Burke, L., Reytar, K., Spalding, M., & Perry, A. (2011). Reefs at risk revisited. Retrieved from World Resources Institute Website: http://www.wri.org/.

Clarke, M. (2015, December 02). President Obama meets with Pacific Island Leaders. *ABC News.* Retrieved from http://www.abc.net.au/news/2015-12-02/president-obama-meets-with-pacific-island-leaders/6992666.

Climate Prediction Centre. (2005). *Frequently asked questions about El Nino and La Nina.* National Centers for Environmental Prediction. Accessed February 28, 2011. http://www.cpc.ncep.noaa.gov/products/analysis_monitoring/ensostuff/ensofaq.shtml#DIFFER.

Community-based Adapatation (Producer) & (Director). (2009). Tofiga o Pili Alii Aau [Documentary]. Samoa: United Nations Development Program.

Connell, J. (2015). Samoa and Tonga: Migration and remittances in the twenty-first century. In W. H. Khonje (Ed.), *Migration and development: Perspectives from small states* (pp. 122–183). London, UK: Commonwealth Secretariat.

CSIRO. (see Australian Bureau of Meterorology).

Daly, M., Poutasi, N., Nelson, F., & Kohlhase, J. (2010). Reducing the climate vulnerability of coastal communities in Samoa. *Journal of International Development, 22,* 265–281. doi:10.1002/jid.1678.

FAO. (1998). Part A: Issues, perspectives, policy and planning processes for integrated coastal area managment. In N. Beus (Ed.), *Integrated coastal area management and agriculture, forestry and fisheries* (pp. 3–84). Rome: Natural Resource Management and Environment.

FAO. (2016, February 22). *FAO puts sustainability on the menu in the world's fastest-growing food sector: Fish.* Retrieved from http://www.fao.org/news/story/en/item/384011/icode/.

Guinotte, J. M., Buddemeier, R. W., & Kleypas, J. A. (2003). Future coral reef habitat marginality: Temporal and spatial effects of climate change in the Pacific basin. *Coral Reef, 22*(4), 551–558.

Hoegh-Guldberg, O., Ortiz, J. C., & Dove, S. (2011). The future of coral reef. *Science, 334*(6062), 1494–1495. doi:10.1126/science.334.6062.1494-b.

IPCC AR5. (2014). Intergovernmental panel on climate change Fifth assessment report. Retrieved from https://www.ipcc.ch/report/ar5/.

Johnson, J., Bell, J., & De Yong, C. (2012). *Priority adapatation to climate change for Pacific fisheries and aquaculture*. Rome, Italy: Food and Agriculture Organization.

Kay, R., & Alder, J. (2005). *Coastal planning and management* (2nd ed.). New York, NY: Taylor & Francis.

Keresoma, L. (2015, January 28). Samoa alerted of strong winds and heavy rain. *Talamua*. Retrieved from http://www.talamua.com/weather-samoa-alerted-of-strong-winds-and-heavy-rain/.

Kleypas, J., Buddemeier, R., Archer, D., Gattuso, J., Landon, C., & Opdyke, B. (1999). Geochemical consequences of increased carbon dioxide in coral reefs. *Science, 284*(5411), 118–120. doi:10.1126/science.284.5411.118.

Kleypas, J., & Gattuso, J. P. (2010). Coral reef. In C. J. Cleveland (Eds.), *Encyclopedia of earth*. Washington, DC: Environmental Information Coalition, National Council for Science and the Environment (First published in the Encyclopedia of Earth October 18, 2006; Last revised March 28, 2010) Retrieved from http://eoearth.org/article/Coral_reef.

Mikami, T., & Shibayama, T. (2015). Tsunami disasters in remote islands: 2009 Samoan and 2010 Mentawai Islands tsunamis. In M. Esteban, H. Takagi, & T. Shibayama (Eds.), *Handbook of coastal disaster mitigation for engineers and planners* (pp. 75–91). Oxford: UK Butterworth-Heinemann (Elsevier).

Morrell, W. J. (2010). *Building resilience to climate change: Pacific food security toolkit*. Rome: Food and Agriculture Organization.

Pelling, M., O'Brien, K., & Matyas, D. (2014). Adaptation and transformation. In J. Birkmann & R. Mechler (Eds.), *Advancing climate change adaptation and risk management* (pp. 113–127). London: Springer.

Puipuiga o le Siosiomaga o Ululaelae. (2011, August). CBA community project assessment. Personal interviews, Crichton, R.

Samoa Bureau of Statistics. (2011). *Population and housing census 2011: Analytical report*. Apia: Government of Samoa.

SBEC. (2010). *Fisheries sector profile*. Retrieved from Small Business Enterprise Center of Samoa Webiste: http://www.sbecsamoa.ws/publications.html.

SPREP. (1999). Samoa. In C. Hunter (Ed.), *Coastal management profiles* (pp. 157–166). Apia, Samoa: SPREP.

Takagi, H., Thao, N. D., Esteban, M., Mikami, T., & Van Cong, L. (2015). Coastal disasters in Vietnam. In M. Esteban, H. Takagi, & T. Shibayama (Eds.), *Handbook of coastal disaster mitigation for engineers and planners* (pp. 235–255). Oxford: Butterworth-Heinemann (Elsevier).

UNCTAD. (2006). *Vulnerability profile of Samoa*. New York, NY: United Nations.

UNDP. (2012). *United Nations development program: Samoa project document*. Apia: United Nations Development Program-Environmental Finance Services.

UNFCCC. (2015). *Conference of the Parties: Twenty-first session. Synthesis report on the aggregate effect of the intended*. Paris: United Nations.

United Nations (2014, January 8). *UN Portal*. Retrived from https://www.un.org/ldcportal/samoa-graduates-from-the-ldc-category/.

Veron, J. E. N., Hoegh-Guldberg, O., Lenton, T. M., Lough, J. M., Obura, D. O., Pearce-Kelly, P., et al. (2009). The coral reef crisis: The critical importance of <350 ppm CO2. *Marine Pollution Bulletin, 58,* 1428–1436.

Wells, S., Ravilious, C., & Corcoran, E. (2006). *In the frontline: Shoreline protection and other ecosystem services from mangroves and coral reefs*. Cambridge, UK: UNEP World Conservation Monitoring Center.

Wilkinson, C., & Brodie, J. (2011). *Catchment management and coral reef consevation*. Townsville: James Cook University.

Yamamoto, L., & Esteban, M. (2014). *Atoll Island States and international law*. Heidelberg: Springer.

Chapter 17
Limits to Capital Works Adaptation in the Coastal Zones and Islands: Lessons for the Pacific

Brendan Mackey and Daniel Ware

Introduction

Adapting to a rapidly changing climate is now recognized as a global goal and an inescapable problem for all nations, communities and organisations (UNFCCC 2015). Climate change impacts global and include: impacts on natural and human systems on all continents and across the oceans; changing precipitation and altered hydrological systems, affecting fresh water quantity and quality; and shifts in wildlife species geographic ranges, seasonal activities, migration patterns, abundances, and interactions. Furthermore, negative impacts of climate change on crop yields have been more common than positive impacts. People who are socially, economically, culturally, politically, institutionally, or otherwise marginalized have been found to be especially vulnerable to climate change and also to some adaptation and mitigation responses, with climate-related hazards exacerbating other stressors, often with negative outcomes for livelihoods, especially for people living in poverty (Field 2014).

While adaptation is now a mandatory response, various intervention options are available for a given problem and the additional problem therefore arises of identifying the most appropriate action. As part of such assessment, it is important to consider the limits of different adaptation interventions. Limits to adaptation come in many shapes and sizes, and are greatly influenced by underlying cultural, social,

B. Mackey (✉) · D. Ware
Griffith Climate Change Response Program, Griffith University,
Gold Coast 4222, QLD, Australia
e-mail: b.mackey@griffith.edu.au

D. Ware
e-mail: d.ware@griffith.edu.au

D. Ware
Centre for Coastal Management, Griffith University,
Gold Coast 4222, QLD, Australia

W. Leal Filho and J. Nalau (eds.), *Limits to Climate Change Adaptation*,
Climate Change Management, https://doi.org/10.1007/978-3-319-64599-5_17

economic, historical, and environmental contexts and processes (Adger et al. 2009; Dow et al. 2013; Klein et al. 2014). Limits can be considered in terms of their efficacy in the face of current and projected climate change impacts. The different modes of adaptation—engineered, ecosystem-based, planning, behavioural change, among others—are implemented with the aim of reducing climate-related risks by addressing the exposure, susceptibility or vulnerability of a community to those risks; assuming community well-being is the goal.

The efficacy of a given adaptation intervention must also be evaluated in the context of ongoing climate change impacts. Unfortunately, even if the world community were to immediately cease use of fossil fuel and end deforestation and degradation, the climate system will continue to be disrupted for millennia. This is the result of both the long lifetime of the atmospheric fraction of a CO_2 pulse (Archer and Brovkin 2008) and the lags in Earth system responses, especially from the capacity of the oceans to store and globally re-distribute heat energy (Hansen et al. 2013). The fact is that even with 100% greenhouse gas mitigation, Earth's climate system is not shifting anytime soon to a new, stable climatic norm. Rather, we are facing ongoing climatic disruption with vast regional differences in the direction and timing of change. It follows that an adaptation intervention, which is appropriate in the short term may become a maladaptation within a few decades. The international community of adaptation scholars and practitioners are only beginning to reflect on the implications of these insights into Earth system dynamics; understanding that has been greatly facilitated by the clarity provided on the issue in the IPCC AR5 Working Group I report (IPCC 2013).

Limits to adaptation can also be considered in terms of the potential perverse outcomes an adaptation intervention generates especially when they result in costs outweighing benefits (Juhola et al. 2016; Noble et al. 2014). Perverse outcomes are (usually but not always) unplanned or unintended negative impacts from an adaptation intervention on assets and social values. In general, maladaptation can be defined as "actions, or inaction that may lead to increased risk of adverse climate-related outcomes, increased vulnerability to climate change, or diminished welfare, now or in the future" (Noble et al. 2014, p. 857). Maladaptive outcomes adversely impact on vulnerability and erode sustainable development (Juhola et al. 2016). Examples of such maladaptive outcomes include adaptation projects that result in ecosystem degradation and biodiversity loss, transfer (i.e., leakage) of climate risk impacts to other locations or sectors, and climate injustices where the burden of costs or harm fall unfairly on those least responsible and most vulnerable. Adaptation interventions, which can be anticipated to generate perverse outcomes therefore face limitations in terms of their social, political environmental or economic desirability and feasibility, when their total costs and benefits are tallied. A further limit is the current limited understanding of climate change impacts, the economic and social costs accruing from impacts and from adaptation options in the small island developing states and territories (Nurse et al. 2014).

In this chapter we focus on the limitations to adaptation of capital works in the coastal zone for Pacific region island states primarily in terms of the potential for perverse outcomes in the areas of coastal processes, ecosystem services, and sustainable livelihoods. We discuss the reasons for the prevalence of capital works as a preferred mode of adaptation. We review case studies of coastal capital works and consider lessons for climate change adaptation in the Pacific coastal zone. We conclude by considering the alternatives provided by working with natural processes and soft-engineered approaches.

Adaptation Challenges in the Coastal Zones and Islands

The universal array of climate risks are all apparent in the Pacific including altered rainfall regimes bringing droughts and floods, increases in the intensity and frequency of tropical cyclones, erosion, along with rising temperatures and heatwaves that will increase coral bleaching and reef degradation, increase in invasive species and human health impacts (Nurse et al. 2014). However, the threats arising from sea-level rise and coastal inundation are given the most media coverage and are at the forefront of public concerns, which is not surprising given that three of the five lowest countries are located in the Pacific (Marshall Islands, Tokelau and Tuvalua; Table 17.1). As expressed in the Paris Agreement (UNFCCC 2015), Small Island Nation States (SIDS) have used the existential threat of rising sea levels to argue persuasively in climate change negotiations for the international community to agree to the goal of holding the increase in the global average temperature to well below 2 °C above pre-industrial levels and pursue efforts to limit the temperature increase to 1.5 °C above pre-industrial levels, recognizing that this would significantly reduce the risks and impacts of climate change (UNFCCC 2015).

Climate change impacts bring both new challenges as well as exacerbating current problems: projected climate change impacts in the Pacific coastal zone are additional to prevailing pressures arising from the expansion of coastal populations, economic activity and settlement growth pressures on coastal and nearshore ecosystems, such as mangroves, marsh, coral reefs, oyster reefs, seagrass beds and barrier islands (Barbier 2015). These pressures are exacerbated in the Pacific by challenges of weak governance and limited capacity in some countries. Ecosystem degradation and biodiversity loss is a major issue in the region with key threatening processes including habitat loss and degradation, invasive species, overexploitation, pollution, and disease (Kingsford et al. 2009). These existing pressures on the sustainability of the coastal zone are well recognized and significant programmes of work are being implemented to address these by governments, communities, NGOs and regional organisations in the Pacific, especially the Secretariat of the Pacific Community (SPC), the Forum Fisheries Agency (FFA), and the Secretariat of the Pacific Regional Environment Program (SPREP).

While many Pacific islands are among the world's poorest nations (e.g., Vanuatu ranked 178th and Tonga 182nd by national GDP; Table 17.1), others have effectively developed country economic status through being legally some category of an OECD nation overseas territory (e.g. USA in the case of American Samoa or France in the case of New Caledonia). In all cases, a priority concern is the welfare

Table 17.1 Pacific island country statistics

Country[a]	Coastline length (km)[b]	Land area (km^2)	CL: LA	% urban	GDP per capita (USD)[d]	Pop.[c]	Max. elevation[e]	Global GDP ranking 2014[f]
American Samoa	116	199	0.6	93	12,584	57,100	966	
Cook Islands	120	237	0.5	74	19,523	14,730	652	
Federated. States of Micronesia	6112	701	8.7	22	3056	102,800		
Fiji	1129	18,333	0.1	51	3757	867,000	1324	156
French Polynesia	2525	4167	0.6	51	22,351	268,270	2241	
Guam	126	541	0.2	94	2708	184,200	406	
Kiribati	1143	811	1.4	54	1442	113,400	81	187
Marshall Islands	370	181	2.0	74	3524	54,880	10	186
Nauru	30	21	1.4	100	11,015	10,840	71	
New Caledonia	2254	18,576	0.1	67	38,100	272,700	1628	
Niue	64	259	0.2	40	12,945	1470	68	
Northern Mariana Island	1482	457	3.2	90	12,239	56,940	965	
Palau	1519	444	3.4	77	13,385	17,950	242	185
PNG	5152	462,840	0.0	13	1931	8,083,700	4509	113
Pitairn Islands	51	47	1.1		37	48		
Samoa	403	2934	0.1	20	4231	187,300	1857	176
Solomon Islands	5313	28,230	0.2	20	1643	642,000	2335	172
Tokelau	101	12	8.4			1160	5	
Tonga	419	749	0.6	24	4280	103,300	1033	182
Tuvalua	24	26	0.9	57	2864	11,010	5	188

(continued)

Table 17.1 (continued)

Country[a]	Coastline length (km)[b]	Land area (km^2)	CL: LA	% urban	GDP per capita (USD)[d]	Pop.[c]	Max. elevation[e]	Global GDP ranking 2014[f]
Vanuatu	2528	12,281	0.2	24	2864	277,500	1877	178
Wallis and Futuna	129	142	0.9		12,399	11,750	524	
Total	31,110	552,188	0.1			11,071,778		

[a]SPC members excluding Australia, France, USA and N.Z
[b]Length of coastline; *Source*: https://www.cia.gov/library/publications/the-world-factbook/fields/print_2060.html
[c]Human population 2015
[d]*Source* SPC pocket statistical summary
[e]*Source* https://en.wikipedia.org/wiki/List_of_elevation_extremes_by_country
[f]*Source* International Monetary Fund World Economic Outlook (October-2015)

of economically disadvantaged rural and urban/peri-urban communities living in the low-elevation coastal zone whose livelihoods are directly threatened by the impacts of climate change on coastal and nearshore ecosystems (Barbier 2015). Recognition of the vulnerability of human settlements and infrastructure in coastal zones, especially in lower income countries with limited resources, have led to calls for adaptation based on regulations and economic incentives to redirect new settlement to better-protected locations and promoting investments in appropriate infrastructure (McGranahan et al. 2007).

The settlement pattern of Pacific islands has distributed communities across vast geographic areas. The spatial distribution and particularly the remoteness of some communities, stretches communications, transport and government service delivery creating significant challenges for adaptation to climate change. Whilst under the current climate these remote communities are largely self-sufficient, the impact of climate change will reduce this with communities increasingly reliant on resources from other areas for food, water, shelter and recovery following extreme events such as such as tropical cyclones such as TC Pam in 2015 (Vanuatu) and TC Winston in 2016 (Fiji).

Potential for Coastal Capital Works in the Pacific

As detailed in Table 17.1, Pacific island countries have a total of 31,110 km of coastline, including countries with greater than 1000 km: Federated States of Micronesia (6112 km); French Polynesia (2525 km); Kiribati (1143 km); New Caledonia (2254); Northern Mariana Island (1482 km); Palau (1519 km); PNG (5152 km); Solomon (5313 km); and Vanuatu (2528 km). A significant proportion of these coastlines are associated with a large number of small islands as indicated by countries with a high ratio of coastline to land area such as the Federated States of Micronesia (8.7) and Palau (3.4) (Table 17.1).

The extent to which Pacific communities are exposed to climate-risks in these coastal zones depends on a myriad of factors including economic status, mode and pattern of human settlement, the kinds of built infrastructure, and the biophysical characteristics of the coastline particular in terms of geomorphology, watershed land cover condition, and sediment transportation processes. Another key factor is the presence or absence of coastal zone policies and programmes and their level of enforcement.

As noted above, Pacific island states are among the world's economically poorest (Fiji ranked 156 lowest GPD, Kiribati 187, Marshall Islands 186, Palau 185, PNG 113, Samoa 176, Solomon Islands 172, Tonga 182, Tuvalua 188, Vanuatu 178) and also the lowest lying (Kiribati highest point 81 m elevation, Marshall Islands 10 m, Tokelau 5 m, Tuvalua 5 m). It follows that due to their economic disadvantage these countries are more likely to have a higher vulnerability to climate-related risks and fewer resources to support adaptation options in terms of central issues related to people, economy and infrastructure/services (Box 17.1).

Box 17.1 Central issues in the vulnerability of human settlements in the Pacific and ket factors to promote effective adaptation (*Source* Vrolijks 1998)

Issues
People · Where people live · How their houses are built · How their economic and social situation can be affected by a disaster · If they have access to a safe place during a disaster
Economy · How can the economic system be disturbed by natural disasters · In what ways does the population depend on these economic activities · Where are the major economic production facilities located · How do they depend on infrastructure and services · What is their construction quality
Infrastructure and Services · In what way do the infrastructure and services contribute to the development of safe settlements · Where are public assembly sites (schools, hospitals, theatres, public markets etc.) located and what is their quality of construction · How is the infrastructure system (water, electricity, access, sanitation, telecommunication) laid out and what is the construction quality of this infrastructure · Where are critical emergency facilities located and are they accessible and safely built.
Key factors for promoting effective human settlement adaptation 1. *Land tenure systems*—affects availability of safe land 2. *Human settlements planning*—proactive guidance to minimise exposure 3. *Housing*—ensuring safe and affordable housing 4. *Rural settlement safety planning*—community adaptation priority actions 5. *Infrastructure development*—integrated development strategies 6. *Economic development*—stable, diverse and resilient sector strategies

As noted by Vrolijks (1998), vulnerability depends on where people live, whether they have access to safe places, how their houses are built, and what their economic conditions are, all of which depend on a range of processes and policies including town planning, squatter policies, land tenure, insurance, economic diversification and others. Successful adaptation interventions therefore must be based on consideration of a broad range of policies and development processes (Noble et al. 2014)

The proportion of rural and urban dwelling communities varies between Pacific countries. PNG, Solomon Islands and Vanuatu with 87, 80 and 76% rural respectively contrast with American Samoa (5%) and Cook Islands (26%). The proportion of coastal dwelling communities also varies considerably. Voyagers who made landfalls on tropical islands of Pacific Oceania were people with maritime skills who typically settled first at sites with sandy substrates on island coasts where boats were easy to beach. Many island inhabitants continued to live along island shorelines during subsequent centuries and millennia (Dickinson 2014). PNG's population is concentrated in the Highlands while others such as Cook Islands are faced with declining outer island populations (UNFPA 2014). For SIDS, the entire island can be properly considered as being "in" the coastal zone. In any case, a substantial proportion of people in the Pacific continue to inhabit coastal human settlements, be they traditional village communities or modern cities though the precise numbers await further research.

What we do know is that every Pacific country's capital city is a port (as is commonly the case throughout the (un-landlocked) world, and that these support the largest populations and are centres of built infrastructure for transportation, trade and communications and are hubs for all business sectors and government activities. Pacific countries are also facing an increase in urbanisation with movement of people to the major cities from outer islands and rural areas. A key issue for adaptation planning in the Pacific is the very high extent to which economic activities rely on key infrastructure systems including wharves, telecommunications, roads and pipelines associated with the coastal zone and the capital city ports in particular: the more reliant, the greater the perceived need to protect these physical assets from climate hazards (Box 17.1). The potential impacts of climate change for the coastal zone are significant globally as well as for the Pacific. Without adaptation, 0.2–4.6% of global population is estimated to be flooded annually in 2100 under 25–123 cm of global mean sea-level rise, with expected annual losses of 0.3–9.3% of global gross domestic product, leading to calls for coastal dikes with global annual investment and maintenance costs at US$ 12–71 billion in 2100 (Hinkel et al. 2014).

Many different types of adaptation options are under consideration for the coastal zone in the Pacific, including ecosystem-based adaptation (Ojea 2015; Reid 2016) and planned retreat (Low Choy et al. 2012). Ecosystem-based approaches are being actively investigated in the region through major aid investments such as PEBACC, a 5-year project funded by the German Government, implemented by the Secretariat of the Pacific Regional Environment Programme (SPREP 2016a). However, there remains strong interest in deploying capital works for adaptation, that is, building and engineering works where construction activity creates a physical asset, designed to reduce the risk associated with a climate hazard now or in the future. An array of capital works are already deployed with the aim of protecting assets from coastal erosions including: vertical or near vertical walls and bulkheads; sand trapping structures such as groynes and causeways; rubble mound structures such as revetments, breakwaters and groynes constructed of hard rock or coral boulders; gabion revetments, baskets and mattress structures; hand-placed coral rubble; stacked sandbags and Longard tubes; and concrete protection units (SOPAC 1994; Environment Agency 2015; Cooper et al. 2016). Thus while perhaps the most commonly recognized type of coastal adaptation capital works is concrete vertical seawalls, related forms of shoreline armouring (i.e., protecting built assets through the construction of hard engineered barriers) proliferate.

The design, construction and delivery cost of capital works as an adaptation strategy is significant particularly for SIDS with limited economic resources. In adopting capital works based adaptation, the role of overseas aid and development to meet the costs must be recognised. Capital works approaches to adaptation simplify the project monitoring and evaluation for OAD funders by creating a tangible asset whose presence can be readily identified at the completion of a project.

Future Sea Level and Climate Change Projections

Limitations of capital works based adaptation in the coastal zone are also highlighted by the growing evidence of the significant impacts arising from a rapidly changing climate. The impacts of climate change on oceans are both direct and indirect. Direct impacts include the warming of ocean surface temperatures, increasing ocean acidification and altered ocean currents. Indirect impacts include rising sea levels from thermal expansion and land ice melt. Coastal inundation reflects the interaction between rising sea levels, king tides and storm surges, along with inland flooding from coastal flowing watersheds. Ocean responses, however, are complex and as discussed below we can expect lagged effects spanning decades to millennia.

The IPPC 5th Assessment Report conclusions regarding coastal systems and low-lying areas were that:

- Due to sea level rise projected throughout the twenty-first century and beyond, coastal systems and low-lying areas will increasingly experience adverse impacts such as submergence, coastal flooding, and coastal erosion;
- The population and assets projected to be exposed to coastal risks as well as human pressures on coastal ecosystems will increase significantly in the coming decades due to population growth, economic development, and urbanization; and
- The relative costs of coastal adaptation vary strongly among and within regions and countries for the twenty-first century. Some low-lying developing countries and small island states are expected to face very high impacts that, in some cases, could have associated damage and adaptation costs of several percentage points of GDP.

IPCC reports, however, are limited to a synthesis of peer-reviewed published science in the preceding years. Since the 2013–2014 5th assessment, a significant body of research has been published investigating the role of Antarctica and the Southern Ocean in potentially accelerating sea level rise this century and beyond based on increased understanding and modelling of processes such as Deconto and

Pollard (2016) who modelled the interactions between ocean temperature and sub-ice-shelf melt rates, hydrofracturing (crevasse penetration in relation to surface liquid water supply), and maximum rates of marine terminating ice-cliff failure; none of which were present in the Earth system models used for the AR5 assessments. The authors concluded that Antarctica has the potential to contribute more than a metre of sea-level rise by 2100 and more than 15 metres by 2500, if emissions continue unabated. Over the course of the twenty-first century, the projections of Deconto and Pollard are commensurate with those of the IPCC AR5 but the former suggest around 1 m rise more than less likely. It follows that existing and new structures built in the coastal zone, along with adaptation interventions such as capital works, need to factor into their location, design and construction the consequences of these projected sea level-related impacts at least for the asset's life-time. Beyond 2100, assuming the vector of Deconto and Pollard projections is of a plausible direction and magnitude, entirely novel adaptation interventions may be needed. Further research is needed, however, in order to draw any specific conclusions as to what these projections may imply for the Pacific.

The Australian Bureau of Meteorology and the CSIRO (2014) have developed climate projections tools and data for the Pacific region along with country specific information. Of particular relevance is their Climate and Oceans Support Program in the Pacific (COSSPaC) which is building tools that can forecast and report on climate, tides and the ocean and is working with stakeholders communicate this information to communities, businesses and Government (Australian Government 2016). Most of the projections described show increases in Western Pacific Monsoon rainfall in a warmer climate, mostly over the wet season, leading to a stronger seasonal rainfall cycle and increases in rainfall within the Inter-Tropical Convergence Zone (ITCZ), particularly in the June to August season, which will amplify the seasonal cycle. El Niño and La Niña events will continue to occur in the future, but projections are uncertain about change in intensity or frequency. While the details of specific projections for rainfall, temperature, drought, waves and other variables vary to some extent between countries, general climate observations and trends across the region indicate:

- On a regional scale, station-based observations show a persistent mean annual warming trend of 0.18 °C since 1961, with most of the warmest years on record in the last two decades. There have been significantly more warm days and nights, and fewer cool days and nights.
- Since 1951, the frequency of warm days and nights has increased more than threefold across the region. Once rare extremes, that used to occur approximately 20 days in a year, are now occurring much more frequently, between 45 and 80 days in a year.
- Rainfall is highly variable, and over the past 30 years, the southwest and northwest Pacific has become wetter and the central Pacific drier. In general, there has been no consistent trend across the region in the long-term for mean and daily extremes of rainfall over the past half century.

- Average annual rainfall will increase with fewer droughts in most areas. Extreme rainfall events that occur once every 20 years on average during 1986–2005 are projected to occur once every 7–10 years by 2090 under a very low emissions scenario, and every 4–6 years by 2090 under a very high emissions scenario.
- Sea surface temperatures have increased, and year-to-year variability is largely due to ENSO. Ocean acidification continues to increase in response to human activities. As the ocean warms, the risk of coral bleaching (recurrence and severity) increases. Extreme temperatures that occur once every 20 years on average are projected to increase in line with average temperatures by up to +2.0–4.0 °C by 2090 under the very high emissions scenario.
- Sea levels have risen, and vary across the Pacific with large-scale climate processes. Extreme sea levels are caused by a combination of long-term sea - level rise from climate change and short-term climate variability factors, such as combined effects of king tides, storm surge and associated wind-wave setup. Wind-driven waves that have influence on coastal regions exhibit strong seasonality and year-to-year variability is largely due to ENSO. The relationship of annual wind-wave properties to ENSO varies regionally.
- Tropical cyclones remain a regular seasonal occurrence with no trend evident in the historical record. However, two category 5 cyclones in 2015 (Pam) and 2016 (Winston), both of which caused extensive damage and deaths in Vanuatu and Fiji, respectively, are stark reminders of the risks from the projected increase in the intensity if not frequency of cyclones.
- Other projected changes include increasing ocean acidification, more frequent and longer lasting coral bleaching, along with changes to wind-driven waves.

Understanding Pacific Coastal Processes

The practice of using capital works to stabile shorelines has developed primarily in the higher latitudes in locations where coastal processes are subject to different influences to those Pacific island states. To identify the limits to the use of capital works as adaptation interventions in the Pacific coastal zone, an understanding of Pacific coastal processes is required. The coastal zones of all regions are highly dynamic environments, which in many circumstances are dominated by natural processes of erosion and accretion along with natural variability in sea levels and wave conditions on seasonal, year-to-year and multi-decadal time scales. However, Pacific island coastal zones have distinctive physical features which must be taken into account when considering adaptation (SOPAC 1994): coral reefs which are a common and often dominant feature; wave conditions which vary significantly in time and location throughout the region; water levels and currents in the coastal zone which are strongly modified by fringing reefs; natural beach material which is commonly and often dominantly supplied from carbonate sources; limitations on construction materials for coastal protection purposes; and mangroves, where they occur, are a prominent and important feature of the coast with recognized value in mitigating storm surges and other kinds of climate-related risks.

As noted, all coastal zones are dynamic and Pacific islands coasts are under constant processes of erosion and accretion with the active beach zone, with heavy erosion following extreme weather events and accretion typically following in due course. Capital works that disrupt these natural processes destroy reefs and mangroves, accentuate erosion, inhibit accretion, increased inundation and ultimately undermine the integrity of the built assets they are designed to protect as well as impacting on the lives of coastal communities. A negative feedback loop is created in an active beach zone by the inclusion of capital works for beach defence that, depending on the design, building and maintenance of these structures, result in the need for more hard barriers of one kind or another as the result of their interactions with the nearshore currents and wave regimes, with these interactions increasing with rising sea levels (SOPAC 1994). The challenge is to identify the circumstances where concrete seawalls are warranted, such as major ports facilities, and where other approaches such as ecosystem-based adaptation might be more feasible (Ojea 2015), along with the design, construction and maintenance guidelines that can optimise their effectiveness and reduce environmental impact.

Capital Works as Climate Change Adaptation

Capital works can be understood as projects that create a physical asset through the commitment of relatively large amounts of human and financial resources. Generally capital works projects involve engineering design and construction and include buildings, roads and bridges, structures or mechanical installations, upgrading or extending existing buildings, and major road works that upgrade existing roads. This definition of adaptation strategy is aligned with that recognised by Biagini et al. (2014) as physical infrastructure within a typology of climate change adaptation projects. Under this approach, physical infrastructure was defined as any new or improved hard physical infrastructure aimed at providing direct or indirect protection from climate hazards (Biagini et al. 2014). Capital works based adaptation strategies share similarities with the concept described by Kay (2012) as ribbon cutting adaptation strategies that operate within short-term political cycles where the motivation behind those sponsoring the adaptation works is that they will personally benefit from the publicity associated with the project's launch. These ribbon cutting strategies manifest in a preference for capital works ahead of other adaptation interventions without new built physical outcomes such as ecosystem-based approaches. An additional factor is the role played by overseas aid provided to Pacific island territories and states by developed countries including USA, N.Z., Australia, Japan, Germany and China. The requirement for these countries to invest relatively large amounts of money in aid projects can create a bias towards larger resource intensive activities, including capital works in the coastal zone. This bias arises in part from the relative ease with which projects can be monitored and evaluated over short time periods. In considering capital works as adaptation strategies it is important to make a distinction between projects whose primary role is adaptation to climate change or natural hazard mitigation such as levees or seawalls and projects whose primary role is to address social issues such as improving access through construction of roads and bridges. While both types of capital works must have regard to climate change projections, our focus here is on the former rather than the latter.

Capital works are delivered across a project lifecycle, which involves design, approvals, construction, operations, maintenance, repair and decommissioning. By replacing decommissioning with upgrade it is possible to create a loop with adaptive qualities. However, just as this connection may enable adaptation it may also create path dependencies, which result in maladaptation.

As an adaptation strategy, the use of capital works seeks to maintain current landuse and settlement patterns and protect associated assets against changing climate conditions and particularly changing magnitude or frequencies of extreme events or natural hazards. The types of capital works projects which are currently under consideration by governments for dealing with climate change are based on existing civil infrastructure types used to provide protection from natural hazards. Table 17.2 provides a summary of capital works structures commonly used in the coastal zone and their main objective in terms of the changes they are intended to bring to natural processes.

From reviewing Table 17.1 it is clear that the objective of each of the capital works structures is to influence physical processes such as inundation and erosion. In fulfilling these objectives capital works will often have a range of intended and

Table 17.2 Summary of capital works structures commonly used in the coastal zone and their main objective in terms of intended changes to natural processes

Structure	Definition	Objective
Breakwaters	Shoreline parallel structures located seaward of the shoreline. Breakwaters may be above or below (artificial reefs) sea level	To reduce erosion by dissipating wave energy before reaching the shoreline
Groynes	Shoreline perpendicular structures connected directly to the shoreline	To reduce erosion by trapping the longshore transport of sediment
Seawalls	Shoreline parallel structures located landward of the shoreline	To hold sediment from sliding into the sea and in doing so to protect land and property from wave impact and erosion
Jetties or Training Walls	Structures at the mouth of a river or water course that create a fixed channel	To restrict the movement of a channel entrance
Dams	Structures across a channel which restrict flow of water	To control the flow of water either for storage or flood control
Levees	An embankment or wall used to control overland water flows	To control flooding by directing the flow of water

unintended consequences. The intended consequences in controlling erosion or inundation is often limited to a location, period of time or magnitude of event to which the structure is exposed. With many engineering solutions, however, come unintended and unknown consequences making it difficult to forecast all potential costs and benefits accruing from such projects and their feasibility under different climate change scenarios (Noble et al. 2014).

Location

Coastal processes are driven by oceanographic and atmospheric phenomena such as pressure, winds, currents, waves and tides. Through these processes locations tend to be interconnected which is recognised by geomorphologists through the concept of coastal or sediment compartments. A sediment compartment is a defined area of coast based on interconnected sediment flows and landforms (Thom 2013). In this way locations within a compartment are said to have a sediment budget, which involves inflows and outflows of sediments. This sediment budget can exist in a variety of states: balanced, positive or negative. The state of the sediment budget will be reflected in the stability of the shoreline, where balanced inflow and outflows result in stable shoreline positions, while negative results in shoreline recession and positive accretion. It is important to emphasise that there are a range of drivers for sediment balances, which are both natural and human influenced. The most recent global review of state of shorelines found that 70% of shorelines are retreating (Bird 1985).

The interconnected nature of coastal processes means that where there is interference from capital works projects there are both local and broader implications, the scale of which is determined by the sediment compartment. As structures perpendicular to the shoreline with the intention to restrict the movement of sand and sediments parallel with the shoreline Groynes highlight this issue. Where these structures are successful in their objective the sediment budget updrift will become increasingly positive resulting in accretion on that side of the structure. While this is likely to be a positive result and the intention of the project, the side effect of this accumulation is that the sediment budget for the downdrift of the Groyne will be increasingly negative resulting in shoreline recession. The term 'Groyne Field' has emerged to describe the tendency of governments to follow an initial groyne designed to control erosion with multiple subsequent groynes to deal with the side effects of the initial groyne.

In addition to these physical consequences, the flow-on consequences of changes to physical locations have important implications for ecosystems and the communities that rely on the services provided by such ecosystems. Shoreline and near-shore capital works occupy the same physical space as highly productive ecosystems such as mangroves, saltmarsh, sandy beach, seagrass and coral reefs. Coastal squeeze describes the loss of habitat that occurs where coastal ecosystems are unable to migrate landwards as sea levels rise where structures such as seawalls fix the shoreline in place.

Time

In examining the intended and unintended implications of capital works the other significant dimension is time. Shorelines are dynamic at varying time scales from daily tidal cycles, seasonal storm frequencies, decadal natural variability and geological time. The construction of structures within the active coastal process anchors the shoreline in place. This is a significant benefit to land use and place based property rights systems which do not readily accommodate the temporally dynamic nature of the shoreline. The fixing of shorelines in place tends to have unintended implications for social and environmental values which result from these benefits to property and land use as well as the need by governments to maximise the returns on investment for such capital works. Shoreline boundaries also have implications for delineation of maritime boundaries.

The value of a given block of land is influenced on the supply side by three factors; characteristics of the lot; characteristics of the location; and permissible use of the lot. Where capital works projects such as seawalls reduce the erosion risk to a property there is a subsequent increase in value and interest from development sector. Where land use adjacent to shoreline capital works structures intensifies this tends to drive a cycle of development and protection works as increased development requires increased protection and in so doing the public good values of an area tend to be increasingly excluded or subject to coastal squeeze as described above.

Engineering Design Process: Design Storm Events

One of the critical challenges in the adoption of capital works adaptation strategies emerges in the design phase of the project lifecycle. The primary influence on the design stage of capital works is the practices of the engineering profession. Engineering is a well established and highly regulated profession with established standards and quality assurance processes which have been developed over centuries and delivered significant societal advances. For the types of capital works projects described in the table above the engineering design process relies on the establishment of the design storm. A design storm is, an event (traditionally precipitation but also applied to waves and storm surge) identified generally from the historical record and used as the design criteria for the capital works project to accommodate (Wenzel and Voorhees 1981).

While there are a range of possible approaches to the identification of a design storm event the requirement for engineering professionals to comply with established standards has implications for their selection. For Australia, most capital works projects are designed for the 0.01 average return interval event so that events exceeding that magnitude may degrade or cause a failure of the structures performance. The return interval varies across locations with flooding defences in the Netherlands, for example, often designed for much lower probability events such as 0.001 ARI and this decision may be as a result of the consequences of events exceeding the design standard.

The engineering process that draws on the design storm event has evolved through a period of human history and in geographical areas where assumptions of climatic stationarity may at the time have been appropriate. Improved understanding of climate change and variability bring this assumption into question particularly in the context of capital works projects which are intended to exist for decades and over which time global climate models project shifting conditions (Kiem and Verdon-Kidd 2013; Verdon-Kidd and Kiem 2015). The engineering discipline has given, until recently, less attention to the benefits associated with soft engineering approaches and ecosystem-based approaches from coastal mitigation in the face of a rapidly changing climate.

As we move into a period of increased climatic uncertainty, the centrality of a design storm event to the engineering design process is an increasing challenge for the use of capital works as an adaptation strategy (Watt and Marsalek 2013; Kuo and Gan 2015). Past experience has shown that the consequences for communities adjacent to capital works defences can be severe when storm events exceed the design standards particularly where those capital works provide residents with a false sense of protection as was the case with both the Hurricane Katrina and Tohoku Earthquake case studies described below.

Hurricane Katrina New Orleans

In August 2005 Hurricane Katrina caused devastating flooding and the loss of more than 1800 lives in the states of Louisiana and Mississippi. Much of the impact of Hurricane Katrina was on New Orleans, a city with a long history and widely recognised as vulnerable to flooding. The development of the city has occurred hand in hand with the implementation of flood mitigation capital works such as levees, drainage networks and land reclamation (Baxter 2014; Gordon and Little 2009; Rogers 2008) so it is a location with a long history of capital works based adaptation to inundation hazards. Prior to the 2005 flood there had not been a major flooding event associated with a Hurricane since 1969.

The flood defence system for New Orleans had traditionally focused on protection of the city from a flood of the Mississippi river. However since the 1990s there had been greater attention to Hurricane driven storm surge inundation risk and the USACE was implementing a program to provide protection for a category 3 Hurricane, which is the magnitude that Hurricane Katrina reached at the time of landfall. The severity of the inundation that occurred within New Orleans has been attributed to the lack of maintenance of the levee defence system at the time of the Hurricane which resulted in numerous levee and floodwall breaches. The geotechnical failure mode that occurred as a result of the combination of high water levels and inadequate maintenance had received limited consideration in disaster planning. The limited consideration of this failure mode contributed to the magnitude of the events impacts as the focus of planning for potential failure modes was on overtopping. This example highlights the limit of capital works approaches which require identification and accommodation of all possible failure modes when this may only be clear in hindsight.

The aftermath of the event was as serious for the residents of New Orleans as the Hurricane itself. The event exposed an almost complete failure of emergency management governance systems with residents unable to be evacuated in time and forced to seek shelter in Football stadium without food and water supplies or sanitary facilities. One of the factors which is felt to have contributed to the failure of governance is complacency driven by the presence of capital works flood protection for most of the city. An important lesson from the event is that structural flood protection is a last rather than first line of defence and early evacuation is critical given the limitations of design event based structures and possibilities of failure modes not considered within engineering design.

The Great Wall of Japan

Capital works are widely used in the coastal zone throughout the world. At least 40 percent of Japan's 22,000-mile coastline is lined with concrete seawalls, breakwaters or other structures meant to protect the country against high waves, typhoons or even tsunamis (Onishi 2011). This is similar to many developed countries; e.g. 32% of the Northern Irish coast is fronted by man-made structures, 44% in England and Wales, and 6% in Scotland (Cooper et al. 2016). The famous Dutch Delta Plan involves the deployment across the Rhine-Meuse-Scheldt delta of a massive array of dams, sluices, locks, dykes, levees and barriers.[1] For European countries, coastal zones dominated by capital works are the norm.

The limitations and risks of vertical concrete seawalls, while long recognized by academia, has been popularised in recent years following the Great East Japan Earthquake which unleashed a massive tsunami. This was the largest earthquake ever recorded in Japan and, according to the Japan Reconstruction Agency, as a result of the disaster, which precipitated the triple disaster of earthquake, tsunami and nuclear accident, more than 15,000 people lost their lives and over 2500 are still officially reported as missing, while a further 6000 suffered injuries. In total, over 470,000 people were evacuated from their homes. As of August 2015, the number of evacuees has decreased to less than 200,000 people, among 70,000 are

[1]http://www.deltawerken.com/Deltaworks/23.html

still in temporary housings. The direct financial damage from the disaster was estimated to be approximately 16.9 trillion yen (US$199 billion), with 10.4 trillion of damage to buildings, 1.3 trillion to lifeline utilities, 2.2 trillion to social infrastructure and 3.0 trillion to other areas as of June 2011. The World Bank estimated that the economic cost could reach up to US$235 billion, making it the costliest natural disaster in world history. It has been widely reported that Japanese authorities plan to build a giant 250-mile long sea barrier to protect its coastline from future devastating tsunamis, at a cost of US$10 billion, with heights reaching up to17 m. A further 14,000 km of Japan's coastline has also been assessed as requiring tsunami protection (Onashi 2011; Dumash 2016).

Extensive and massive vertical concrete sea walls come with an array of problems and limitations including: destruction of coastal ecosystems that provide sustainable livelihoods for local communities; blocking of inland flood waters and debris; and increased future risks by encouraging rebuilding of settlements and infrastructure in zones still exposed to over-topping or catastrophic collapse of walls by extreme events. Nonetheless, the funding playing field appears to remain tipped in favour of capital works for coastal zone defence with major donor's familiarity with capital works perhaps facilitating their prioritisation in overseas development aid investments for Pacific island territories and states.

As noted above, many of the developed countries who are major donors to the Green Climate Fund,[2] and other international sources of finance intended for mitigation and adaptation projects in developing countries, are familiar with capital works as these remain the dominant mode of defending coastal infrastructure and human settlements. Cooper et al. (2016) suggested the following reasons for this: ignorance on the part of the public and managers of the implications of sea defences for coastal ecosystems; structuring of decision-making processes such that some form of hard defence is the only plausible outcome; active lobbying for engineered interventions by the engineering profession; poor planning decisions that permit construction in high risk zones; a higher priority being afforded to private property than maintenance of the communal coastal resource; and an inability to contemplate large scale removal of infrastructure to less vulnerable locations. There is also the social pressure evident in many governments to maintain the status quo rather than embark on a radical change in risk management strategies. Such approaches are common where policy problems are defined as routine issues that can be dealt with by minor modifications to existing management procedures (Nalau and Handmer 2015). There can also be community pressure, particularly after extreme weather events, for governments to be seen to be 'doing something' which can provide further pressure towards engineering works. Finally, we cannot ignore the ribbon cutting opportunities of capital works and the positive publicity such events afford local politicians and donors.

[2]See http://www.greenclimate.fund/contributions/pledge-tracker.

Conclusions

Given projected climate change impacts including sea level rise and the increasing intensity of tropical storms and the precipitation cycle, along with the concentration in the coastal zone of major infrastructure including port facilities, airports, road, and urban centres, the pressure for capital works adaptation responses in the Pacific is set to magnify in the coming decade. Capital works as coastal defences therefore warrant special attention especially in terms of their economic, social and environmental costs and perverse outcomes in the medium to longer term, in addition to anticipated short term benefits.

Admittedly, high construction and maintenance costs pose a financial limit for capital works based adaptation in the least developed countries of the Pacific. Such adaptation limits can also be technological (not all technological options are available), social (capital works are opposed by communities and/or decision makers) and environmental (capital works do not fit in with environmental context as an option). However, the financial limit will be somewhat addressed in the coming years given that the international community has committed to a goal of mobilising climate finance funds of US$100 billion per year every year from 2020 with the aim of achieving a balance between adaptation and mitigation, taking into account country-driven strategies, and the priorities and needs of developing country Parties, especially those that are particularly vulnerable to the adverse effects of climate change and have significant capacity constraints, such as the least developed countries and small island developing States.[3]

It is now widely accepted that there are three major strategies to adaptation in the coastal zone: retreat, accommodate, or defend (Low Choy et al. 2012). To this list we can add a fourth proactive strategy, namely, *exclude*, which refers to zoning land so that valuable assets are not located in high risks locations in the first place. Given the naturally dynamic nature of the coastal zone, a general planning principle would be to take such a precautionary approach unless there are special circumstances, such as port facilities, that warrant expensive capital works-based defence strategies.

There is now a substantial body of work documenting projects and investigations into adaptation options and their effectiveness, at least in the short term including cost-benefit analyses comparing engineered and ecosystem-based approaches (Rao 2014). The Pacific Adaptation to Climate Change (PACC) Programme, for example, which began in 2009, is working to enhance adaptive capacity on the ground, and is driving the mainstreaming of climate risks into national development planning and activities. Working in 14 Pacific island countries, the Programme is *demonstrating best-practice adaptation* in three key climate-sensitive areas: coastal zone management, food security and food production, and water resources management. Each country is hosting a pilot project in one of these theme areas to demonstrate how climate change adaptation can work on the ground (SPREP

[3]Details of the Green Climate Fund can be found at http://www.greenclimate.fund/home.

2016b). This growing body of evidence should be draw upon to inform adaptation decisions in the region, particular given that the factors noted in Box 17.1, which are central to the vulnerability of human settlements in the Pacific, suggest that Pacific island territories and states will place higher value on nature based coastal protection schemes compared to economically developed countries. OECD countries are less directly dependent on local ecosystem services and in any case have more financial resources available to support and maintain capital works interventions.

A critical aspect of human-forced climate change that has yet to fully resonate with the public and decision makers is the fact that even if we are successful in reducing anthropogenic greenhouse gas emissions to the level required to limit global warming to well below 2 °C, as per the Paris Agreement commitment, the global climate system will not be shifting to a single, stable new state. Rather, the climate system will continue to be perturbed for millennia. This is the consequence of lags and the highly variable rates of response in different components of the Earth system responses, such as those associated with sea level rise noted. Adaptation therefore will also need to follow an adaptive management approach of monitoring changing impacts, evaluating the efficacy of adaptation responses, and modifying adaptive strategies and options as necessary. While the challenge of mitigating anthropogenic greenhouse gas emissions has a definite time horizon,[4] unfortunately, climate change adaptation will be required "forever" and therefore must be mainstreamed into planning and decision making in all sectors and levels of government, and especially for decisions regarding long term infrastructure investments in the coastal zone.

Mainstreaming adaptation into planning and decision, however, is hampered by the prospects of continuous and ongoing climatic disruption as this means the assumption of stationarity, i.e., that the relationships inferred from historical data remain valid into the future, no longer holds. This means, among other things, that engineering specifications based on climatic norms will no longer be a valid approach per se and a far more nimble approach based on an adaptive management will be required.

Finally, we note again that there are circumstances where capital works are the most appropriate adaptation response, vertical concrete seawalls are needed, and the consequences for ecosystems and sustainable livelihoods must be accepted. However, the key research problem is to identify the precise circumstances under which use of capital works are an appropriate and cost effective coastal defence strategy, those where a soft-engineering approach such as beach nourishment is preferable, and the situations where an ecosystem-based approach is the most effective and to be recommended. Ultimately, therefore, Pacific island territories and states will need a mix of adaptation strategies for the coastal zone that draw

---

[4]To have a greater than 50% chance of limiting global warming to 2 °C below the pre-industrial global surface average, requires total anthropogenic greenhouse emissions are reduced by about 40% from 1990 levels by 2030 and then to around zero by about 2080. See *IPCC 5th Assessment Report, Working Group III Mitigation of climate change*; http://www.ipcc.ch/report/ar5/wg3/.

upon ecosystem-based adaptation, soft engineering use of natural processes, along well designed and maintained capital works.

Acknowledgements This publication draws upon the results of research undertaken by the Griffith University Ecoadapt Pacific research project which is funded by the University and an anonymous charitable trust.

References

Adger, W., Dessai, S., Goulden, M., Hulme, M., Lorenzoni, I., Nelson, D., et al. (2009). Are there social limits to adaptation to climate change? *Climatic Change, 93*(3), 335–354.

Archer, D., & Brovkin, V. (2008). The millennial atmospheric lifetime of anthropogenic CO_2. *Climatic Change, 90,* 283–297.

Australian Government. (2016). Climate and Oceans Support Program in the Pacific (COSPaC). Bureau of Meteorology and the CSIRO. http://cosppac.bom.gov.au/.

Barbier, E. B. (2015). *Climate change impacts on rural poverty in low-elevation coastal zones.* Shock waves: Managing the impacts of climate change on poverty. Background Paper. Policy Research Working Paper 7475. Development Economics Climate Change Cross-Cutting Solutions Area, World Bank.

Baxter, V. (2014). Rent, real estate, and flood mitigation in New Orleans East. *Antipode, 46*(4), 1014–1031. doi:10.1111/anti.12080.

Biagini, B., et al. (2014). A typology of adaptation actions: A global look at climate adaptation actions financed through the Global Environment Facility. *Global Environmental Change, 25* (1), 97–108. doi:10.1016/j.gloenvcha.2014.01.003.

Bird, E. C. F. (1985). *Coastline changes: A global review* (p. 219). Chichester: Wiley.

Bureau and Meteorology and Commonwealth Scientific and Industrial Research Organisation (CSIRO). (2014). *Climate variability, extremes and change in the Western Tropical Pacific. Pacific: New Science and Updated Country Reports 2014.* Australia Climate Change Science and Adaptation Planning Program.

Cooper, J. A. G., O'Connor, M. C., & McIvor, S. (2016). Coastal defences versus coastal ecosystems: A regional appraisal. *Marine Policy.* doi:10.1016/j.marpol.2016.02.021.

Deconto, R. M., & Pollard, D. (2016). Contribution of Antarctica to past and future sea-level rise. *Nature.* doi:10.1038/nature17145.

Dickinson, W. R. (2014). Beach ridges as favored locales for human settlement on pacific islands. *Geoarchaeology, 29,* 249–267. doi:10.1002/gea.21476.

Dow, K., Berkhout, F., Preston, B. L., Klein, R. J. T., Midgley, G., & Shaw, M. R. (2013). Limits to adaptation. *Nature Climate Change, 3*(4), 305–307.

Dumas, D. (Published: March 7, 2016 8:17AM). The Great Wall of Japan divides a country still reeling from 2011's earthquake. *The Sydney Morning Herald.* http://www.smh.com.au/action/printArticle?id=1006558175.

Environment Agency. (2015). *Cost estimation for coastal protection—Summary of evidence.* Report –SC080039/R7. Environment Agency, Horizon House, Bristol. www.environment-agency.gov.uk.

Field, C. B., Barros, V. R., Mach, K. J., Mastrandrea, M. D., van Aalst, M., Adger, W. N., Arent, D. J., Barnett, J., Betts, R., Bilir, T. E., Birkmann, J., Carmin, J., Chadee, D. D., Challinor, A. J., Chatterjee, M., Cramer, W., Davidson, D. J., Estrada, Y. O., Gattuso, J. -P., Hijioka, Y., Hoegh-Guldberg, O., Huang, H. Q., Insarov, G. E., Jones, R. N., Kovats, R. S., Romero-Lankao, P., Larsen, J. N., Losada, I. J., Marengo, J. A., McLean, R. F., Mearns, L. O., Mechler, R., Morton, J. F., Niang, I., Oki, T., Olwoch, J. M., Opondo, M., Poloczanska, E. S., Pörtner, H. -O., Redsteer, M. H., Reisinger, A., Revi, A., Schmidt, D. N., Shaw, M. R.,

Solecki, W., Stone, D. A., Stone, J. M. R., Strzepek, K. M., Suarez, A. G., Tschakert, P., Valentini, R., Vicuña, S., Villamizar, A., Vincent, K. E., Warren, R., White, L. L., Wilbanks, T. J., Wong, P. P., & Yohe, G. W. (2014). Technical summary. In C. B. Field, V. R. Barros, D. J. Dokken, K. J. Mach, M. D. Mastrandrea, T. E. Bilir, M. Chatterjee, K. L. Ebi, Y. O. Estrada, R. C. Genova, B. Girma, E. S. Kissel, A. N. Levy, S. MacCracken, P. R. Mastrandrea, & L. L. White (Eds.), *Climate change 2014: Impacts, adaptation, and vulnerability. Part A: Global and sectoral aspects. Contribution of Working Group II to the Fifth Assessment Report of the Intergovernmental Panel on Climate Change* (pp. 35–94). Cambridge University Press: Cambridge, United Kingdom and New York, NY, USA.

Gordon, P., & Little, R. G. (2009). Rethinking flood protection. *Public Works Management & Policy, 14*(1), 37–54.

Hansen, J., Kharecha, P., Sato, M., Masson-delmotte, V., Ackerman, F., Beerling, D. J., et al. (2013). Assessing "dangerous climate change": Required reduction of carbon emissions to protect young people, future generations and nature. *PLoS ONE*. doi:10.1371/journal.pone.0081648.

Hinkel, J., Lincke, D., Vafeidis, A. T., Perrette, M., Nicholls, R. J., Tol, R. S. J., et al. (2014). Coastal flood damage and adaptation costs under 21st century sea-level rise. *Proceedings of the National Academy of Sciences of the United States of America, 111*(9), 3292–3297. doi:10.1073/pnas.1222469111.

IPCC. (2013). Climate change 2013: The physical science basis. In T. F. Stocker, D. Qin, Plattner, G.-K. Tignor, M. Allen, S. K. Boschung, J. Nauels, A. Xia, Y. Bex, V. & P. M. Midgley (Eds.), *Contribution of Working Group I to the Fifth Assessment Report of the Intergovernmental Panel on Climate Change*. Cambridge, United Kingdom and New York, NY: Cambridge University Press.

Juhola, S., Glaas, E., Linner, B. O., & Neset, T. S. (2016). Redefining maladaptation. *Environmental Science & Policy, 55*, 135–140.

Kay, R. C. (2012). Adaptation by ribbon cutting: Time to understand where the scissors are kept. *Climate and Development, 4*(2), 75–77.

Kiem, A. S., & Verdon-Kidd, D. C. (2013). The importance of understanding drivers of hydroclimatic variability for robust flood risk planning in the coastal zone. *Australian Journal of Water Resources, 17*(2), 126–134.

Kingsford, R. T., Watson, J. E. M., Lundquist, C. J., Venter, O., Hughes, L., Johnston, E. L., et al. (2009). Major conservation policy issues for biodiversity in oceania. *Conservation Biology, 23*(4), 834–840. doi:10.1111/j.1523-1739.2009.01287.x.

Klein, R. J. T., Midgley, G. F., Preston, B. L., Alam, M., Berkhout, F. G. H., Dow, K., et al. (2014). Adaptation opportunities, constraints and limits. In C. B. Field, V. R. Barros, D. J. Dokken, K. J. Mach, M. D. Mastrandrea, T. E. Bilir, M. Chatterjee, K. L. Ebi, Y. O. Estrada, R. C. Genova, B. Girma, E. S. Kissel, A. N. Levy, S. MacCracken, P. R. Mastrandrea, & L. L. White (Eds.), *Climate change 2014: Impacts, adaptation, and vulnerability. Part A: Global and sectoral aspects. Contribution of the Working Group II to the Fifth Assessment Report of the Intergovernmental Panel on Climate Change* (pp. 899–943). Cambridge, New York: Cambridge University Press.

Kuo, C., & Gan, T. (2015). Risk of exceeding extreme design storm events under possible impact of climate change. *Journal of Hydrologic Engineering, 20*(12), 4015038. doi:10.1061/(ASCE)HE.1943-5584.0001228.

Low Choy, D., Serrao-Neumann, S., Crick, F., Schuch, G., Sanò, M., van Staden, R., et al. (2012). *Adaptation options for human settlements in South East Queensland—Main report*. The South East Queensland Climate Adaptation Research Initiative, Griffith University. https://www.terranova.org.au/repository/seqcari/final-seqcari-adaptation-options-main-report.

McGranahan, G., Balk, D., & Anderson, B. (2007). The rising tide: Assessing the risks of climate change and human settlements in low elevation coastal zones. *Environment & Urbanization, 19*(1), 17–37. doi:10.1177/0956247807076960.

Nalau, J., & Handmer, J. (2015). When is transformation a viable policy alternative? *Environmental Science & Policy, 54*, 349–356.

Noble, I. R., Huq, S., Anokhin, Y. A., Carmin, J., Goudou, D., Lansigan, F. P., et al. (2014). Adaptation needs and options. In V. R. Barros, C. B. Field, D. J. Dokken, M. D. Mastrandrea, K. J. Mach, T. E. Bilir, M. Chatterjee, K. L. Ebi, Y. O. Estrada, R. C. Genova, B. Girma, E. S. Kissel, A. N. Levy, S. MacCracken, P. R. Mastrandrea, & L. L. White (Eds.), *Climate change 2014: Impacts, adaptation, and vulnerability. Part A: Global and sectoral aspects. Contribution of Working Group II to the Fifth Assessment Report of the Intergovernmental Panel on Climate Change* (pp. 833–868). Cambridge, New York: Cambridge University Press.

Nurse, L. A., McLean, R. F., Agard, J., Briguglio, L. P., Duvat-Magnan, V., Pelesikoti, N., et al. (2014). Small islands. In V. R. Barros, C. B. Field, D. J. Dokken, M. D. Mastrandrea, K. J. Mach, T. E. Bilir, M. Chatterjee, K. L. Ebi, Y. O. Estrada, R. C. Genova, B. Girma, E. S. Kissel, A. N. Levy, S. MacCracken, P. R. Mastrandrea, & L. L. White (Eds.), *Climate change 2014: Impacts, adaptation, and vulnerability. Part B: Regional aspects. Contribution of Working Group II to the Fifth Assessment Report of the Intergovernmental Panel on Climate Change* (pp. 1613–1654). Cambridge, New York: Cambridge University Press.

Ojea, E. (2015). Challenges for mainstreaming ecosystem-based adaptation into the international climate agenda. *Current Opinion in Environmental Sustainability, 14,* 41–48.

Onishi, Norimitsu (Published: March 13, 2011). Seawalls offered little protection against tsunami's crushing waves. *The New York Times.* http://www.nytimes.com/2011/03/14/world/asia/14seawalls.html?pagewanted=all&_r=0.

Rao, N. S. (2014). *An economic analysis of ecosystem-based adaptation and engineering options for climate change adaptation in Lami Town, Republic of the Fiji Islands.* The Secretariat of the Pacific Regional Environment Programme (SPREP), Apia, Samoa. A technical report providing an overview of an economic analysis (least-cost analysis and cost-benefit analysis) of adaptation options, including EBA, in Fiji.

Reid, H. (2016). Ecosystem- and community-based adaptation: Learning from community-based natural resource management. *Climate and Development, 8*(1), 4–9.

Rogers, J. D. (2008). Development of the New Orleans Flood Protection System prior to Hurricane Katrina. *Journal of Geotechnical and Geoenvironmental Engineering, 134*(5), 602–617.

SOPAC. (1994). *Coastal protection in the South Pacific.* SOPAC Technical Report 190 SOPAC SECRETARIAT in association with SPREP.

SPREP. (2016a). PEBACC: Pacific ecosystem-based adaptation in the Pacific. Secretariat of the Pacific Regional Environment Program; https://www.sprep.org/pebacc.

SPREP. (2016b). The Pacific adaptation to climate change (PACC) Programme. Secretariat of the Pacific Regional Environment Program; https://www.sprep.org/pacc.

Thom, B. (2013). *Coastal compartments project: Summary for policy makers.* Department of Environment and Energy: Canberra, ACT.

UNFCCC. (2015). Adoption of the Paris Agreement conference of the parties twenty-first session Paris, 30 November to 11 December 2015. United Nations Framework Convention on Climate Change. https://unfccc.int/resource/docs/2015/cop21/eng/l09r01.pdf.

UNFPA. (2014). *Population and development profiles: Pacific Island countries.* Suva: United Nations Population Fund, Pacific Sub-Regional Office.

Verdon-Kidd, D. C., & Kiem, A. S. (2015). Non-stationarity in annual maxima rainfall across Australia—Implications for intensity–frequency–duration (IFD) relationships. *Hydrology and Earth System Sciences Discussions, 12*(3), 3449–3475. http://www.hydrol-earth-syst-sci-discuss.net/12/3449/2015/.

Vrolijks, J. (1998). *Human settlements development and disaster risks in Pacific Island countries.* The South Pacific Programme Office of the UN Department Of Humanitarian Affairs and UN Department of Economic and Social Affairs.

Watt, E., & Marsalek, J. (2013). Critical review of the evolution of the design storm event concept. *Canadian Journal of Civil Engineering, 40*(2), 105–113.

Wenzel, H. G., & Voorhees, M. L. (1981). *An evaluation of the urban design storm concept.* University of Illinois: Water Resources Center.

Chapter 18
A 'Cost Barrier' Perspective to Adaptation on a Coral Triangle Initiative (CTI) and Mangrove Rehabilitation Projects (MRP) in Solomon Islands

Michael Otoara Ha'apio, Walter Leal Filho and Morgan Wairiu

Introduction

Climate change is one of the major challenges facing Pacific Island Countries and Territories face (PICTs) (Fig. 18.1), including the Solomon Islands (Kelman and West 2009). It threatens not only people's livelihood and well-being, but also the viability of isolated insular communities and rural households. Like many other PICTs, the Solomon Islands has been identified as one of most vulnerable countries in respect of the adverse impacts of climate change (Barnett 2011). The country's high vulnerability stems from two many factors:

(a) Geographical position: as a country composed by many low lying islands, quite remote, vulnerable to sea level rise and coastal erosion; and
(b) Socio-economic elements: as a developing country with a low GDP (US$ 1.819), it is excessively dependent on foreign aid.

In addition, the fact that the majority of the population lives within 1.5 km from the coastline and the high poverty rate, increases its vulnerability even further.

The country has a population of half a million, of which some 85% live in rural households that rely on subsistence activities from agriculture, forestry and marine resources (Albert et al. 2010). The dependency of the villagers on marine resources has strained the productivity of the coral reefs and destroyed most of the reef habitats due to over-harvesting through the practice of illegal fishing techniques and other unconventional methods over the years.

M. O. Ha'apio (✉) · Morgan Wairiu
Pacific Centre for Environment and Sustainable Development,
The University of the South Pacific, Suva, Fiji
e-mail: mhaapio@gmail.com

W. Leal Filho
School of Science and the Environment, Manchester Metropolitan
University, Chester Street, Manchester M1 5GD, UK

W. Leal Filho and J. Nalau (eds.), *Limits to Climate Change Adaptation*,
Climate Change Management, https://doi.org/10.1007/978-3-319-64599-5_18

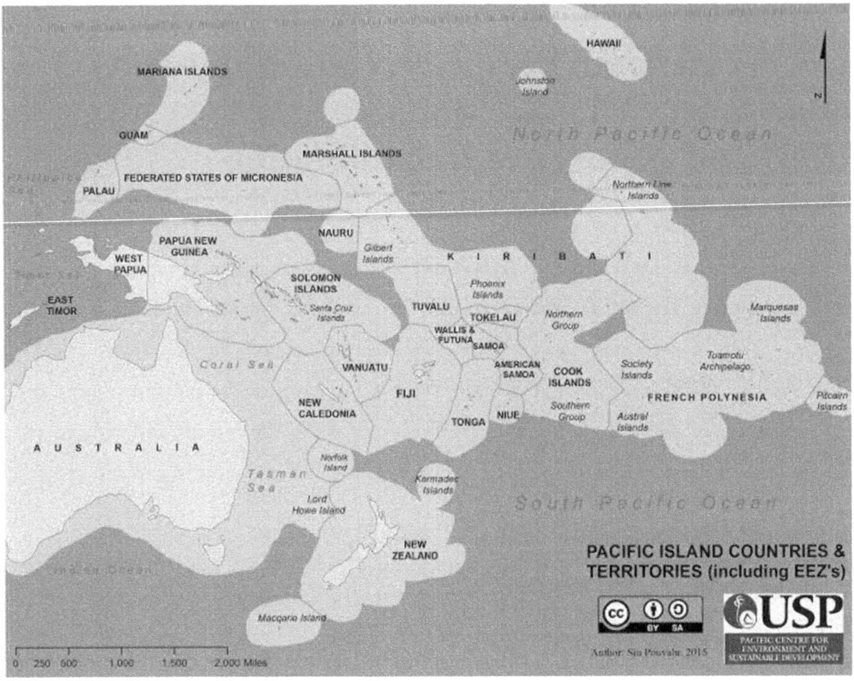

Fig. 18.1 Schematic map of the South Pacific showing the location of Solomon Islands in the Pacific context

The high vulnerability of the Solomon Islands to climate change creates an opportunity to explore the situation of rural households and how government strategies may help to mitigate and adapt to climate change impacts. The Solomon Islands government, international donor partners and agencies have spent more than US$1.1billion (MECDM[1] 2008) in the last 8 years in initiatives aiming to integrate Climate Change[2] in developing plans and budgets, capacity building and implementation of adaptation projects to increase the resilience to climate change. International donor agencies such as the US Agency for International Development (USAID), the Global Environment Facility (GEF) and the Asian Development Bank (ADB) have had a major role in assisting Solomon Islands in addressing climate change risks. The country is yet to explore other funding sources e.g. Green Climate Fund (MECDM 2008).

This study focuses on three rural coastal communities in Solomon Islands. It looks at costs and benefits that participants received from these conservation initiatives. It especially looked at the costs as barriers which deter participants from adapting effectively to the negative impact of climate change at these communities

[1]Ministry of Environment, Climate Change, Disaster Management and Meteorology.

[2]Climate Change—The change in weather pattern over longer period of time.

and how these villagers go about overcoming these constraints. The costs include financial costs, opportunity costs, social costs and other costs, which communities encountered during conservation of mangrove and coral reefs rehabilitations.

Description of Mechanisms for Adaptation

The Coral Triangle Initiative

Coral Triangle Initiative (CTI) is an example of a partnership between six countries —the Philippines, Malaysia, Indonesia, Timor-Leste, Papua New Guinea and Solomon Islands employing adaptation practices—in Southeast Asia and the Pacific Triangle (Clifton 2009). The main purpose of the CTI is to assist coastal communities to rehabilitate and protect their coral reefs for the use of future generations. It does not explicitly define food security as one of its goals, despite this being the wider aim used to justify conservation action (Foale et al. 2013).

The Solomon Islands component of the CTI was implemented across the country in more than 50 villages in the rural communities in 2008, when the MECDM established CTI projects in selected communities in Malaita, Guadalcanal and Western Provinces, with the main objective of conserving coral reefs and marine ecologies, by emphasising economic development and building resilience to climate change (Lawrence 2012). Selected communities in these three provinces have worked with MECDM and implementing agencies such as World Fish and World Wide Fund for nature in educating and assisting communities to protect and conserve their coral reefs. The villagers through the CTI have identified parts of their coral reefs where they were not allowed to fish and have established policies and preventive measures to enable the fish and other marine resources to replenish. The general connection between conservation and building resilience to climate change is that when the CTI participants conserve their coral reefs, it may increase their level of household income through sales of fish and other marine resources which will eventually increase their adaptive capacity. The increased adaptive capacity will prevent them from over-depending on those resources for their survival (Sulu and Boso 2011). These will in turn assist the villagers to overcome associated costs as barriers to effective adaptation.

Mangrove Rehabilitation Project

The second adaptive measure is the establishment of mangrove rehabilitation programs (MRP), assisted by government and some agencies. In this study, the MRP was a private initiative by the villagers. Mangroves protect villages from storms, surges and winds, and also provide avenues where the many marine inhabitants can replenish and regenerate. Mangroves provide a buffer against hurricanes and protect inland areas from tidal surges, and as plants, they help to hold the soil, preventing soil erosion (Albert and Schwarz 2013). The mangrove products also provide the villagers with food as an important component of their livelihood. Replanting mangroves has played a significant role in conserving habitats and reservoirs of biodiversity, home of many species including (it is hoped) some as yet undiscovered at this site (Maczulak 2009). Conservation of mangroves means some of the villagers especially at Oibola will not able to sell mangrove trees for "fire wood and building materials" thus cutting away a source to an immediate income need. Villagers must therefore find alternate income to supplement this source.

Marine Protected Area

The International Union for Conservation of Nature (IUCN) defines an MPA (Dudley 2008) as "a clearly defined geographical space, recognised, dedicated and managed, through legal or other effective means, to achieve the long-term conservation of nature with associated ecosystem services and cultural values". It includes a wider variety of governance types (including community-based areas) and includes, but is not limited to, no-take areas. Furthermore, these MPAs are often regarded as the cornerstone of coastal and marine resource management and are considered essential strategies in integrated coastal and fisheries management (Nichols 1999).

The establishment of MPAs has proven to attract the interest of tourists who come to dive and do research at these sites. For example, in the Western Province of Solomon Islands, the introduction of MPAs in the surrounding reefs increased the number of home stays at the site because of such tourists. Further, conservation increased the value of fish stock allowing the villagers to fish, obtain income and in turn invest in businesses that will diversify their dependency on marine resources and, thereby, reduce the barriers to effective adaptation to climate change (Hannesson 2002).

Importance of MPAs at the National Level

Over decades, the agriculture sector has been the main source of livelihood to most of the villagers within the country. That supremacy has now been challenged by marine resources, particularly the coral reefs (Pauku and Lapo 2008). Because of this high dependency and the deteriorating state of coral reefs, the Government has secured financial assistance from foreign aid donors to ensure that the coral resources are revived and rehabilitated for future generations' use and benefits (Sore 2010). As part of that strategy, the CTI was used to source funding from aid donor partners with the aim of re-enforcing better management of the coral reefs and establishing policies including MPAs, and how to manage and harvest the resources sustainably.

Historically most MPAs are focused on potential population- and ecosystem-scale benefits during establishment. These benefits include: increased biomass; spill over of fish from the protected area into the allocated areas for fishing; larval export of protected species; and reduced habitat loss. Little consideration was given to the actual (social and financial) costs of most of the coral rehabilitation, including MPA establishment, around the region (Carneiro 2011). In the Solomon Islands the number of MPAs has increased from 21 MPAs in 2008 to 52 in 2013, with MPAs now covering 2802 km^2 (Personal communication, Senior Fisheries officer Mr. Peter Kenilorea[3]).

Costs as Barriers to Adaptation to Climate Change

Costs are known to be one of barriers to climate change adaptation and include a variety of elements among planning and implementation (Klein et al. 2014). This study considers costs as one of the factors or conditions that constrain the ability for effective adaptation as response to climate change and are largely insurmountable (McCarthy 2001). These costs become barriers to adaptation when communities are unable to meet their livelihood needs and faced with the adaptation options where the benefits are longer term rather than short term. In such situation communities try to maximize benefits at the expense of adaptation. On the other hand, cost barriers are the conditions that render adaptation difficult as a response to climate change (Nielsen and Reenberg 2010). Yet, these are often mutable (Adger et al. 2009).

Materials and Methods

In order to understand the role played by costs as barriers to adaptation, the study built a baseline of socioeconomic information at household level in selected rural communities (villages) where conservation initiatives have already been implemented. The study considers the design and implementation of an exploratory sample. The use of secondary sources of information from government agencies, non-government organisations (NGOs), the literature, and international cooperation agencies complements the field data. After collecting the necessary information, the study conducted an exploratory and descriptive analysis of the information. A special focus was on the costs and benefits associated with the implementation of conservation projects at village level.

Given the sample nature of the instrument used, this was not designed to be a probabilistic sample. The main purpose of this sample was to explore the

[3]Peter Kenilorea is the Senior Marine officer responsible for the MPA within the Ministry of Fisheries and Marine resources.

relationship that might exist between the socioeconomic attributes of villagers and the conservation initiatives already implemented. There was also a study conducted in the Western Province of Solomon Islands which had adopted this similar approach, '*Building social and ecological resilience to climate change in Roviana, Solomon Islands*' to find the socio-economic factors influencing the customary marine tenure in the respective communities In their study they found that

> For generations communities in the Western Pacific have employed a range of resource management techniques (including periodic reef closures, gear restrictions, entry limitations, and the protection of spawning aggregations) to limit marine resource use. Localized control over marine resources, commonly known as customary marine tenure (CMT), is the legal and cultural foundation for many of these practices. Because of their perceived potential to meet both conservation and community goals, these traditional resource management techniques are being revitalized by communities (Cinner 2005, p. 1).

In this research the aim was to learn more, through collecting primary data, about the perceptions and expectations of villagers with respect to the conservation projects. Although this was not a probabilistic sample, it might be considered as representative of the specific situation of each study site selected.

The study chose only one initiative per province, i.e., either a CTI or an MRP. Moreover, because of time and budget constraints the study selected the sites with better access, that is, that are closer to main ports or townships (Fig. 18.2). The study used three measurement instruments: (1) a survey at household level, (2) a

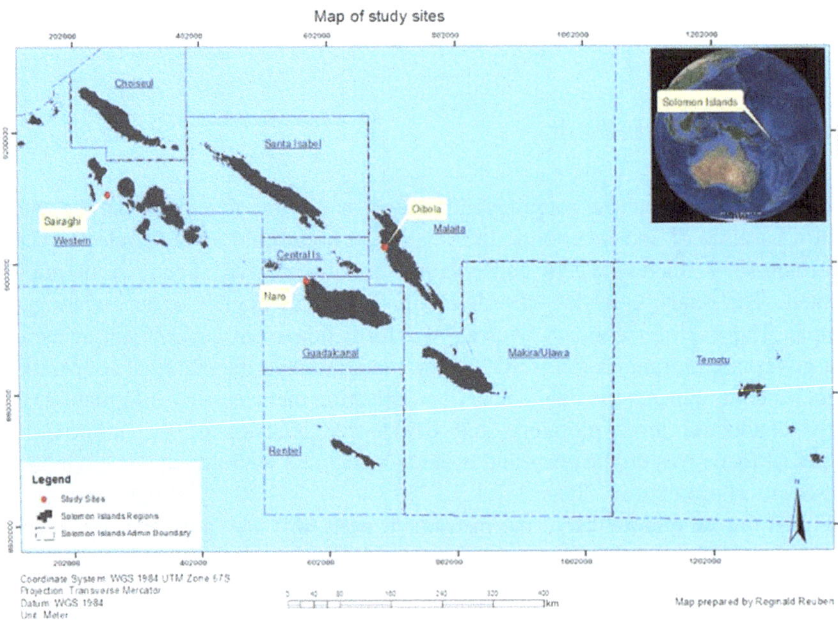

Fig. 18.2 Map of the three main study sites in the Solomon Islands: *1* Sairagi. *2* Naro. *3* Oibola villages

semi-structured survey at village level and, (3) a focus group questionnaire, including village experts.

Household Survey

A household survey was used as the instrument for obtaining respondents' perceptions about the benefits and costs they incurred from CTI and MPR projects. This baseline information collected through the household survey questionnaires helped to gauge the level of benefits they have gained 'with or without' the project. Some of the main variables in the household questionnaires included: level of education, level of income, level of their community participation, number of household members, and the alternative business activities they participated in besides fishing or farming in the respective rural community.

Semi-structured Questionnaires

Semi-structured questionnaires are usually conversational and informal in tone (Longhurst 2003). The research team utilised this instrument to interview various members of the communities in individual and group-interviews. The intention was to interview key respondents[4] within the communities individually while the rest of the villagers were grouped[5] in peers e.g., the youths, women, men and some villager elders together to get their collective opinions.

The research team started by explaining the motive of our interview; first by explaining the nature of CTI and MRP projects. The CTI includes the creation of a marine protected area in the fishing grounds close to the village, the MPA bans fishing activities but the MPA's objective is to improve the quality and quantity of fishing resources. The respondents were then asked whether they are aware of climate change, its causes and consequences on the communities. The research team also linked the theme of questions with the importance of CTI and MRP conservation initiatives at these sites. A focus was given to costs as these could prevent the villagers from effectively adapting to the impacts of climate change.

Some of the general questions asked were as follows: In your understanding what are some of the negative impacts of climate change on the community? Do you think these projects have a positive or negative impact on the communities? What are some of the costs that the establishment of projects has caused to your livelihoods? What are some of the measures you employ to overcome these costs as barriers? What could the community do without the project initiative to adapt to the

---

[4]People who hold responsibilities, such as village organizers, chiefs and religious leaders etc.

[5]One group of interviewees ranges from 6 to 8 people.

impact of climate change, e.g. increasing sea-level rise, decline in fish catches etc.? What are the benefits of the CTI and MRP initiatives to your family unit? Why are CTI and MRP important to the communities?

Special Focus Group Interviews

Aside from the respondents to the semi-structured questionnaires, the research team interviewed another category of people during our trips to these sites. These were mainly fishermen aged from 25 to 60 years old. Examples of questions discussed with them are indicative but roughly included (1) Cost of fuel to new fishing ground, (2) Are there any benefits to the villagers? (3) How long do you (fishermen) spend on fishing? (4) What types of fishing methods did your community (fishermen) practice? (5) On the types of transport used for fishing (6) On the radio communication for weather warning? (7) Safety at sea (8) sea worthiness of the fishing boat.

Data Collection Technique and Time Frame

The principal researcher of this study collected the data with the three research field assistants. The engagement of the research field assistants from the respective sites was necessary since they knew the communities and could relate well with respondents during the interview process. Additionally the research field assistants knew the protocols and how to approach community leaders at respective site. The fieldwork was undertaken from 2 May to 8 June, 2013 in the respective conservation sites.

110 survey respondents and 40 semi-structured questionnaires were collected as part of this study. In the special group interviews the fishermen in particular were asked about costs and benefits: for example, how much cost they incurred when going to their new fishing grounds and in doing so what were some costs that acted as barriers in the course of adaptation to climate change. Some of the data around costs and benefits came also from the *semi-structured* questionnaires, which formed part of the household survey issued to the villagers. The questionnaires were asked in Solomon Islands' pidgin (a creole language and the lingua franca of Solomon Islands) and when respondents needed more clarifications, these were explained in the local language. The researcher and the field assistants' then interpreted and translated the responses into English, which was the official language the research team used in the research. Both the principal researcher and research assistants assisted in distributing the questionnaires and also take leading roles in interviewing in the semi-questionnaires.

Results and Discussion

The answers produced a variety of categories on costs versus benefits of establishing the projects at these three sites (Table 18.1). The variables used to obtain information included costs incurred by fishermen in travelling to new fishing areas, what forms of transport they use to the fishing sites, the type of boat they use, what were costs of other sources of income such as tourism sites etc., damage to the eco-system, time spent travelling to the fishing ground, risk faced in travelling to new fishing ground, traveling equipment e.g., radio and etc.

Benefits

It was challenging for project designers to ascertain the exact amounts in dollars against the likely costs and the benefits of the project, but an estimated figure would provide decision makers with a platform to select the best alternative investment option. Benefits may increase or diminish over time depending on the nature of the project. In this study, the focus of discussion narrowed down to cover only three main user groups of benefits and costs as identified from the conservation programs during the field trip. The first benefit was received by extractive users, the second by non-extractive users, and the third category by those in charge of management and implementers of the projects.

Table 18.1 Cost and benefits of coral reef and mangrove rehabilitation program

Categories	Benefits	Costs
• Extractive uses	• Increase in number of catches • Reduced variation in catches • Improved catch mix	• Decrease in catch • Congestion on the fishing grounds • Users' conflicts • High costs associated with choice of fishing location • Increase in safety risk
• Non-extractive uses (e.g. divers, ecotourism, and existence value)	• Maintain species diversity • Greater habitat complexity and diversity • High density level	• Damage to marine ecosystem • Loss of traditional fishing community
• Management	• Scientific knowledge • Hedge against uncertain stock • Assessments, skills and educational opportunities	• Increase in monitoring and enforcement costs • Opportunity costs (forgone economic opportunities, e.g. oil, gas, and mineral exploration and non-bio prospecting

Source Livelihood Survey 2013

Extractive Users

The extractive users referred to in this context are the project participants (fishermen) who use more than one technique to catch fish in these designated protected areas. According to the survey respondents, extractive users, who represent 38% of the respondents from the three sites, stated that there were increases in the size and number of species of fish surrounding the project sites. This was expected as the rehabilitation and replenishment of the coral reefs served to attract various types of fish to the surrounding coral reefs. Study findings of both Roberts et al. (2001) and Halpern and Warner (2002) support this expectation by the village participants. Thus the villagers had realised that establishing conservation projects benefited the community when they were allowed to fish after several years of abstinence from it. About 56% of Naro respondents also confirmed that the introduction of the CTI project enabled villagers' catch to be consistent, compared to the pre-CTI project period, when the villagers had to travel further before finding the next fishing ground to catch enough for either household consumption or commercial purposes.

More than 77% of the respondents from the three sites have confirmed that the fishermen spent less time to catch some varieties of fish compared to the pre-CTI period. Contrary to Oibola, where fishermen travelled long distances to their next fishing ground, at Naro they save a lot of time fishing adjacent to the project site. From households' perspective, less fishing time implies more time to attend to village communal activities such as church activities, education, health and other alternative livelihood activities such as farming, casual employment or tourism.

As it was expressed by the village chief Cornelius of Naro community, prior to the CTI project implementation more of their farming time was allocated to fishing. About 40% of the respondents did not practise any farming; however, the remaining 60% dedicated between 0 and 12 months a year to farm activities (Table 18.2). The constraint this has caused to the villagers was that less time people spend in their farms, thus allowing them to spend longer time for fishing and at times because of poor fishing boats and gears during cyclones and storm surges fishermen lost their lives at sea.

Those 77% furthermore confirmed that the mix of fish catches had also improved after the implementation of the CTI projects. According to Roberts et al. (2001) the establishment of MPA benefits the marine ecology and increases the varieties of fish

Table 18.2 Time spent on farms by respondents

Description (months)	Percentage
0	40
0–3	31
3–6	13
6–9	7
9–12	9
Total	100

Source Household Survey 2013

within the protected areas. The Naro community spokesman, Chief Cornelius Vulu expressed the opinion that prior to the implementation of the CTI project, the community was unable to raise enough finance to complete their church building, a project that had dragged on for more than 20 years. The community members had tried all available means to raise funds towards completion of the church project without success. It was only after the introduction of the CTI project that the community managed to raise enough funds from fishing, during the permitted times at the MPAs to make progress on the construction of the church building. During those past 5 years of the moratorium they were allowed to fish in the protected area fewer than eight times. The community church building was now completed and dedicated during a recent field trip in 2015.

The communal achievement that villagers gained from these projects motivated them to cooperate in their efforts to ensure these conservation initiatives were successful. This common achievement has bred unity amongst the villagers to cooperate in their efforts and ensure this current project is successful. Such communal achievement strengthened the bonds among the villagers, which is a social benefit to the community.

Non-extractive Users

There was also a group among the interviewees that was classified as non-extractive users at the selected sites. These non-extractive users have adopted some fishing techniques and methods that are selective and aimed at catching enough fish and not exploiting them during the processes. One of the benefits that respondents had experienced with this type of fishing was maintaining the various species of fish within the CTI project implemented areas compared to the pre-CTI and MRP implementation period. The fishermen claimed that the number of fish both in species and varieties had also increased at these sites. They perceived that there was also an increase in the habitat organisms living in the conservation implemented areas. The fishermen further felt the increase meant that there was complexity in the number of living organisms, which was seen as advantageous to the biodiversity surrounding these MPA implemented sites.

In addition, another benefit that the respondents experienced was the increase in small business entrepreneurship such as ecotourism in the surrounding MPA (CTI) designated areas. According to Chief Stanley Hebala[6] of Sairaghi, over 15 small businesses were established during the past 5 years in the area (2008–2012). This benefit had never been experienced at this growth rate before at this site. He stated that the introduction of the CTI project in this site has prompted villagers to invest in alternative revenue generating businesses. These micro-businesses that were established surrounding the coral reef protected area acted as positive

[6]Chief Stanley Hebala is the one of the key informants at the Sairaghi site.

externalities to the surrounding communities and greatly relieved the pressure villagers were putting on the marine resources. This has diversified the sources of income for the community members at these three selected sites from depending solely on marine and agricultural produce as their primary source of income and as such has helped them to overcome one of the barriers to climate change adaptation at these sites such as the "barrier of limited alternate source of income at these sites".

Management

In terms of management benefits to the villagers and implementing agents, these conservation initiatives have provided opportunities for the management (especially locals) to learn some basic scientific knowledge about the technical nature of these projects. The project site managers have admitted having very little knowledge in general about marine science skills and knowledge prior to working for the projects. After working for the CTI projects they were quite familiar with the basic skills and knowledge of rehabilitating the coral reefs and enabling that knowledge to grow and pass on to the villagers as the project progressed. For example, Chief Cornelius Vulu of Naro community has confirmed learning the skills on how to manage an MPA from a village community perspective. He further explained this task required some technical skills and knowledge if this activity is to be performed effectively. However, as benefit of the CTI project implemented sites, it was discovered that the project coordinators were able to perform effectively this form of skills learned through the CTI project implementation for the benefit of the communities.

The project site managers have access to various educational training opportunities during the course of the project to foster better management of their respective project sites. This knowledge opportunity was important as it acted as the basis for educating the villagers and managing the project sites sustainably into the future. Some of the villagers were unable to support the projects because of their limited knowledge about the CTI and MRP objectives within the country.

Costs

These environment conservation designers have also taken into account the factors that will have negative impacts on the communities. In most cases these were seen as constraints that limits effective adaptation of these environment conservation programs. The general rule of thumb is that the expected benefits from the projects must outweigh the associated costs and expenses that might be incurred during the establishment and implementation of the project. Costs of CTI projects were explored under the same three categories as benefits: extractive and non-extractive users and management (see Table 18.1).

Extractive Users

According to some respondents there was a huge decrease in the number of catches in the selected fishing areas. This was a direct cost incurred from limiting the fishing ground available to villagers closer to the project site. Besides this, the MPAs were governed by policies and regulations preventing fishermen or villagers from fishing at any time during the year. In one of the project sites this has caused the extractive users to resort to illegal fishing techniques. According to chief Benjamin of Oibola community, the extractive users often resorted to illegal fishing techniques and engaged in fishing methods and techniques that would destroy the corals and the marine ecosystems. These illegal fishing techniques were regarded as barriers to adaptation because they damage the coral reefs and destroy the marine ecosystem for generations a head.

Congestion and over-fishing in the allocated fishing ground was another consequence of MPAs. This was seen as a barrier to adaptation because it destroyed the unity and harmony in the village when people of same tribe disputed their ownership of the protected site. Sairaghi project site experienced a similar situation in the late 2000s. A disagreement between two clans on fishing rights in an allocated fishing ground triggered an argument on who rightfully owned the surrounding coral reefs. The disagreement on these rights had later negatively affected the support and management of the coral reef rehabilitation and replanting in the area during that period and led to the failure of the effort at coral management and MPA establishment. Similar incidents of disagreement were also found among the villagers at the Naro and Oibola communities (Table 18.3).

The majority of respondents have agreed with the establishment and general management of the conservation projects at these selected sites (71%). However, 13% declared that there was disagreement because of employment issues and 10% believed that their main cause of disagreement was due to a leadership issue. A further 6% stated that land disputes were the most significant factor that was responsible for most of the disagreement and disunity amongst the villagers about the establishment of these projects.

Furthermore, one of the costs evident amongst the villagers was the safety risk of fishermen travelling to these fishing grounds, especially under bad weather, unsafe fishing boats, poor communication system with no safety gears on boat. In some earlier incidents, fishermen from these locations got lost out in the high seas after travelling long hours to some of these fishing grounds. For example, the Oibola

Table 18.3 Reasons for dispute over the coral triangle initiative implemented areas

Description	Percentage
Employment issues	13
Land ownership	6
Leadership issues	10
No incident of land dispute	71
Total	100

Source Household Survey 2013

community fishermen had to travel on every fishing trip into open sea and were exposed to a higher risk of losing their lives. The Naro community fishermen had to travel across to Gela in the Central Island province for their next fishing ground. The other option for the Naro community villagers was to buy their fish from the central market which was an expensive option to the villagers. The above points were compounded as constraints to adaptation and compliances to the conservation programs.

Non-extractive Users

Non-extractive users used part of the coral reefs to establish ecotourism development, paid beaches and other tourism related activities along the beaches and the surrounding reefs, diversifying the alternative sources of income. However, such ventures could destroy the ecosystem and the marine ecology of the marine resources in the long term, particularly the coral reefs in the surrounding area. This was evidenced along the beaches on the west side of Sairaghi. Chief Stanley Hebala stated that since the establishment of these facilities; users of the beaches have been destroying and disturbing the coherent existence of the marine ecology at these allocated tourism sites. As the number of tourists increases, so does the villagers' revenue (short-term benefit) but at the same it increases the chances of tourists and other users damaging the ecology, which will be felt over a considerable length of time into the future.

Management (Costs)

The establishment of the CTI projects within the country has brought higher direct and indirect costs to the management teams of all the project sites throughout the provinces. The MECDM, the supervising ministry, incurred additional costs by employing additional staff to oversee the management and administration of the CTI project nationwide. On the operational level, after the establishment of the CTI project along the western side of Gizo, the coral reef management team has brought in additional expenses such as training and enforcement and time and monetary costs for Sairaghi project sites communities. Chief Hebala from Sairaghi confirmed that some villagers and potential investors displayed huge interest in exploration of the coral reef areas for sea mining and other mineral potential but they have now decided against this, with the intention of conserving the coral reefs for present and future generations.

Most of the respondents believed that the conservation projects have been good and beneficial to the communities. The respondents believe that effective and sustainable management of the projects is important to the long-term viability and beneficial to the people of these selected sites (Table 18.4).

Table 18.4 Reported benefits of the CTI and MRP at the three sites

Description	Percentage
The community respondents dislike the projects	17
Future generations benefit from the resources	20
Community survival (ecologically)	27
Community survival (economically)	31
The project may benefit if effectively implemented	5
Total	100

Source Household Survey 2013

Table 18.5 Decision making in favour of establishment of the CTI and MRP

Description	Percentage
Youth representatives	29
Mothers' group representatives	17
Senior NGO officers	39
Senior government officers	15
Total	100

Source Household Survey 2013

The CTI has also brought in multiple opportunities for the community members to gain employment and assume leadership roles within the community over the management of the projects (van Beukering et al. 2007). This was evidenced with the Naro and Oibola communities' coral reef and mangrove rehabilitation project sites. These started out as community based coral reef conservation and demonstration sites, and the villagers have organised themselves and allocated certain individuals with responsibility as full-time caretakers over the project sites for the projects' duration. The communities also set up coral triangle initiative village based committees with the objective of better managing the coral triangle initiatives and mangrove rehabilitation programs at the selected sites. As was learned during the field trip, women and youths are now active members of these committees.

In terms of decision making, with the introduction of CTI, external institutions were often the decision-makers for the community to apply and negotiate for the project. This was never the idea of the community leaders, and such projects as this often were unsuccessful and failed to live up to expectations. Thus, the study noted from the research that the community participation by different demography of villagers in the community project is important for its long-term success. The communities' latest inclusion of females, youths, church leaders, community leaders and project coordinators into the decision making and general governance of the projects is vital for their long-term success (Table 18.5).

Furthermore, a short-term benefit from this project has been that the community members were also given financial assistance and cash benefits while working during the establishment of these MPAs and the coral reefs rehabilitation at the initial stages. These economic activities have broadened the cash inflow into the communities and thus assisted in improving the livelihood of the people living in

Table 18.6 Main sources of income for the three project sites

Description	Percentage
Money from non-agricultural shops	22
Income from fishing and marine resources	41
Income from wages	16
Income from own farm	21
Total	100

Source: Houshold Survey 2013

the surrounding areas of these three project sites. According to a similar study undertaken previously on the Arnavon community in the Choiseul Province, the introduction of a similar project has economically impacted their livelihoods (van Beukering et al. 2007). The employment opportunities that the members of these three communities received became their sources of income for the duration of the projects. Although some communities' members did not engage in longer-term employment contracts for the duration of the project, the CTI indirectly contributed significantly, contributing as part of their sources of income and cash flow within these respective sites through the sale of fish and other marine based products.

It was also evidenced from the research project that the general income from agricultural sources has been on a downward trend over the last 5 years, the period of the study. On the other hand, there was a positive surge in the marine resources as the primary source of income to the respondents of Sairaghi and Oibola communities. 41% of the respondents also supported this perception, which was shared by the interviewees, that villagers in general are now heavily dependent on marine products as their livelihood (Table 18.6).

Conclusions

One of the conclusions of this research is that conservation of coral reefs and mangrove rehabilitation are vital for the food security and livelihood sustenance of the people at these project sites. The conservation of these habitats not only provides a means to replenish and rejuvenate the marine ecology, but also improve the livelihood, provide barriers against storms, which minimize wave impacts from storms such as cyclones on the communities. At the three project sites the positive impact that these three environment conservation programs have on the communities are undeniable. Unfortunately, because of both human and climate change impacts on coral reefs, these ecosystems are becoming increasingly degraded and in need of conservation even more than ever. The biggest threats include overfishing, destructive fishing practices, and sedimentation and pollution from land-based source including acidification, coral bleaching, and diseases. The conservation measures now in place are aimed at reversing the impediments it has on the people and community at large. These measures are based on the assumption that people are now depending on the conservation programs as a means to secure the viability of these ecosystems in the future.

However, during the conservation programs communities have to overcome some basic costs as barriers to the adaptation in their endeavour. For example, while the MPAs and MPRs are in place these communities' residents often risk their lives in high seas on unsafe fishing boats with poor lifesaving equipment just to catch enough for their business or family use. These costs act as constraints and barriers to effective adaptation and therefore the government must provide alternative income source to diversify the villagers' dependency on marine resources (especially fishing) by helping them to over some of the barriers for effective adaptation.

Policy Recommendations

The findings of this study provide an opportunity to make three policy recommendations.

(a) Firstly, environment conservation programs are vital for the sustenance ecosystem which rural communities depend on for their livelihood. As such, in order for these communities to participate faithfully in these projects, the government or aid donors must provide alternate livelihood endeavors to ensure these villagers cope with the impact of climate change. A variety of alternative livelihood micro-projects such as piggery, poultry, sawing programs with women and others should be implemented. These will enable the communities to depend less on marine resources, leading to a lesser pressure on their use.
(b) The government should investigate ways of establishing a micro-finance scheme to allow fishermen to invest in safe and quality fishing assets. This recommendation includes the establishment of better and reliable weather forecasting systems and other relevant climate services.
(c) Finally, further in-depth studies are needed in order to move towards an improved characterization of adaptation and to identify the most suitable means to overcome the costs and barriers to adaptation.

Acknowledgements This research was supported by the University of the South Pacific particularly the Pacific Center for Environment and Sustainable Development; the Research Office and Government of Solomon Islands. The lead author is thankful to his supervisors who provided expert advise that greatly assisted the paper.

References

Adger, W. N., Dessai, S., Goulden, M., Hulme, M., Lorenzoni, I., Nelson, D. R., & Wreford, A. (2009). Are there social limits to adaptation to climate change? *Climatic Change, 93*(3–4), 335–354.

Albert, S., Grinham, A., Bythell, J., Olds, A., Schwarz, A., Abernethy, K., et al. (2010). *Building social and ecological resilience to climate change in Roviana, Solomon Islands*. Brisbane: University of Queensland.
Albert, J. A., & Schwarz, A. J. (2013). *Mangrove management in Solomon Islands: Case studies from Malaita Province*. Penang, Malaysia: CGIAR, Research Program on Aquatic Agricultural Systems.
Barnett, J. (2011). Dangerous climate change in the Pacific Islands: Food production and food security. *Regional Environmental Change, 11*(1), 229–237.
Carneiro, G. (2011). Marine management for human development: A review of two decades of scholarly evidence. *Marine Policy, 35*(3), 351–362.
Cinner, J. (2005). Socioeconomic factors influencing customary marine tenure in the Indo-Pacific. *Ecology and Society, 10*(1), 36.
Clifton, J. (2009). Science, funding and participation: Key issues for marine protected area networks and the Coral Triangle Initiative. *Environmental Conservation, 36*(02), 91–96.
Dudley, N. (Ed.). (2008). *Guidelines for applying protected area management categories*. Gland: IUCN.
Foale, S., Adhuri, D., Aliño, P., Allison, E. H., Andrew, N., Cohen, P., et al. (2013). Food security and the coral triangle initiative. *Marine Policy, 38*, 174–183.
Halpern, B. S., & Warner, R. R. (2002). Marine reserves have rapid and lasting effects. *Ecology Letters, 5*(3), 361–366.
Hannesson, R. (2002). The economics of marine reserves. *Natural Resource Modelling, 15*(3), 273–290.
Kelman, I., & West, J. J. (2009). Climate change and small islands development states, a critical review. *Ecological and Environmental Anthropology, 5*, 1–16.
Klein, R. J. T., Midgley, G. F., Preston, B. L., Alam, M., Berkhout, F. G. H., Dow, K., et al. (2014). Adaptation opportunities, constraints, and limits. In: C. B. Field, V. R. Barros, D. J. Dokken, K. J. Mach, M. D. Mastrandrea, T. E. Bilir, M. Chatterjee, K. L. Ebi, Y. O. Estrada, R. C. Genova, B. Girma, E. S. Kissel, A. N. Levy, S. MacCracken, P. R. Mastrandrea, & L. L. White (Eds.), *Climate change 2014: Impacts, adaptation, and vulnerability. Part A: Global and Sectoral Aspects. Contribution of Working Group II to the fifth assessment report of the Intergovernmental Panel on Climate Change* (pp. 899–943). Cambridge, United Kingdom and New York, NY, USA: Cambridge University Press.
Lawrence, A. (2012). *Blue Carbon: A new concept for reducing the impact of climate change by conservation coastal systems in the Coral Triangle*. World Wide Fund Nature (WWWF).
Longhurst, R. (2003). Semi-structured interviews and focus groups. In: N. Clifford, & G. Valentine (Eds.),*Key Methods in Geography* (pp. 117–132). London: SAGE Publications.
Maczulak, A. E. (2009). *Conservation: Protecting our plant resources*. New York: Conservation InfoBase Publishing.
McCarthy, J. J. (2001). *Climate change 2001: Impacts, adaptation, and vulnerability: Contribution of Working Group II to the third assessment report of the Intergovernmental Panel on Climate Change*. Cambridge: Cambridge University Press.
MECDM Solomon Islands National Adaptation Plan of Action. (2008). *Ministry of Environment, Climate change, Disaster Management and Meteorogy*. Honiara, Solomon Islands.
Nichols, K. (1999). Coming to terms with "integrated coastal management": Problems of meaning and method in a new arena of resource regulation. *The Professional Geographer, 51*(3), 388–399.
Nielsen, J. Ø., & Reenberg, A. (2010). Cultural barriers to climate change adaptation: A case study from Northern Burkina Faso. *Global Environmental Change, 20*(1), 142–152.
Pauku, L. R., & Lapo, W. (2008). *The Solomon Islands national biodiversity strategy and action plan*. Honiara: Maraghoto Holdings Company Limited.
Roberts, C. M., Bohnsack, J. A., Gell, F., Hawkins, J. P., & Goodridge, R. (2001). Effects of marine reserves on adjacent fisheries. *Science, 294*(5548), 1920–1923.

Sore, R. (2010). *Solomon Islands National Plan of Action: Coral triangle initiative on coral reefs, fisheries and food security*. Honiara: Ministry of Environment, Disaster Management and Meteorology (MECDM).

Sulu, R., & Boso, D. (2011). *State of the Coral Triangle Report of the Solomon Islands*. Honiara: National CTI Coordinating Committee.

van Beukering, V. P. J. H., Scherl, L. M., Sultanian, E., & Lisher C. (2007). *Arnavon community marine conservation area (Solomon Islands). Case Study 2. The role of marine protected areas in contributing to poverty reduction*. Honiara, Solomon Islands.

Chapter 19
Customary Land and Climate Change Induced Relocation: A Case Study of Vunidogoloa Village, Vanua Levu, Fiji

Dhrishna Charan, Manpreet Kaur and Priyatma Singh

Introduction

The South Pacific being the hub of climate change associated environmental and social developments is irrefutably one of the world's most predisposed regions when it comes to the climate and weather induced disasters (Boege 2011). Particularly susceptible are the several of the low-lying coral islands (Nunn 2012). The livelihoods of majority of the Pacific Islanders which revolve around the Pacific Ocean are being acutely affected due to rising sea levels, increased coastal erosion, inundation, flooding and salinization of coastal aquifers (Ferris et al. 2011). For several of the communities in the South Pacific, adaptation has become an immediate necessity for survival. The pressing need to acclimatize to climate change adversities has escalated over the last couple of years and the issue of climate change taking its toll on many island nations has surfaced in recent discourses (Barnett and Campbell 2010).

On the onset, Fiji's marine and coastal ecosystems endow considerable physical, financial, societal, ecological and cultural benefits to approximately half of the country's estimated 902,964 population (Govan 2009). Yet, the repercussions of climate change on the coastal ecosystems are threatening the way of life of the coastal inhabitants and for the residents of Vunidogoloa in the province of Cakaudrove in Vanua Levu, relocation has emerged as a reality for more than three decades. In February 2014, the village was the first in Fiji to reposition; moving

D. Charan · P. Singh
Department of Science, The University of Fiji, Saweni, Lautoka, Fiji
e-mail: drishnac@unifiji.ac.fj

P. Singh
e-mail: priyatmas@unifiji.ac.fj

M. Kaur (✉)
Department of Language and Literature, The University of Fiji, Saweni, Lautoka, Fiji
e-mail: manpreetk@unifiji.ac.fj

W. Leal Filho and J. Nalau (eds.), *Limits to Climate Change Adaptation*,
Climate Change Management, https://doi.org/10.1007/978-3-319-64599-5_19

2 km inland after years of inundation, storm surges, coastal abrasion and unwarranted flooding had made their village susceptible to the impacts of climate change (United Nations Office for the Coordination of Humanitarian Affairs 2014). The traditional responses of disaster relief were no longer protecting the village community despite thousands of dollars spent on the construction of sea walls. Community relocation was the only cogent solution to safeguard the inhabitants of Vunidogoloa (Edwards 2012, p. 3). Conversely, this was an enormously emotional and harrowing headway for the villagers especially since they had to retreat from their customary land which has been part of their culture and identity for their entire life.

Relocation may be the last resort but also one of the best adaptation responses for several of the coastal Fijian villages currently facing similar tribulations as Vunidogoloa (Rubelli 2015). This also indicates that quite a few of these vulnerable villagers will be experiencing similar limitations as faced by the people of Vunidogoloa. Some of the drawbacks are the availability of land for settlement, governance and funding and perhaps the most intricate of all is the traditional and emotional place attachment. Disputes over land rights as well as loss of social and communal cohesion will highly likely create some of unconstructive effects of population relocations (Ferris et al. 2011). According to Wewerinke (2013), the cultural identity of the people is likely to be impeded by the loss of customary land that is anticipated to occur as a result of climate change. A report by Nurse et al. (2014a, b) explains that barriers to taking action have also been attributed to endogenous factors such as traditional values and awareness.

In many indigenous communities access to land depends on membership in a specific clan. For the iTaukei (indigenous Fijians) the ownership of land is vested in the *mataqali* (Fijian clan or landowning unit) (Fonmanu et al. 2003). Land offers not only livelihood but it is also the source of the traditional and spiritual wellbeing for many of the island communities. This is why despite the distressed situation on the islands there are still people who do not want to relocate (Boege 2011). Generation gap also influences the decision to relocate. In the Vunidogoloa resettlement case, it was particularly the elderly who did not want to move, while members of the younger generation were keen to move.

Developing countries also have a major limitation in capacity making adaptation difficult. Limitations include both human capacity and financial resources. The lack of funding available in various forms and difficulties in accessing the funds, which are available, represents a major barrier for adaptation, particularly for local community action (United Nations Framework Convention on Climate Change 2006). Climate-induced population displacement entails a governance and policy framework that can holistically respond to communities challenged with harsh impacts of climate change. Lack of proper awareness and institutional capacity also limits adaptation process (Amundsen et al. 2010). Relocation of Vunidogoloa village provides an opportunity to address the multiple societal issues to foster long-term sustainability in the process of relocating communities.

The document on Peninsula Principles on Climate Displacement within States (2013) forms a preliminary guiding framework and premise for policy and lawmakers, based on current international law. Myriad doctrines such as community engagement and consent, provision of affordable housing, land solutions, basic services and economic opportunities to those affected, have been experiential in Vunidogoloa.

This study provides a synopsis of some of the fundamental challenges encountered by the people of Vunidogloa village from the inception till the completion of the entire resettlement. This paper considers the limitations of adaptation as exemplified by testimonies from some key people of Vunidogoloa village and the government administrators. In particular, it accentuates the intricacies surrounding the socio-cultural aspects of the relocation process. The loss of Fijians customary land that is projected to occur as a result of climate change could plausible impede with their cultural identity and associated climate induced repositioning.

The challenges to relocating the community manifestly exhibits the exigency for new policies and procedures that specifically respond to climate induced relocation. This paper concludes by proposing some strategies that can be applied to accomplish an improved transition that suits closely knit Fijian communities as a whole and also cares for the various socio-cultural facets that embrace the community.

Aim and Method

Many forecasts have been made to predict the number of climate change induced migration. The International Organization for Migration (2014) estimates the number of environmental migrants to vary by a factor of 40, ranging from 25 million to a billion by the year 2050. In the past, more than 15 million people have been estimated to have been displaced due to natural disasters annually around the globe. These numbers are projected to increase significantly due to the increasing risk brought upon by climate change (International Displacement Monitoring Centre 2014). The government of Fiji, in recognition of the eminent dangers and urgency of climate induced relocation, is currently engaged in a design and consultation process of establishing a relocation guideline to assist communities forced to migrate due to climate change impacts. The vulnerability assessments carried out by the Fijian government have identified as many as 830 communities that are at risk from climate change with 45 of these being recommended for relocation (Fiji Broadcasting Corporation 2015). The few communities that have fully or partially relocated have depended on government assistance to carry out the relocation.

This article examines Vunidogoloa village relocation with respect to the challenges faced by the villagers, essentially in terms of their attachment to the land, cultural and community cohesion and governance and funding. The sentiments surrounding relocation are complex and as such questionnaires were employed to gain an insight into the way people of Vunidogoloa perceive climate change and its impacts. The questionnaire comprised of a range of questions related to the topic of this research.

Interviews using the questionnaire were conducted as means of obtaining qualitative data from the villagers and key informants. In-depth semi-structured key informant interviews were undertaken with government officials involved in this relocation (n = 10). Structured interviews were also conducted with villagers (n = 20). The interviews were taken in English and iTaukei language and focused on eliciting information on the experiences of the villagers during the relocation and the contribution from the government during the entire process. Communication in iTaukei language was carried out and interpreted by a translator who was engaged to assist the interviewers.

Focus of Vunidogola Relocation

The Vunidogoloa village is located in Cakaudrove Province in Fiji's second largest, northern island of Vanua Levu. The original site of the village was located on a tract of land overlooking the Natawa Bay in the rural town of Savusavu with the houses sited only a few meters inland from the coast. A report by Nurse et al. (2014a, b) predicts with certainty the acceleration in sea level rise for the smaller island countries in the Pacific and forecasts severe sea floods and erosion for low-lying coastal areas. This holds true for the previous site of the Vunidogoloa village where heavy rain and high tides continuously coalesced causing inundation and salt water intrusion, rendering local community gardens extremely difficult to manage and causing widespread damage with floodwater incessantly threatening the safety and health of the villagers (McNamara and Des Combes 2015). Following abortive attempts to adapt by building seawalls, which easily succumbed to raging waves, in 2014 the villagers moved two kilometres inland from their original village to a new site, which fell within the customary land boundaries of the community. The villagers named this new site as *Kenani*—a Fijian word for the biblical word *Canaan for* "promised land".

Undoubtedly, the issue of relocation is a sad predicament facing many communities around the world. More so, many villagers like those in Vunidogoloa in their "promised land" are increasingly and innocently being subjected to complex problems such as forced land abandonment and likely socio-environmental changes while having a very minute contribution to anthropogenic climate change and associated sea level rise. The paper by Mitchell et al. (2015) proposes a joint response by the government and community to ensure the success of such relocations. The government of Fiji largely organized the Vunidogoloa relocation as an integrated action across different government ministries.

Dynamics of Land, Identity and Adaptation

Irrefutably, Fijians have always shared a special relationship with land, regarding it as a foundation of their identity, a place that defines them as the rightful owners of their land, closely associated to the natives' wealth, status and placement in their respective *mataqali's*. Hence, land is revered nowhere more fervently than here at home in Fiji. For the natives of Fiji, land is viewed as being more than a resource, with islanders possessing an instinctive and spiritual attachment to their land (Edwards 2014a, b). An iTaukei community refers to their land as *vanua*, a term which unites the concepts of personhood and land ownership as an inseparable entity where the ownership of an area of *vanua* is translated to mean the land area one is identified with.

Majority of land in Fiji is owned in trust by natives with the remainder being state land and freehold (Boydell 2001). Currently, 87% of the land is owned by traditional Fijian *mataqali's* and according to the laws of the country, this land is prohibited from being sold, exchanged or sub-let by the *mataqali*. The iTaukei Land Trust Board (TLTB) oversees all the indigenous land and is also responsible for liaising with various *mataqali's*. Depending on the size, a village can have several land-owning *mataqali's* within its boundaries. As a result of such strong land laws, the Fijians remain a proud race, forming strong spiritual and family connections with the land (Boydell and Reddy 2000), which is an integral component of Fijians' communal life style.

Boydell and Shah (2003) compacts this relationship fittingly by suggesting that a customary land belongs to a communal stewardship rather than an individual with the *mataqali* entrusted with the responsibility of taking care of the land for the spirits of one's ancestors, use of one's life and protection and to ensure sustainability of the land for one's descendants. Due to these reasons, relocating becomes psychologically stressful for the entire village. They collectively view the process of relocation as agonizing, poignant and one that is robbing them of their communal identity. This shared identity of belonging to a *mataqali* and people of a respective *vanua* is a pride that they uphold as landowners and landowning units in their village. Hence, acquiring alternative land for relocating is a primary obstacle and this is even true for communities where such land is available as villagers habitually associate leaving their ancestral homes as potentially losing their identity and becoming a displaced, landless people (Edwards 2014a, b).

Such is the overwhelming case of the Vunidogoloa village, which moved to a land owned by the same *mataqali*. Since the community moved to a land within its boundaries, the transition was smoother than what the villagers would have experienced if a suitable resettlement site was not available. Noteworthy is the fact that even though the relocation was within the vicinity of the existing place, the prospect of leaving their place of birth, the very land that defines them and solidifies their identity, has a pessimistic bearing on the villagers.

The unwillingness to relocate was echoed by a villager who blatantly responded, "We were trying to adopt by our own so that we don't have to leave our land and each time the sea came to our doorsteps, we moved a little away from it until it became so worse that we knew we had to relocate" (Vunidogoloa village interviewee, personal communication, 28 January 2016, Savusavu, Fiji). Sovaraki (2014) quoted a village spokesman of Vunidogoloa who revealed that plans to relocate had begun in 1956 but due to the reluctance of the elders to leave their ancestral grounds/boundaries, they had to wait till 2006 to start the process of relocating. The hold of the land was so great that it bound the villagers for almost half a century, delaying their process of adaptation till the sea burst its bounds. Moving away from their land inevitably affects an individual's physiological and mental health, particularly in the context of disaster recovery and extreme weather events (Frumkin et al. 2008). Social networks and community connections are likely to be fractured as a result of climate induce displacement.

The prime contributing factor of the Vunidogoloa relocation stemmed from the constant salt-water inundation, which coupled with the consequent relocation to another land, cogenerated significant disruptions to the social and mental health of the villagers. The emotional distress and anxiety of leaving their ancestral land was expressed by a villager who vividly recalled, "Initially relocating was not an option to us at all but climate change came like an enemy that chased us away by taking our land, taking our food, taking everything" (Vunidogoloa village interviewee, personal communication, 28 January 2016, Savusavu, Fiji). The connection to the land and the environment, traditions and customs associated with it are very profound for a Fijian community, forming part of their identity. Relocation due to rising sea level is likely to cause communities to have some physiological impacts while they try to re-build their lives in a new location after their upheaval from the ancestral lands.

The Cultural-Spiritual Dimension of Resettlement

The findings also recount the emotional ordeal, which obscures the process of relocation. As recollected by one of the villagers in an interview, "Movement to new site was very painful and upsetting" (Vunidogoloa village interviewee, personal communication, 25 November 2015, Savusavu, Fiji). This is a precise indication of the level of attachment and sentimental value upheld by people who are in the process of relocation. As shared by the villagers, it was a difficult situation to leave the old site where they had lived all their lives, and to go through the harrowing decision to disentomb and shift the remains of their ancestors to a new burial site. In an interview with another village representative, it was noted, "We didn't want to leave the cemetery where it was, to be washed away, so the church arranged for the burial site to be moved. Sadly the first burial at the new site was that of a still-born child which is interpreted as a bad omen from our ancestors" (Vunidogoloa village interviewee, personal communication, 28 January 2016, Savusavu, Fiji).

The Fiji Police Force facilitated the relocation process and in a conversation with one of the police officers, it was recorded that, "the spiritual connection to the land was so strong that some families were adamant not to move despite being fully aware of the severity of climate change impacts" (Fiji Police Force interviewee, personal communication, 28 January 2016, Savusavu, Fiji). All the villagers interviewed during field study indicated that climate change and the need to relocate had always been a topic of discussion at community gatherings, church assemblies, and even at homes amongst family members and such deliberations had often ended with sundry responses from the people. It seemed that the youngsters were willing to relocate while the elders of the village could not come to terms with having to uproot an entire community and relocate—it meant detachment from their identities, parents and cultural ambiance. In an interview with an administrator from Cakaudrove Provincial Council, it was recorded, "People were not ready and discussions concerning relocation took several years" (Cakaudrove Provincial Council interviewee, personal communication, 25 November 2015, Savusavu, Fiji).

For the people of Vunidolgoa, relocation would not have eventuated if it was not utterly necessary. The old site was linked to some exceptional and momentous nostalgic significance. The day of actual relocation was the most difficult moment in the lives of the people of Vunidogoloa. A representative of the Vunidogoloa Women's group described the movement, "as if a funeral procession was underway" (Vunidogoloa village interviewee, personal communication, 27 January 2016, Savusavu, Fiji). Although, the relocation for Vunidogoloa village was planned and executed with the consent of the villagers the entire process from its inception to completion created a profound spiritual predicament. An increased feeling of alienation and stress associated with relocation may pose threat to the health, and the overall welfare of the people (Fresque-Baxter and Armitage 2012).

The Vunidogoloa relocation is the first successful project of its kind in the South Pacific and for Fiji. However, in an interview with an administrator of Ministry of iTaukei Affairs it was revealed that climate change adaptation through relocation may not be so easy to achieve for many of the other vulnerable Fijian villages (Ministry of iTaukei Affairs interviewee, personal communication, 26 November 2015, Suva, Fiji). Piazza and Bolalevu (2014) elucidate that the most important factor is the willpower of the people especially since the cultural and spiritual significance that indigenous people attach to their lands and territories goes far beyond any monetary or productive value or even the value of their life.

In the case of another village, Vunisavisavi, also located in the Cakaudrove Province, the relocation is impeded by traditional obligation of the villagers to the Tui Cakau (High Chief of Cakaudrove Province in Fiji). Silaitoga (2016) quoted a village spokesman of Vunisavisavi village who revealed that the cultural obligation towards the *Tui Cakau* was an ancestral tradition and upheld great values. The villagers believe in customary punishment succeeding relocation, which will affect their livelihood at the new location. The residents of Vunisavisavi treasure their land and associated cultural values so much that despite being strongly advised to relocate, the villagers are not willing to budge (Frontline Truths 2015).

In an interview with a Vunidogoloa villager representing the chiefly clan, it was highlighted that there is a need to develop the faith of the people. He mentioned "detachment from customary land is heartbreaking and it is important to move people together with their church and faith to make relocation a success" (Vunidogloa interviewee, personal communication, 29 January 2016, Savusavu, Fiji). People's opinion of climate change and its impacts is greatly influenced by the church, which can at times act as a significant barrier to adaptive capacity (Kuruppu and Willie 2015). The Pacific Conference of Churches, village church leaders and the village chiefs play a pivotal role in ensuring that the traditional values are integral during and after the repositioning process. For indigenous societies, every aspect of climate change is mediated by culture. Community cohesion and place attachment are also key elements in sustaining indigenous societies actions against climate change (Adger et al. 2013).

Adaptation Limits: From a Governance and Funding Perspective

The initial consultations between the village headman and the Fijian Government officials occurred in year 2006. It was not until year 2012 that some events started to transpire. During this period the villagers waited tensely for the government to reciprocate (United Nations Office for the Coordination of Humanitarian Affairs 2014). In the 2012 national budget, the Fijian Government sanctioned an allocation of FJD $1.0 million to the Ministry of Provincial Development and National Disaster Office for Disaster Risk Reduction measures. The total for Vunidogoloa relocation project cost is estimated at around FJD $980,000 of which FJD $740,000 was government contribution while an approximate sum of $240,000 was subsidized by the community in the form of timber used for construction (Cakaudrove Provincial Council interviewee, personal communication, 29 January 2016, Savusavu, Fiji). Timber was provided from their forest reserves. Relocation is undeniably costly in many ways and is often considered as a last resort for vulnerable communities (McNamara and Des Combes 2015). Community contribution is essential as it eases the financial burden endured by the government. However, several coastal Fijian communities have limited resources and may not be able to make significant contributions, which will further exhaust government funding.

The legislative issues coupled with community consultations also complicate the relocation process and results in delays. This concern was noted in an interview with a Fijian government official, "It is not easy for government to relocate a community, as witnessed in the case of Vunidogoloa, the process takes a lot of time as a lot of government departments are engaged, and there is a lot of paper work" (Fiji Government interviewee, personal communication, 28 January 2016, Savusavu, Fiji). Similar concern was noted in another interview with an iTaukei provincial council official, "It takes a lot of time to consult with iTaukei administrators and ensure that indigenous protocols are observed" (Cakaudrove Provincial Council interviewee, personal communication, 26 November 2015, Savusavu, Fiji). There is an immediate need not only for Fiji but for other Small Island Developing States to reinforce synchronization between various departments responsible for climate change adaptation and disaster management (Kuruppu and Willie 2015).

Fiji's current climate change policy does not specifically address relocation as an adaptation measure. There is no clear mandate to accommodate the relocation process and therefore, a national relocation guideline is being conscripted, however, it is still a work in progress (Pareti 2015). There is no clear indication as to much longer it would take to establish the guideline, which must support a factual and participatory approach to planning and execution of relocation (McNamara and Des Combes 2015). A lack of legal and policy frameworks, procedural and human resources and financing for relocation impedes progress of the guideline (Wilson 2015). Customary land plays a fundamental role in identity, way of life, communal unity and source of revenue for Pacific islanders, and forced detachment from the land is a sensitive issue, which further complicates relocation policies. Although relocation is considered to be adaptive response to climate change, many Pacific Islanders oppose the notion of resettlement and this has significant repercussions for national policy that supports relocation as an adaptive strategy (Campbell and Warrick 2014).

For the people of Vunidogoloa, the preferred new site was available about 2 km inland on higher ground and was owned by the same *mataqali* who owned the old site. This avoided any land issues that would have emerged if additional customary land were to be acquired from a different *mataqali* (Mitchell et al. 2015). In Fiji, majority land is owned by different *mataqali's* and while some are prepared to accommodate new occupants at a small cost, some may still engage in extensive negotiations and sizeable remuneration (Wilson 2015). Cultural obligations also play a significant role in shaping people's perceptions of climate risks at the local level in the Pacific and a thorough relocation policy would have to adequately address this issue. This concern was noted in an interview with one of the Vunidogoloa villager, "government must consider relocating people together with their church". The reluctance of Vunisavisavi villagers to move to new site is also an example of a cultural issue that can hinder adaptation. Several government ministries provided their input in the relocation process. Some of the authorities provided funding while some provided significant resources for the provision of essential services and new livelihoods. The multi-sectoral approach is vital in maintaining the socio-economic status of the relocated families and must be prioritised in the policy framework.

National governments do not have the capacity to offer displaced people infinite financial support (Edwards 2013, p. 2). Developing nations also experience complications in accessing global climate change funds. The prerequisites to acquiring such funds are sometimes difficult to achieve and as a result, developing nations fail to benefit from them. The process of accreditation is a lengthy one, which is prompting some developing nations to opt for alternative route. The Fijian government managed to access some grant from the Green Climate Fund through a partnership with the Asian Development Bank, which has already attained accreditation (Lo 2016). The Fijian government's relocation policy must emphasize on the concept of "relocation with dignity", since relocation has countless strings attached to it for the person, family or community who has or need to be relocated.

Community Cohesion

Fijian settlement is extensively identified as a closely-knit community due to its communal way of living. This is ingrained in the minds of the people who in many Fijian villages around the country practice and even today are unable to view themselves as a separate entity from their *mataqali*. During the field survey almost all the interviewees stressed that the villagers deal with issues collectively and not as individual units. One villager shared the reason for Vunidogoloa settlement taking so much time to finally relocate, "Our people of the *Vanua* view this place as *kece* (everyones) and as a community we deal with problems together" (Vunidogloa interviewee, personal communication, 25 November 2015, Savusavu, Fiji). In recognition of the important role that community social protection plays in Fiji, efforts to relocate the people to improve targeting, inclusiveness and governance of community-level is a huge challenge here in Fiji. Due to the nature of family units, a strong bonded Pacific community with its cultural roots defining them as one people makes it rather complicated to talk communities for relocation (Ferris et al. 2011).

In addition, according to one villager interviewed during field study, he echoed that "Elders of Vunidogoloa were really against relocation that the younger generation decided not to move unless all the elders within certain age category finally died" (Vunidogloa interviewee, personal communication, 29 January 2016, Savusavu, Fiji). The respect and solidarity displayed by the younger generation to not hurt the feelings of the elders is noteworthy. The support system in place in this village is a testimony of community unity and to what extent these common people can support each other during times of tribulation. Together these ideologies of shared community and one people of the *vanua* can buffer these communities against vulnerabilities such as climate change and its impacts. For many years the villagers lived at the same place that continued to threaten their livelihoods, rip them off their prospect of progress, only to live in harmony at their place of birth, a place that redefines them as a unique clan. This is one reason it took them so long to relocate even fully understanding the threat their current place of residence posed on their lives and their families.

Awareness and Monitoring

The Pacific Island countries (PICs) are projected to be one of the first to experience the impacts of climate change. This calls for a proactive approach by the PICs governments to create awareness amongst its citizens to ensure that they are equipped to tackle the looming dangers. In the Pacific Islands, however, the reactive approach as compared to a more proactive approach, is more common as the awareness follows after the first hand experiences with the effects of climate change (Ferris et al. 2011). This is clearly demonstrated by the relocated villagers of Vunidogoloa who were initially caught unaware that the receding shore and the advance of the sea waves were signs of climate change. This led to a period of inactivity with little effort by the villagers to adopt using soft adaptation techniques before the seawall construction and the subsequent relocation were finally ensued after 40 years of first detection in the 1950s. Awareness and community outreach programs by civil organisations and the government is of paramount importance as such incentives spark behavioural changes in local people that enhance social resilience and enable them to better adapt to the impacts of climate change. Capacity building has been highlighted as a priority in developing countries to ensure access to information, knowledge and participatory principles (Kumamoto and Mills 2012). In the case of Fiji Islands, this becomes even more imperative since the national government is usually the contact point and the recipient of international adaptation financing. Studying the case of the Vunidogoloa relocation has made it clear that awareness of the cause of their inundation issues was essential in advancing the relocation and in enhancing the adaptive capacity of the villagers. Government monitoring of communities that are particularly vulnerable and raising awareness at the initial detection of vulnerability is critical to ensure that community members actively search for options to adapt rather than be caught unaware until the full effects of climate change come to bear on them.

Way Forward: Lessons Learnt

Fijian government has identified a number of communities that need to be relocated in the next decade and is also developing a national relocation guideline. An official relocation policy document which encompasses all the legislative issues and planning framework needs to be meticulously prepared in consultation with all the relevant stakeholders including research and academic institutes. This is to ensure that climate change induced relocation is aligned to national strategic plan and government primacies. Relocation as a climate change adaptation strategy must be considered as a last resort particularly because it involves an increased cost that encumbrances the national budget. Communities differ in the level of vulnerability and a systematic preliminary assessment must identify the most appropriate adaptation strategies for each community to reduce their vulnerability.

For the Vunidogoloa community the call for support from government to consider relocating the village was genuine and necessary. Procuring land for relocation may prove to be problematic since majority of the land in Fiji is customary and protected by the laws of the country. The bonding to land is also strongly embedded in the lifestyle of the Fijian community and a national relocation plan must be able to subtly tackle this issue. Culture, communal unity, and traditional obligations to ancestors play a colossal role in deliberations relating to relocation and in extreme cases even prevents communities to move. It is thus essential to incorporate sociocultural parameters in a relocation policy. In-depth consultations and community outreach programs with vulnerable villagers' necessitates for them to fathom that customary livelihoods may be enhanced through relocation than by staying back and guarding an existing settlement. Socio-economic quandary embraces even greater sway in decisions related to community resettlement and for this reason it is imperative to adopt a cross-sectoral and a participatory approach that is inclusive of environmental consultants, pertinent government and non-government officials, church leaders, community residents, social researchers, academics and local community specialists.

Conclusion

The findings from the survey unveil some valuable qualitative insights into people's perception of climate change and also establish the fact that relationship between land and culture is inherent in the context of climate induced relocation. The Vunidogoloa relocation being the first effective relocation project in Fiji, and the South Pacific sets the premise for future climate change induced relocations by recognizing a list of potential complications that Fijian government must address in their national relocation policy. In relocating vulnerable communities, it becomes fundamental to contemplate not only the financial and bureaucratic conditions surrounding their relocation, but also the numerous socio-cultural features that encompass the community. Fijian communities are diverse and entrenched in cultural notions of the environment, which as realized through this study can be a significant socio-cultural limit to climate change adaptation. Therefore, relocation must be conducted in a holistic manner that befits communities and conserves quality of life. Indubitably, it is essential to note that many such communities embrace distinctive sets of traditional knowledge and abilities that provide opportunities for adaptation both by staying put and by relocating.

References

Adger, W. N., Barnett, J., Brown, K., Marshall, N., & O'Brien, K. (2013). Cultural dimensions of climate change impacts and adaptation. *Nature Climate Change, 3*(2), 112–117.

Amundsen, H., Berglund, F., & Westskog, H. (2010). Overcoming barriers to climate change adaptation—A question of multilevel governance? *Environment and Planning C: Government and Policy, 28*(2), 276–289.

Barnett, J., & Campbell, J. (2010). *Climate change and small island states: Power, knowledge, and the South Pacific*. London: Earthscan.

Boege, V. (2011). *Challenges and pitfalls of resettlement measures: Experiences in the Pacific Region*. Bielefeld: Center on Migration, Citizenship and Development Working Paper Series.

Boydell, S. (2001). *Philosophical perceptions of Pacific property—Land as a Communal Asset in Fiji*, viewed 20 February 2016, http://prres.net/Papers/Boydell_Philosophical_Perceptions_of_Pacific_Property.pdf.

Boydell, S., & Reddy, W. (2000). Contemporary land tenure issues in the Republic of Fiji. In *6th Pacific Rim Real Estate Society (PRRES) Conference* (pp. 23–27).

Boydell, S., & Shah, K. (2003). The international association for the study of common property. In *Second Pacific Regional Meeting*, 7–9 September. Brisbane.

Campbell, J., & Warrick, O. (2014). *Climate change and migration issues in the Pacific, Report - United Nations Economic and Social Commission for Asia and the Pacific*, viewed December 10, 2015. http://www.unescap.org/sites/default/files/Climate-Change-and-Migration-Issues-in-the-Pacific.pdf.

Edwards, J. (2012). *Location, location, relocation.* Fiji Newsletter, April–May, viewed December 5, 2016. http://www.methodist.org.uk/downloads/wcr-julia-edwards-fiji-newsletter-aprilmay2012.pdf.

Edwards, J. (2013). *Financing the future wave of relocations,* viewed December 5, 2015. http://www.methodist.org.uk/media/757401/wcr-pcc-relocation-policy-fiji-jan13.pdf.

Edwards, J. (2014). Climate justice and the Pacific conference of churches. Moving the relocation agenda in the Pacific. In *Sustainable alternatives for poverty reduction and eco-justice: Volume 1* (2nd ed, 1, p. 209).

Edwards, J. (2014). *Land linked to cultural heritage and identity*, viewed December 5, 2015. http://www.unocha.org/top-stories/all-stories/fiji-building-resilience-face-climate-change.

Ferris, E., Cernea, M. M., & Petz, D. (2011). *On the front line of climate change and displacement: Learning from and with Pacific Island Countries*. Washington, D.C.: Brookings Institution.

Fonmanu, K. R., Ting, L., & Williamson, I. P. (2003). Dispute resolution for customary lands: Some lessons from Fiji. *Survey Review, 37*(289), 177–189.

Fresque-Baxter, J. A., & Armitage, D. (2012). Place identity and climate change adaptation: A synthesis and framework for understanding. *Wiley Interdisciplinary Reviews: Climate Change, 3*(3), 251–266.

Frontline Truths by the Pacific Climate Warriors 2016, Frontline Truths—Fiji, viewed January 29, 2016. http://350pacificclimatewarriors.tumblr.com/post/122267652905/george-nacewa-fiji-islands.

Frumkin, H., Hess, J., Luber, G., Malilay, J., & McGeehin, M. (2008). Climate change: The public health response. *American Journal of Public Health, 98*(3), 435–445.

Govan, H. (2009). Achieving the potential of locally managed marine areas in the South Pacific. *SPC Traditional Marine Resource Management and Knowledge Information Bulletin, 25,* 16–25.

Kumamoto, M., & Mills, A. (2012). What African countries perceive to be adaptation priorities: Results from 20 countries in the Africa adaptation programme. *Climate and Development, 4*(4), 265–274.

Kuruppu, N., & Willie, R. (2015). Barriers to reducing climate enhanced disaster risks in least developed Country-Small Islands through anticipatory adaptation. *Weather and Climate Extremes, 7,* 72–83.

Lo, J. (2016). *Why can't poor countries access the climate finance they were promised?* The Guardian, February 15, viewed March 10, 2016. http://www.theguardian.com/global-development-professionals-network/2016/feb/15.

McNamara, K. E., & Des Combes, H. J. (2015). Planning for community relocations due to climate change in Fiji. *International Journal of Disaster Risk Science, 6*(3), 315–319.

Mitchell, D., des Combes, H. J., Myers, M., & McEvoy, D. (2015). Addressing land issues in disaster risk management in the Pacific island countries. *Land Tenure Journal, 1*, 12–15.

Nunn, P. (2012). Climate change and Pacific Island Countries. *Asia-Pacific Human Development Report Background Papers Series 2012/07.*

Nurse, L. A., McLean, R. F., Agard, J., Briguglio, L. P., Duvat-Magnan, V., Pelesikoti, N., et al.. (2014). Small islands. In *Climate change 2014: Impacts, adaptation, and vulnerability. Part B: Regional aspects. Contribution of Working Group II to the Fifth Assessment Report of the Intergovernmental Panel on Climate Change*, p. 1640.

Nurse, L. A., McLean, R. F., Agard, J., Briguglio, L. P., Duvat-Magnan, V., Pelesikoti, N., Tompkins, E., et al. (2014). Small islands. In *Climate change 2014: Impacts, adaptation, and vulnerability. Part B: Regional aspects. Contribution of Working Group II to the Fifth Assessment Report of the Intergovernmental Panel on Climate Change*, p. 1613.

Pareti, S. (2015). *Fiji shares its 'do's' and 'don'ts' on relocation*, viewed February 15, 2016. https://www.sprep.org/climate-change/fiji-shares-its-dos-and-donts-on-relocation.

Peninsula Principles on Climate Displacement within States. (2013). viewed February 13, 2016. http://displacementsolutions.org/wp-content/uploads/2014/12/Peninsula-Principles.pdf.

Piazza, B., & Bolalevu, A. (2014). *Rising sea levels have prompted the Fiji government to relocate one of the country's most vulnerable seaside communities.* World News Australia Radio, radio transcript, viewed 5 March 2016.

Rubelli, E. (2015). *Escaping the waves: A Fijian village relocates.* The Sydney Morning Herald, October 3, viewed December 12, 2015. http://www.smh.com.au/good-weekend/escaping-the-waves-a-fijian-villages-forced-relocation.

Sovaraki, A. (2014). *Vunidogoloa relocation mooted in the 50s.* Fiji Sun, August 20, viewed December 15, 2015. http://fijisun.com.fj/2014/08/20/vunidogoloa-relocation-mooted-in-the-50s/.

United Nations Office for the Coordination of Humanitarian Affairs. (2014). *Fiji: Building resilience face climate change*, viewed February 18, 2016. http://www.unocha.org/top-stories/all-stories/fiji-building-resilience-face-climate-change.

United Nations Framework Convention on Climate Change. (2006). *Climate change: Impacts, vulnerabilities and adaptation in developing countries* report, viewed March 01, 2016. https://unfccc.int/resource/docs/publications/impacts.pdf.

Wewerinke, M. (2013). *A right to enjoy culture in face of climate change: Implications for "Climate Migrants".* Cambridge: Centre of Governance and Human Rights, University of Cambridge.

Wilson, C. (2015). Fiji leads Pacific region on Climate adaptation efforts: Vunidogoloa: An exercise in inclusive adaptation, viewed on February 15, 2016. http://www.ipsnews.net/2014/05/fiji-leads-pacific-region-climate-adaptation-efforts/.

Chapter 20
Limits to Adapting to Climate Change Through Relocations in Papua-New Guinea and Fiji

Dalila Gharbaoui and Julia Blocher

Introduction

For generations Pacific Island communities' traditional adaptive strategies, culture and practices have included retreating from coastal zones following extreme environmental events. One can point to a number of cases of significant out-migration, as well as environmentally-induced partial and staggered community relocations, which exist outside of "normal" migratory patterns. Furthermore, many traditional risk management and response strategies have been lost in the post-colonial era, due in part to today's pre-eminence of "modern" strategies. This loss also applies to strategies of risk-sharing with traditional trading and kinship partners, who are now found across artificial international borders. It leaves exposed communities and specific vulnerable groups with fewer capacities to respond to extreme weather events and the (gradual but permanent and assured) loss of habitable and arable land (as in the case of land subsidence, volcanic eruptions, as well as, for low-lying coastal areas in particular, sea level rise, coastal erosion, saltwater intrusion, and frequent or recurrent meteorological hazards). The result may be the loss of shared social and cultural identities, spaces and meanings; the creation of bifurcated, altered or hybrid identities. Those who migrate find themselves in

D. Gharbaoui (✉) · J. Blocher
Hugo Observatory for Environmental Migration University of Liège,
Bâtiment 31 - Boîte 45 - 7, Boulevard du Rectorat, 4000 Liège, Belgium
e-mail: dgharbaoui@doct.ulg.ac.be; dalila.gharbaoui@pg.canterbury.ac.nz

J. Blocher
e-mail: blocher@unu.edu

D. Gharbaoui
Macmillan Brown Center for Pacific Studies (MBC), University of Canterbury,
Private Bag 4800, Christchurch 8140, New Zealand

J. Blocher
United Nations University, DC2-2062, 2 U.N. Plaza, New York, NY 10017, USA

W. Leal Filho and J. Nalau (eds.), *Limits to Climate Change Adaptation*,
Climate Change Management, https://doi.org/10.1007/978-3-319-64599-5_20

tension with those who remain, and are treated as outsiders or deserters. Some migrants remain anchored to the physical source of a shared heritage, contributing to barriers to integrate elsewhere and perpetuating deep attachment to a literal home that may be disappearing.

These concerns are of particular importance for community members governed by customary leadership and land ownership structures. The adverse effects of climate change are likely to increasingly incite islanders to migrate to cope with threats to their livelihoods. Risks associated with relocations are well established, and it should therefore be considered as a last resort strategy (Barnett and Webber 2010; UNHCR 2014; López-Carr and Marter-Kenyon 2015). The relocation of whole communities may be more likely to preserve community cohesion and cultural integrity, implementation of relocation plans is challenging and costly (Bronen and Chapin 2013; López-Carr and Marter-Kenyon 2015). This paper explores divergent examples of environmentally-induced community relocations in Papua New Guinea and Fiji that have shown that loss of culture are unavoidable results of relocation if customary land tenure is not considered at very early stage at the planned relocation process. In term of definition, there is no internationally agreed terminology on human mobility in the context of climate change. However, we have decided to use 'planned relocation', as defined in Paragraph 14(f) of the Cancun Climate Change Adaptation Framework, as the "planned process of settling persons or groups of persons to a new location".[1] Good governance and best practice addressing limits to adaptation should include this dimension. There is a major gap in the understanding of the limits to adapting to climate change through community relocation (Campbell et al. 2005; UNHCR 2014; Gemenne and Blocher 2016, 2017). The issue has received limited scholarly attention as compared to migration and displacement in response to climate change (López-Carr and Marter-Kenyon 2015). The paper also aims to identify the limits to adaptation in relation to the loss of land and cultural cohesion taking Papua New Guinea as case study and Fiji as an example.

Results from the study of Papua New Guinea community relocation derive from desk research, discussions with key informants and practitioners in Port Moresby, Madang and in Europe—in particular, from local staff of the International Organization for Migration (IOM)—and observations during site visits, all of which were made in the framework of research for the Migration, Environment and Climate Change: Evidence for Policy (MECLEP) programme. One of the authors served as a research advisor in this programme, which was funded by the European Union and led by IOM. Within the framework of the six-country project, a key informant interview grid was developed and intended to be adapted to the context of each case study. Interviews derived from these interview grid/guidelines, as well

---

[1]This form of human mobility is one of the three forms of population's movement in the context of climate change with "displacement" understood as the primarily forced movement of persons and "migration" as primarily voluntary movement of persons. *In* UNFCCC, supra note 9, paras.14 (b)(c)(h). Report of the Conference of the Parties on its eighteenth session, held in Doha from 26 November to 8 December 2012, Decision 3/CP.18.

as the surveys developed in the frame of the MECLEP project methodology,[2] were carried out by research partners in care centers populated by Manam islanders known as Potsdam, Assuramba and Mangem in Madang province outside of Bogia in Papua New Guinea, and on Manam island itself. The author had the opportunity to visit *in situ* and participate in some of the interviews. Research for the Papua New Guinea (hereafter PNG) case study of the MECLEP project culminated in an assessment report based on fieldwork in May 2015 led by research partners.[3] The analysis presented in this chapter is different and complementary to that of assessment report. The example of customary land tenure and community relocation in Fiji aims to illustrate another Pacific country that will urgently need to address the limits to adapting to climate change through relocation by anticipating the implications of land loss and cultural cohesion on relocatees' vulnerability.

This chapter draws a number of conclusions that could be beneficial to evidence-based policy making on planned relocations and resettlements in the context of climate change and its impacts. It is the intention of the authors that this piece of work provides a different angle on the substantial foundation of local knowledge and practices in regards to climate change adaptation, and apply it to current scholarly debates. The discussion below furthermore makes a unique contribution to the literature as a rare example of a comprehensive work comparing different areas under the rule of customary land tenure, although we make little attempt to directly compare very disparate contexts. The two case study countries chosen, Fiji and PNG, approach the question from different vantage points and, in so doing, expose complementary dynamics that provide insight into the longer term feasibility of relocations in the context of deeply culturally embedded customary land tenure regimes.

A key argument presented below is that traditional channels—with the support of the government, at the numerous relevant levels, and impartial land experts—must be central in the process of land negotiation, if there are to be positive relocation outcomes. Sustainable relocation is more likely achieved while land-based tensions related to loss of cultural heritage are likely to be limited and better managed when customary land owners and chiefs are at the first line of negotiations over land with hosting communities and chiefs (c.f. Gharbaoui and Blocher 2016a). We suggest that inserting customary leadership and governance structures (i.e. land owners and chiefs) at the centre of the relocation process may provide a promising avenue towards overcoming challenges associated with loss of land and cultural heritage.

---

[2]For more information about the methods used, please see the MECLEP website and: Gemenne and Blocher (2016, 2017). The interview grids were meant to complement a household survey, which was carried out in five countries but not in PNG.

[3]John Connell, University of Sydney and Nancy Lutkehaus, University of Southern California, to whom the authors owe sincere thanks. Any errors or analysis here should not be attributed to them. The full assessment report will ultimately be published by IOM (http://www.environmentalmigration.iom.int/migration-environment-and-climate-change-evidence-policy-meclep).

Lessons from PNG provide an interesting counterbalance to the research carried out in Fiji, primarily due to the trajectory of approach. In order to fully represent the role of mobility (broadly encompassing displacement, migration, relocation and resettlement) as a strategy to adapt to the nefarious impacts of climate change, researchers must integrate the impacts of movement on the adaptive capacities of both the communities of "origin" and communities of "destination" in their assessments (Gemenne and Blocher 2016, 2017). This paper provides practical insights to substantiate this conceptual foundation. The research carried out in PNG was primarily from the viewpoint of communities originating from Manam Island, over ten years after their evacuation due to a volcanic eruption and ensuing displacement. While this case supports the thesis that integrating traditional leadership structures into relocation and resettlement planning is key, it furthermore shows clearly the limits to this approach under certain conditions. The research was primarily aimed at understanding what adaptive measures have the Manam islanders been more or less able to employ successfully in the context of environmental stress before and since their relocation and displacement. To explore this question requires comparing the strategies of the islanders, a diverse collection of community groups themselves, to those of the "host" community (customary land owners), as well as those that were transferred or are shared.

The example of Fiji provides a reflection firstly on how customary land tenure systems and legitimate traditional authorities are considered throughout the relocation planning and implementation processes. Secondly, it seeks to propose a reflection on how to optimize land governance and tenure security and create the conditions for a more inclusive environmental relocation framework adapted to the local customary land tenure governance mechanisms.

As a pre-condition for our primary recommendation, we posit that customary authorities and institutions are legitimate governance actors. In the Pacific region in particular, these leadership structures and traditions form their own governance mechanisms that may be external to those of the "modern" state and legal systems (FAO 2012; Tobin 2008; Farran 2011). Strategies addressing climate change adaptation in the Pacific should include both state-based governance mechanisms combined with customary non-state institutions. In order to combine those two forms of governance, it is necessary to include traditional authorities in the relocation decision-making process. This cannot be done without a deep respect for their view of the world, a profound understanding of how they represent the climate and migration within their belief systems and how traditional knowledge directly addresses those questions.

Limits to Adaptation and Cultural Cohesion in the Pacific

In this chapter, "limits to adaptation" will be defined as: "the point at which an actor's objectives or system's needs cannot be secured from intolerable risks through adaptive actions" (Klein et al. 2014). The paper explores the limits to

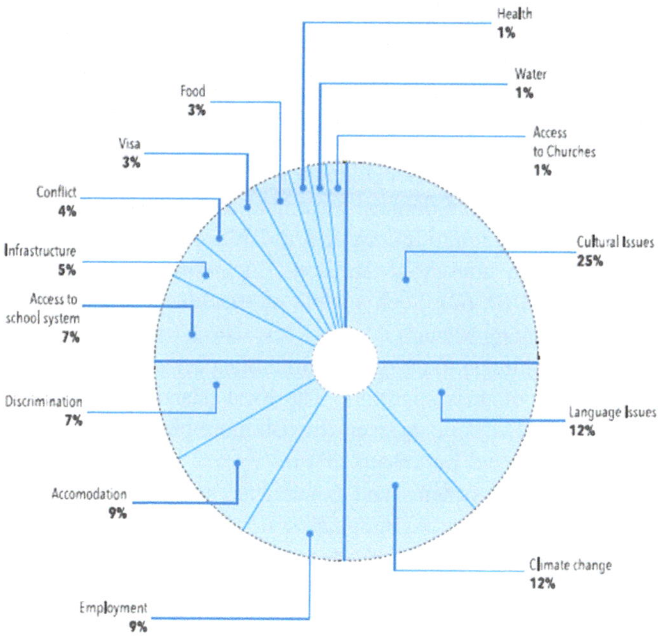

Fig. 20.1 Main challenges faced by migrants when arriving in new communities (Corendea et al. 2015: 12)

adaptation through relocation when an "unacceptable measure of adaptive effort is required, to maintain societal objectives," referring to cultural cohesion among such objectives (Klein et al. 2014). For Pacific Islanders, a cultural component must be central to adaptation efforts. This was reaffirmed in the 2008 Niue Declaration on Climate Change, which emphasizes that the peoples of the Pacific have the strong desire to live in their own countries if possible, in order to contribute to the preservation of their social and cultural identities (Nansen Initiative 2013).[4]

The research project "Promoting human security and minimizing conflict associated with forced migration in the Pacific region," led by the Pacific Islands Forum Secretariat, revealed some of the main challenges that people forced to migrate report experiencing when arriving in new communities. In selected case studies in the Pacific (Fig. 20.1), 25% of the challenges reported are cultural issues (the largest portion) and 12% are climate change-related (Corendea et al. 2015: 12).

Pacific communities, who must choose to relocate in order to adapt to climate change, are facing the challenges associated with the loss of their traditional lands embedded by their cultural cohesion. The loss of sense of place and cultural identity associated with migration (Adger et al. 2009) could limit adaptation to climate

---

[4]The Nansen Initiative consultation on the Pacific: https://www.nanseninitiative.org/pacific-consultations-intergovernmental/.

change. Preston and Stafford Smith observed in this context: "the feasibility of transformational adaptation may therefore be dependent in part on whether it results in outcomes that are perceived to be positive versus negative" (Preston and Stafford Smith 2009). John Campbell, in his study of historical cases of community relocation in the Pacific, also underlined that "the sense of loss [associated with among other things relocation from traditional lands] is especially pronounced in the wake of environmental disasters that damage local land and resources" (Campbell et al. 2005). Within Pacific communities, culture and place attachment are among the most decisive factors in household decisions, including the decision to migrate (Barnett and Webber 2010: 62). Such a depth of attachment to place and to land is evident in PNG and among Manam interviewees, who after over ten years of living in another location continued to carry out burial practices on their home island, the physical resting place of their ancestors. The local islanders interviewed insisted that the coconuts will always be sweeter, the fish more plentiful, and the soil richer on their island (Connell and Lutkehaus 2016).

Customary land tenure systems and governance structures are sometimes presented as constraining factors. As a social system undermined by contemporary government-supported economic and social developments in the context of a globalized world, customary tenure presents barriers to development, cultural and economic integration, citizenship and, implicitly, to the ability of customarily based communities to adapt to 'modern' life, its challenges and its opportunities (Fien et al. 2000: 173). Customary land systems are associated to community cultural cohesion and social norms could therefore potentially be constraints to adaptation in the Pacific region; constraints to adaptation defined as "factors that make it harder to plan and implement adaptation actions. {…}. Types of constraints to adaptation include {…} social norms, identity, place attachment, beliefs {…}" (Klein et al. 2014: 923).

On the other hand, customary relationships and systems may offer flexibility and solutions to face contemporary challenges. Customary partners provide safe refuge and a support system in case need, in the aftermath of a natural hazard, for example. Land tenure systems may also provide greater flexibility to approach land negotiations when occurring within the same system of tenure, by embedding traditional chiefs and landowners at the centre of the decision-making process (Gharbaoui and Blocher 2016a). Considering the positive aspects of migration as 'way of life' for Pacific nations as stated in the 2008 Niue Declaration, includes also protecting community resilience that is part of the Pacific cultural heritage. Opportunities to preserve cultural cohesion despite fundamental changes exist when disequilibrium of societal structures adapt by coming back to equilibrium through self-regulation mechanisms as part of cultural heritage (Campbell 2010).

As we demonstrate below, a central constraint in decisions to resettle is strongly linked to the potential for land-disputes at the new site (Mitchell et al. 2016: 60). Participatory approaches to relocation planning are central to overcome this challenge and "disputants are more likely to engage positively with dispute resolution processes, when they have participated in their design" (Constantino and Merchant 1996: 66). Hybrid systems based on both Western and traditional approaches of conflict prevention in Melanesia are increasingly recognised and encouraged to engage with conflict management. Research in conflict management systems is mostly not inclusive of customary processes but mainly uses Western style processes, however, scholars suggest that "customary and Western systems can work together to effectively manage land-related conflict" (Loode et al. 2009: 63).

Pacific governments have shown reluctance to choose customary land as selected site for planned relocation in order to avoid associated land-based tensions (Nansen 2013).[5] However, most of the land available for relocation in most Pacific countries such as Fiji is customary-based. Land ownership in Fiji is 88% customarily-based, and 97% customarily based in PNG (GoPNG 2007; AusAus 2008). 52% of the population of Fiji lives in urban areas that are mostly coastal, while 11.4% of the total land area below 5 m of elevation. 11% of the population is living in those areas where elevation is below 5 m (UN Habitat 2015). In PNG 61% of the population live within 100 km of the coastline (ADB 2012), while the population living in (mainly coastal) urban areas is 12.6%, one of the lowest rates in the region (United Nations 2014: 152; UNDP 2014). PNG has four large islands and 600 smaller ones (GoPNG 2012). Approximately one fifth of the land in PNG is subject to regular inundation, as low-lying areas are increasingly experiencing frequent storm surges due to sea-level rise (Australian Bureau of Meteorology and CSIRO 2014: 220).

We argue below that it is possible to reduce limits to adaptation and risks of post-relocation localized conflicts[6] associated with land and vulnerabilities linked to the erosion of the cultural cohesion. Successful relocations must integrate, *inter alia*: criteria for societal well-being, livelihood or subsistence activities, community cohesion and cultural integrity (Bronen and Chapin 2013; Gharbaoui and Blocher 2016a). To do this, customary authorities, impartial local land experts and affected communities, as key stakeholders in land negotiation, should have a leading role in decision-making related to the relocation process. They should therefore be legitimate and central stakeholders in the development of the national relocation guidelines developed at the State level.

Why Customary Land Boundaries Matter for Manam Islanders: Living in Displacement Over a Decade After Disaster

Among Pacific islands the limits to adaptation are a concern of significant weight in PNG, where nearly all natural hazards co-exist, most of which will be exacerbated by climate change. PNG's volatile environment and geographic location on the

---

[5]https://www.nanseninitiative.org/pacific-consultations-intergovernmental/.

[6]Unless indicated otherwise, we use this term to mean (small) localized conflicts or tensions among or between community members.

so-called Pacific Ring of Fire (Ramakrishna and Bang 2014: 68) makes its people vulnerable to many hazards, such as volcanic eruptions, tsunamis, earthquakes, tropical cyclones, floods, landslides, El Niño and droughts (NDC 2015; Australian Bureau of Meteorology and CSIRO 2014: 220; Government of Papua New Guinea 2010: 13). According to the Internal Displacement Monitoring Center (IDMC), a total of 151,000 people were newly displaced by disasters in PNG between 2008 and 2013, two thirds of which were due to natural hazards. Another 50,000 people were reportedly internally displaced by conflict and violence during the same period, further contributing to a cycle of vulnerability to shocks and stresses (Ibid).

The rise of sophisticated disaster risk management and response strategies have, among other elements of contemporary and globalized life, replaced or eroded traditional community-based strategies to protect from the threats of physical harm and social disruption posed by natural hazard (Campbell 2006). In a country where of all the land area, only 3% is outside of customary use ownership (Mugambwa 2007; GoPNG 2007), and where the 'legitimacy' of the state and the preponderance of Western economic systems are relatively new developments (Fien et al. 2000), attachment to land can be all-important.

After eleven years living in protracted displacement away from their ancestral lands, a small but critical brick has been laid in the foundation of the future for the displaced Manam people (Fig. 20.2). This group of communities from different villages of Manam Island, located in the northeast of PNG, have been living in protracted displacement in government-supported "care centres" on the mainland in Bogia district since they were evacuated due to an impending volcanic eruption in 2004–2005. In April 2016, much to the satisfaction of local and provincial leaders alike, the Manam Resettlement Bill was passed in PNG's parliament (Shisei 2016). The law authorizes the legal basis for the main authority in Bogia District charged to seek viable options and use government funds for a resettlement plan for the Manam groups.

The Manam people have suffered through persistent joblessness, place disarticulation from their ancestral lands, physical and mental ailments, and repeated conflict with the customary land owners of the disused colonial-era plantations on which the "care centres" were established (Nalu 2011a; IDMC 2014; Connell and Lutkehaus 2016). Facing economic inactivity, loss of access to resources, disconnection from communal, disarticulation from their land, and an erosion of traditional communal rites and ceremonies, many Manam Islanders have become pessimistic about their situation and antagonistic towards the government (Connell and Lutkehaus 2016, 2017). These sentiments are heightened by the continued lack of viable alternatives to the care centres and conflicts with the owners of the customary land on which they currently live, with whom they had previously had unpleasant but overall minor exchanges with youth members of the Manam community. While the Manam islanders—as traditional trading partners—had previously been hosted by this community for short periods of time in reponses to minor hazards befalling the island, the mainland community was unprepared for a long-term occupation, particularly as the group swelled to potentially 15,000–20,000 residents (IDMC 2014; Connell and Lutkehaus 2016, 2017).

Fig. 20.2 Young Manam men in a care centre boil fish, yams and local greens in coconut milk for their visitors using traditional cookware as others look on. Traditionally women and girls prepare meals. In recent years, according to Manam residents, they have had more difficulty fishing and are often finding smaller and smaller fish between the mainland and Manam Island. This is likely to be a result of overfishing as well as the destruction of corals due to dynamite fishing, cyanide fishing, and the harvesting of coral to make powdered lime (which is chewed with betel nut, a popular stimulant). (*Photo* © Julia Blocher 2015)

Customary land systems are more than about geographical and marine space; they regulate the sustainable usage of resources (AusAID 2008) as well as, by extension, a shared norm to regulate interpersonal relationships, to share living space, and regulate collective local practices (Thaman 2000: 139). Customary land can thus catalyse land conflicts and social conflicts (Weir et al. 2010: 10), representing one of the main limits to adaptation in the region. In the case of the displaced Manam population, existence across land boundaries amounted to suffering. They are required to deal with restrictions on land, water, forest resources and marine access and must return to Manam to harvest timber and other resources, and to bury their dead. Access issues may contribute to the poor sanitation, hygiene, health and nutrition that is evident among the Manam and their children—some of whom have severe skin disease (such as tinea corporis) and visible signs of malnutrition (kwashiorkor)—as well as limited improvements to acceptable housing and energy. As one Manam resident stated "I can't get materials to build a new house. The posts are slowly rotting, the walls are falling apart. The owners of the land don't allow us to use their land to get wood or roofing" (Waide 2014). On the other hand, observers noted that as early as 2005 the Manam people was causing tension with local communities over land use issues (UNOHCHR 2010: 13), and in 2007 landowners were "planning a fight to claim more land as [they] have given up on the government" (Mercer and Kelman 2010: 419). Villages near the Asuramba care centre asked their Member of Parliament to evict the Manam islanders in 2006 because of security concerns (UNOHCHR 2010). Several clashes ensued; a clash at Tobenam in 2008 resulted in the centre burning to the ground and two fatalities (Connell and Lutkehaus 2016).

For years the provincial officials, committed to resolve the situation, had been unable to procure suitable customary land on which a township with basic services could be founded. Finally they lost the federal approval to pursue the resettlement when a sunset provision on the previous resettlement bill halted its effect in 2012 (Radio New Zealand 2006). As another resident stated;

> The situation in the care centres is getting worse. We want a quicker resettlement process because government is spending a lot of money in humanitarian relief rather than development… the displaced Manam people will continue to be 'reluctant' and 'stubborn' and already some have expressed a desire to return to their ancestral land and die there. We have suffered too long; we are like refugees on our own land, forgotten and neglected… We don't know who to turn to. (quoted in Connell and Lutkehaus 2016)

The Chairman of the Manam Island Council of Chiefs stated in 2012 that the majority of people wanted to move away from the care centres as soon as possible and were receptive to the proposed move to Andarum (Radio Australia 2012), a largely inaccessible area 30–40 km inland. Moving the population farther from the Island and fishing waters from whence they make their livelihoods and to which they connect "home" comes with a number of drawbacks economic activity, social and cultural cohesion, and identity. The alternatives—staying in the care centres or returning to a volcanic island the government has designated as uninhabitable and discontinued services to, and which again erupted in July 2015—may be worse in other respects.

Eroding of 'Traditional' Methods of Resilience

Across the Pacific, indigenous disaster risk reduction and disaster management practices have largely been superseded by 'modern' strategies (Campbell 2006), which are often funded and carried out by international actors. The population of Manam Island was no exception. Previously, the communities on Manam had widespread knowledge of risk exposure integrated into the repeated and normal communications. They employed rudimentary—but effective—warning signals such as sounding an alarm through a conch shell, church bell or *Garamut* (a slit drum they would trade for).

In previous times of crisis, the Manam people could often rely on their traditional trade and kinship Murik[7] partners on the mainland (*taoa*). Many of them

[7]The Murik reside around the Murik lakes region, along the north coast of PNG in the East Sepik Province in the Sepik (river) estuary.

were to the west of the mouth of the Sepik river and thus across contemporary administrative borders. These bonds were strengthened by marriage, hospitality towards one another, trade ties and reciprocal exchanges of pigs and *galip* nuts (also known as Tahitian almonds) (Lutkehaus 2016). An additional insurance strategy reserved for the members of status was to move to and purchase land in the Wewak area of East Sepik Province, another area nearby and across provincial boundaries.

International protection and assistance actors have not to-date adequately addressed the effects wrought by the erosion of customary partnerships and indigenous strategies in increasing community vulnerability to shocks. In the case of the Manam people (Fig. 20.3), for example, the erosion of traditional measures meant that volcanic eruption presented an increasingly uncertain level of risk to life and well-being (Connell and Lutkehaus 2016).

Fig. 20.3 A home in the Potsdam Care Centre in, Madang Provice, PNG. Most of the houses in the care centres for the Manam communities were built with the help of NGOs, as the Manam people have been largely forbidden by the mainland community from felling timber. (*Photo* © Julia Blocher 2015)

How Customary Relationships in PNG Facilitates Risk Response

Individual's rights to land and community inclusion are protected by customary law, but are more largely contextualised via the collective rights of kinship groups (Thaman 2000). Under customary principles, all people in PNG are considered landowners, however, a general perception persists that new arrivals can be considered squatters (RICS 2016). People constrained to self-resettle following displacement often find themselves in informal settlements, usually on state land, unoccupied customary land, or freehold land (such as church owned land). This puts populations at risk of land disputes, hazards and isolation. State-led allocations land may result in violent disputes, given that the customary landholders are likely to have social and political relationships to the land, and have residual ownership legitimacy and rights to land across generations. However, not only can customary relationships facilitate temporary relocations or long-term resettlements of disaster-affected peoples; they are the crux of such strategies.

Although there are no reliable estimates, as many as a quarter of the Manam population joined their *taoa* hosts on the mainland in the immediate aftermath of the 2004 eruption. This significant contribution to the protection and assistance to the plight of the Manam is unrepresented in most reporting around the Manam displacement, which has generally accompanied the NGO and government activity centred around the care centres at Bogia. More research would be needed to assess whether these community-based relocations are more or less successful than government- or NGO-supported efforts. According to Connell and Lutkehaus (2016: 50) "it is generally perceived by both the Provincial Government and Manam islanders that any formal resettlement solution must occur within the province, which rules out the possibility of a formal migration to East Sepik where *taoa* partners exist (and where some have gone individually and purchased land)".

The tension between community-based strategies and the pre-eminence of post-colonial administrative boundaries is important—as is the communities' evolving perception of to whom they can turn to for protection and assistance. This reality belies an evolving relationship and concept of responsibility and care at multiple levels: between the Manam islanders and their government, between the Manam communities and their partner communities, and between the islanders and the international responders.

It is uncertain whether these community-led and customary protection mechanisms remain viable today. Ultimately, the problem of relocation for the Manam may have been less a result of relocating away from their *taoa* partners, but the fact that they are in protracted displacement in itself, whether they were displaced to the land of *taoa* partners or not (Nalu 2011a; Connell and Lutkehaus 2016). The vulnerabilities and lack of viable options for these potential settlers, perceived as outsiders to the mainland community despite sharing nationality, raises issues of ambiguity, identity and citizenship (Connell and Lutkehaus 2017). In an era of modern and post-colonial ascendancy, the strength of these reciprocal relationships

—forged through the bonds of tradition and trust—have been difficult to maintain. These relationships may be of less consequence as the use of modern currency expands. Evidence of this can also be seen in the decline of inter-community pig exchange (*buleka*) and gifts of shelled nuts. The act of shelling the nut and presenting them to trading partners, ensuring they will be eaten immediately, is a symbolic gesture that reaffirms the social relationships that make life possible (Lutkehaus 2016). Population growth, recurrent hazards and other factors contributing to declining wealth also meant the Manam communities had fewer resources to invest in generosity and hospitality, including towards *taoa* partners. In recent years, guests are not greeted with baskets of *kangari* nuts as they once were (Lutkehaus 2016).

Re-establishing Community Based Protection for Future Displacement Events

An influx of international NGOs in the post-colonial era as well as improved implementation and funding of the Disaster Management Act and the National disaster mitigation policy by the National Disaster Centre (NDC) have contributed to an better implementation, monitoring, evaluation and management of disaster mitigation (IRIN 2013). In line with the global trend, in PNG evacuation effectiveness has improved (Johnson 2013: 187) and fewer causalities and deaths have resulted from disasters (EM-DAT 2015). This is partially because volcanic eruptions give discernable warnings with enough time to act, whether by modern instruments or traditional detection strategies (smoke and ground rumblings, for example). This is often not the case for landslides and earthquakes, which combined account for 89.5% of disaster mortalities between 1990 and 2015, according to the EM-DAT database. Between 1900 and 1960, 3506 people died from four major volcanic events, compared to nine people since 1950 (EM-DAT 2015).

In the case of the Manam people, government-assisted evacuation from the Island occurred rapidly between November and January of 2004–2005 and saved lives. This is compared to the previous eruption in 1996, during which thirteen people were killed instantly when they were engulfed in a pyroclastic flow near Budua (c.f. Connell and Lutkehaus 2016). Modern volcanologist early warning and evacuation systems, as well as the confidence of the people in and compliance with government-mandates measures, have proven their worth (Mulina et al. 2011); most of the Manam Islanders were reportedly able to evacuate on time.[8]

Despite the efforts of the authorities, among the Manam communities and others around the Pacific, 'traditional' disaster risk reduction and management strategies

[8]Even with a successful evacuation, five people tragically died. Most died of respiratory complications and from drinking ash-contaminated water; one person was killed when struck by large fragments of volcanic rock.

are being lost—and with them, practices of cultural and communal significance. Furthermore, these strategies remain important when resource-poor or remote areas do not have or lose access to modern amenities, communications, and international funding needed for large-scale projects. For example, the use of mobile phones on Manam Island, charged with solar batteries, is not universal and cannot always be relied on. After ten years of waiting for the massive government and international mobilization that would be needed to resettle 10,000–20,000 people,[9] the Manam communities developed largely inadequate ineffective autonomous adaptation measures. Preserving traditions and indigenous knowledge may therefore not merely be a question of social cohesion and preservation of a rich cultural history; it is also a matter of risk exposure.

Customary Land Tenure and Post-Relocation Vulnerabilities: The Example of Fiji

Fiji, spread over 332 islands, is highly threatened by climate change through change in intensity and frequency of environmental disasters and accelerated sea-level rise (Gravelle and Mimura 2008). Most urban areas in Fiji are located along the coastline where communities will be increasingly vulnerable to rising sea levels and coastal erosion. Flooding is becoming a persistent issue mostly due to the intensity and frequency of tropical cyclones. Storm intensification is becoming increasingly visible across the Pacific. On February 20, 2016, the Cyclone Winston (Fig. 20.4), classified as a "Category 5 tropical cyclone"[10] was recorded as the strongest storm in history for Fiji and one of the most powerful storm in the Pacific Islands region, more intense than Cyclone Pam in 2015 and Cyclone Zoe in 2002 (Shultz et al. 2016: 41).

The consequences of Cyclone Winston on communities already threatened by the intruding sea became an increasingly important issue in Fiji. The cyclone made communities more vulnerable and exposed, forcing many to leave to safer grounds. In the aftermath of the cyclone Winston, at least 63 villages were considered as «in need to relocate» following the assessments carried out by the iTaukei Affairs Board, Lands and Mineral Resources Department, Water Authority of Fiji and Fiji Roads Authority officials (Bolatagici 2016).

This event reaffirms the necessity of considering relocations within the scope of disaster risk reduction efforts. Previous disasters, particularly if there is a lack of anticipated relocation planning, drives post-disaster vulnerability and exposure. The resilience of coastal populations, hit by recurrent rapid-onset hazards such as typhoons and slow-onset events such as drought and land degradation, can be

[9]Government-used estimates of 20,000 are derived from the media may not be reliable; Conell & Lutkehaus estimate the total population figure far lower at around 12,000.

[10]On the Saffir-Simpson hurricane rating scale.

Fig. 20.4 Cyclone Winston impacting the Republic of Fiji, February 20, 2016 (NASA 2016)

eroded over time. Planning for sustainable relocations in a proactive manner reduces the human and financial costs involved, but it also helps reduce socio-economic factors contributing to vulnerability as well as to mediate exposure to natural hazards which can culminate in life-threating disaster events (Fig. 20.5).

For its recovery from Cyclone Winston, Fiji followed the Sendai Framework for Disaster Risk Reduction 2015–2030. This voluntary non-binding instrument recognizes that the State has the primary role to reduce disaster risk, but it also emphasizes that this responsibility should be shared by the local government and local communities as stated in Article 19.(f): "While the enabling, guiding and coordinating role of national and federal State Governments remain essential, it is necessary to empower local authorities and local communities to reduce disaster risk, including through resources, incentives and decision-making responsibilities, as appropriate" (Sendai Framework for Disaster Reduction 2015–2030).[11] The central role of local authorities and customary chiefs in relocation planning was clearly underlined by an interviewee from the Ministry of Land in Fiji when discussing an on-going relocation caused by Cyclone Winston in Koro Islands: "The Native Lands Commission (NLC) holds a key role in the relocation process as demarcation of land and maps that are key to avoid conflicts are based on land ownership knowledge. Customary chiefs and land owners holds an equal chair in

[11]The Sendai Framework for Disaster Reduction 2015–2030 is the successor instrument to the Hyogo Framework for Action (HFA) 2005–2015: Building the Resilience of Nations and Communities to Disasters In http://www.unisdr.org/files/43291_sendaiframeworkfordrren.pdf.

Fig. 20.5 Manam Island, still spouting smoke regularly, is visible from the beach at Potsdam care centre in Madang Province. The island serves as the cultural, spiritual, and economic centre of gravity for the Manam communities. The inhabitants of the care centres travel there by canoe to bury their dead, to fish, harvest coconuts, and gather timber. (*Photo* © Julia Blocher 2015)

the NLC" (Fiji Ministry of Lands interviewee, personal communication, 30 April 2016, Christchurch, New Zealand).

Relocation governance frameworks should also include mechanisms allowing hosting and affected communities as well as their traditional leaders to have access to reliable information, in order to enable participation in the decision-making process in an optimal manner (Gharbaoui and Blocher 2016b). Efforts from decision-makers should also target awareness-raising on the adverse effects of climate change and induced relocations among exposed or affected communities and their leadership authorities. Challenges such as translating "climate change" into different Fijian languages and dialects should not be neglected (Janif et al. 2016). The relocation process should also include Faith-Based Organization as much as possible as they play a crucial role in raising awareness on climate change and in convincing communities to be empowered in relocation planning and not passive to the divine will. FBOs can also play an important role in relocation planning as key informant about cultural and spiritual features associated with communities' land (Ministry of Churches interviewee, 18 February 2016, Wellington).

Involvement of local researchers and local land experts with their input in framing the agenda in terms of relocation planning should be further prioritised, encouraged and considered at the decision-making level in order to address cultural cohesion, loss of cultural heritage and associated tensions limiting adaptation through relocation (Federation of International Surveyors interviewee, June 2016). Western researchers and institutions are still taking the lead on climate-induced migration. It is crucial that traditional knowledge as well as locally-based expertise is valued when planning relocation as perceptions of the environment and environmental change have cultural elements (Gharbaoui and Blocher 2016b).

Despite having no mention of relocation in their 2012 climate change policy (McNamara and Des Combes 2015), the Fiji government has been recently showing leadership in planning relocation through the national relocation guidelines that are still currently being developed. Peter Emberson, Director of Climate Change at Fiji Ministry of Foreign Affairs, underlined at the annual Pacific Climate Change Roundtable (PCCR) (15 May 2015, Apia, Samoa), that a participatory approach in relocation planning inclusive of technical specialists, social scientist and local community experts needed to be inserted into the national relocation guidelines (Pareti *in press*).[12] However, the guidelines "{…} do not appear to be done in consultation with any potential affected communities" (McNamara 2015: 318).

Strategies addressing climate change adaptation in the Pacific should include both state-based governance mechanisms combined with customary non-state institutions. In order to combine those two forms of governance, it is necessary to include traditional authorities and affected communities to the decision-making process on relocation.

Post-relocation vulnerability associated to land-based conflicts, loss of customary land systems and associated culture need to be considered by decision-makers when planning for relocation as adaptation to climate change in Fiji. It is crucial that more efforts and strategies are developed by the government to ensure the participatory and leading role of customary authorities, local landowners and land experts in the relocation planning and along the relocation process from conception to finalization including monitoring and follow-up. Flexible approaches to land governance are key for dealing with security of tenure. Developing policies optimizing registration of customary lands would be an example of measure that would help to protect the land rights of communities forced to move as result of climate change. Unsecure tenure and issues around customary land titles and ownership including unclear villages boundaries demarcation can increase the vulnerability of relocatees and provide fertile ground for post-relocation land-based conflicts.

Conclusions

In modern-day governance of Pacific Islands and Territories, the administrative boundaries established by nation-states rarely correspond to those of customarily owned and demarcated land. Insecurity of land is often very high following displacement, while land grabbing and eviction has been shown to among the primary fears of affected peoples in the region (RICS 2016). Large-scale and long-term funding required for a project such as the Manam resettlement may not always be accessible.

[12]https://www.sprep.org/climate-change/fiji-shares-its-dos-and-donts-on-relocation.

Clear government policies that combine both modern and traditional approaches, informed by regular and extensive consultation with affected communities, are paramount. Before disasters, risk reduction and management strategies must target not only physical safety but must also place emphasis on community cohesion and the preservation of Pacific people's identity, culture, social and belief systems. Following and in anticipation of displacement risk, protection and assistance strategies should be sensitive to land tenure boundaries. In many cases, they should be the main borders of relevance when planning for local integration, temporary relocation of communities, and longer-term resettlement. In addition to avoiding loss of cultural heritage as well as conflicts associated to land ownership, this direction will imply much-needed sensitivity to community-based leadership and governance structures necessary for buy-in and the long-term success of any solution to displacement.

In the context of the Pacific, and potentially other customary land tenure systems, protecting and assistance policies should place emphasis on community cohesion and the preservation coherent with customary systems. Given the existence of customary boundaries that cut across modern day states, regional cooperation may be a key fit-for-purpose approach. UN-HABITAT and others in the international community have developed guidance on frameworks for land governance that can be turned to, as they consider the combination of the spatial, legal and institutional frameworks and include key principles such as Good land governance preventing bureaucratic barriers.

Participation is paramount. Locally-based expertise and traditional knowledge are valuable sources of information to develop strategies to address displacement, or, in the case of many Pacific communities, to inform planned relocations and resettlements. Flexible approaches to land governance and the concept of borders are key.

References

Adger, W. N., Dessai, S., Goulden, M., Hulme, M., Lorenzoni, I., Nelson, D. R., Otto Naess, L., Wolf, J., and Wreford, A. (2009). Are there social limits to adaptation to climate change? *Climatic Change, 93*(3–4), 335–354.

Asian Development Bank Regional Coopertion on Knowledge Management, Policy and Institutional Support of the Coral Triangle Initiative. (2012). http://www.coraltriangleinitiative.org/sites/default/files/resources/PNG%20SCTR_web%20copy.pdf.

AusAID. (2008). Making land work—Volume one: Reconciling customary land and development in the Pacific. Canberra: AusAID http://www.ausaid.gov.au/publications/pdf/MLW_VolumeOne_Bookmarked.pdf.

Barnett, J., Campbell, J. (2010) Climate change and small island states:power, knowledge and the South Pacific.Earthscan, London.

Bolatagici, L. (2016). After assessments, many more Fiji Villages Slated For Relocation. Pacific Islands Development Program, East-West Center With Support From Center for Pacific Islands Studies, University of Hawai'i http://www.pireport.org/articles/2016/04/21/after-assessments-many-more-fiji-villages-slated-relocation.

Bronen, R., & Chapin, F. S. (2013). Adaptive governance and institutional strategies for climate-induced community relocations in Alaska. *PNAS, 110*(23), 9320–9325.

Byamugisha, F. (2013). Sharing Africa's land for shared prosperity. A program to scale up reforms and investments. World Bank, Washington. http://www.scribd.com/doc/144377306/Securing-Africa-s-Land-for-Shared-Prosperity.

Campbell, J. R., Goldsmith, M., & Koshy, K. (2005). Community relocation as an option for adaptation to the effects of climate change and climate variability in Pacific Island Countries (PICs) Kobe: Asia-Pacific Network for Global Change Research (APN).

Campbell, J. R. (2006). Traditional disaster reduction in Pacific Island communities. *GNS Science Report, 2006*(38), 46.

Campbell, J. (2010). An overview of natural hazard planning in the Pacific Island Region. *The Australasian Journal of Disaster and Trauma Studies*, ISSN: 1174–4707, Vol. 2010–1 http://www.massey.ac.nz/~trauma/issues/2002-2/mcdowell.htm.

Connell, J., & Lutkehaus, N. (2016). *Another Manam? The forced migration of the population of Manam Island, Papua New Guinea; Due to Volcanic Eruptions 2004/2005*. Geneva: IOM.

Connell, J., & Lutkehaus, N. (2017). Environmental refugees? A tale of two resettlement projects in coastal Papua New Guinea. *Australian Geographer.*

Constantino, C., & Merchant, C. (1996) Designing Conflict Management Systems: A Guide to Creating Productiveand Healthy Organizations, Jossey-Bass Publishers, London.

Corendea, C., Bello, V., & Bryar, T. (2015). Pacific research project: Promoting human security and minimizing conflict associated with forced migration in the Pacific region. Pacific Islands Forum Secretariat, UNU-GCM, UNU-EHS. https://collections.unu.edu/view/UNU:3171.

EM-DAT. (2015). The OFDA/CRED—International Disaster Database. Brussels: Université catholique de Louvain http://www.emdat.be. For a disaster to be entered into the database at least one of the following criteria must be fulfilled: Ten (10) or more people reported killed; a hundred (100) or more people reported affected; a declaration of a state of emergency; a call for international assistance.

Farran, S. (2011). Navigating between different land tenure and land laws in pacific island states. *Journal of LegalPluralism 64.* http://commission-on-legal-pluralism.com/volumes/64/farran-art.pdf.

FAO. (2012). Voluntary guidelines on the responsible Governance of Tenure of Land, Fisheries and Forests in the Context of Food Security. Rome. http://www.fao.org/docrep/016/i2801e/i2801e.pdf.

Fien, J., Sykes, H., & Yencken, D. (2000). *Environment, education and society in the Asia-Pacific: Local traditions and global discourses*. London: Routledge Advances in Asia-Pacific Studies.

FIG/GLTN. (2010). The social tenure domain model. FIG Publication No. 52, FIG Office, Copenhagen, Denmark. http://www.fig.net/pub/figpub/pub52/figpub52.htm.

Gemenne, F., & Blocher, J. (2016). How can migration support adaptation? Different options to test the migration-adaptation nexus Migration, Environment and Climate Change: Working Paper Series 1/2016 IOM, Geneva.

Gemenne, F., & Blocher, J. M. (2017). How can migration serve adaptation to climate change? Challenges to fleshing out a policy ideal. *The Geographic Journal*, pp. 1–12. http://onlinelibrary.wiley.com/doi/10.1111/geoj.12205/abstract.

Gharbaoui, D., & Blocher, J. (2016a). The reason land matters: Relocation as adaptation to climate change in Fiji Islands. In Milan, A., Cascone, N., Schraven, B., & Warner, K. (Eds.), *Migration, risk management, and climate change—Evidence and policy responses.*

Gharbaoui, D., & Blocher, J. (2016b). Climate change, planned relocation and land tenure in Fiji and Papua New Guinea. Macmillan Brown Centre for Pacific Studies, Pacific Policy Brief 2016/7, Policy Brief Series, University of Canterbury.

Government of Papua New Guinea (GoPNG). (2007). The national land development taskforce report: Land administration, land dispute settlement, and customary land development. National Research Institute, Monograph 39, Port Moresby.

GoPNG (2012). Systems, process and institutions. PNG Decent Country Program. Monograph. Department of Labour and Industrial Relations, Port Moresby. Papua New Guinea Vision 2050, National Strategic Plan Taskforce. Papua New Guinea's Strategic Program for Climate Resilience. Recognised Seasonal Employer Policy: Inter-Agency Understanding Papua New Guinea.

Gravelle, G., & Mimura, N. (2008). *Sustainability Science 3*: 171. http://link.springer.com/article/10.1007/s11625-008-0052-2?no-access=true.

IRIN News. (2013). Papua New Guinea government improves disaster preparedness funding. http://www.irinnews.org/report/98433/papua-new-guinea-government-improves-disaster-preparedness-funding.

Janif, S. Z., Nunn, P. D., Geraghty, P., Aalbersberg, W., Thomas, F. R., & Camailakeba, M. (2016). Value of traditional oral narratives in building climate-change resilience: insights from rural communities in Fiji. *Ecology and Society, 21*(2), 7. www.ecologyandsociety.org/vol21/iss2/art7/#Relocation.

Johnson, R. W. (2013). *Fire mountains of the Islands: A history of volcanic eruptions and disaster management in Papua New Guinea and the Solomon Islands*. Canberra: Australian National University Press.

Klein, R. J. T., Midgley, G. F., Preston, B. L., Alam, M., Berkhout, F. G. H., Dow, K., & Shaw, M. R. (2014). Adaptation opportunities, constraints and limits. Climate Change 2014: Impacts, adaptation, and vulnerability. In Field, C. B., Barros, V. R., & Dokken, D. J. et al. Part A: Global and sectoral aspects. *Contribution of the Working Group II to the Fifth Assessment Report of the Intergovernmental Panel on Climate Change* (pp. 899–943). Cambridge: Cambridge University Press.

Loode, S., Nolan, A., Brown, A., & Clements, K. (2009). Conflict management processes for land-related conflict Suva, Fiji: Pacific Islands Forum Secretariat.http://www.forumsec.org/resources/uploads/attachments/documents/LMCM%204_1%20COMPLETE.pdf.

López-Carr, D., & Marter-Kenyon, J. (2015). Human adaptation: Manage climate-induced resettlement. *Nature, 517,* 7534.

Lutkehaus, N. C. (2016). Finishing Kapui's Name: Birth, death and the reproduction of Manam Society, Papua New Guinea. In D. Lipset & E. K. Silverman (Eds.), *Mortuary dialogues: Death ritual and the reproduction of moral community in Pacific Modernities*. New York, Oxford: Barghahn.

McNamara, K. E., & Des Combes, H. J. (2015). *International Journal of Disaster Risk Science, 6,* 315.

Mitchell, D., Orcherton, D., Numbasa, G., & Mc Evoy, D., (2016). The implications of land issues for climate resilient informal settlements in Fiji and Papua New Guinea, RICS, 2016 http://www.rics.org/Global/Land_Issues_Fiji_PNG_300316_dwl_aa.pdf.

Mugambwa J. (2007). A comparative analysis of land tenure law reform in Uganda and Papua New Guinea. *Journal of South Pacific Law, 11*(1), 7–8.

Mulina, K., Sukua, J., & Tibong, H. (2011). Report on Madang Volcanic Hazards Awareness Program. 6–29 September 2011, Rabaul Vulcanological Observatory.

Nalu, M. (2011a). The tragedy of Manam Islanders … Refugees in their own country. http://malumnalu.blogspot.com/.

Nansen Initiative (2013) "Human mobility, natural disasters and climate change in the Pacific": Report from the Nansen Initiative Pacific Regional Consultation 21–24 May 2013, Rarotonga, Cook Islands.

Preston, B. L., Stafford-Smith, M. (2009) Framing vulnerability and adaptive capacity assessment: Discussion paper. CSIRO climate adaptation flagship working paper no. 2. http://www.csiro.au/org/ClimateAdaptationFlagship.html

Radio New Zealand. (2006). PNG government commended for establishing Manam resettlement authority http://www.radionz.co.nz/international/pacific-news/163830/png-government-commended-for-establishing-manam-resettlement-authority.

Radio Australia. (2012). Manam islanders displaced by volcano to re-settle http://www.radioaustralia.net.au/international/radio/program/pacific-beat/manam-islanders-displaced-by-volcano-to-resettle/1020584.

Ramakrishna, A., & Bang, S. (2014). The Impacts of Climate Change on Food Security and Management in Papua New Guinea. In Ganpat, W.G, & Isaac, W.A.P. (Eds) Impacts of Climate Change on Food Security in Small Island Developing States, 67–68. Hershey: IGI Global.

Report for Royal Institution of Chartered Surveyors (RICS). (2016). *The implications of land issues for climate resilient informal settlements in Fiji and Papua New Guinea*. London: RICS.

Sendai Framework for Disaster Risk Reduction 2015–2030 (Sendai Framework). (2015). United Nations Office for Disaster Risk Reduction (UNISDR), Geneva, Switzerland http://www.unisdr.org/files/43291_sendaiframeworkfordrren.pdf.

Shisei, R. (2016). Manam resettlement bill passed' EM TV News http://www.emtv.com.pg/article.aspx?slug=Manam-Resettlement-Bill-Passed—2&.

Shultz, J. M., Cohen, M. A., Hermosilla, S., Espinel, Z., & McLean, A. (2016). Disaster risk reduction and sustainable development for small island developing states. *Disaster Health, 3* (1), 32–44. doi:10.1080/21665044.2016.1173443.

Thaman, K. H. (2000). Cultural rights: A personal perspective. In Wilson, M., & Hunt, P., & UNESCO (Eds.), *Culture, rights and cultural rights: perspectives from the South Pacific : the proceedings of a colloquium organised by the Unesco Office for Pacific Member States and the Centre for New Zealand Jurisprudence (School of Law, University of Waikato, New Zealand) October 1998*. Huia, Wellington, N.Z.

Tobin, B. (2008). The Role of Customary Law in access and benefit-sharing and traditional knowledge governance: Perspectives from Andean and Pacific Island Countries, UNU-WIPO. www.wipo.int/export/sites/www/tk/en/resources/pdf/customary_law_abs_tk.pdf. 2016. In Customary Law and Traditional Knowledge, Background Brief 7, p.1. http://www.wipo.int/edocs/pubdocs/en/wipo_pub_tk_7.pdf.

UN-HABITAT, GLTN. (2008). Secure land rights for all. https://www.responsibleagroinvestment.org/sites/responsibleagroinvestment.org/files/Secure%20land%20rights%20for%20all-UN%20HABITAT.pdf.

UN-Habitat. (2015). *Urbanization and climate change in small Island developing states.*

United Nations Office of the High Commissioner for Refugees (UNHCR). (2014). *Planned relocations, disasters and climate change: Consolidating good practices and preparing for the future, background document*. Geneva: UNHCR.

United Nations Office of the High Commissioner for Human Rights (UNOHCHR), Regional Office for the Pacific. (2011). Protecting the Human Rights of Internally Displaced Persons in Natural Disasters: Challenges in the Pacific. Discussion Paper, April, Suva.

Waide, S. (2014) *10 years after the eruption*. www.PNGBlogs.com/2014.

Chapter 21
Atoll Habitability Thresholds

Mark H. N. Stege

Introduction

The tide supplies cadence in one's day-to-day life on an atoll. When it rises, it signals the temporary closure of reef flat walkways and less strenuous bucketing of fresh well water, among other benign characteristics of atoll life. Now however, owing to global climate changes including 0.6–1.2 m of sea level rise by the end of this century (IPCC 2013: 1140), atoll community leaders like this author are beginning to view the diurnal occurrence of an encroaching tide differently. Rather than benign, it is becoming more and more evocative of a difficult conversation in which every atoll dweller will inevitably be obliged to participate on some level: a conversation on climate-induced migration.

When former Marshall Islands President Christopher Loeak stoically rejected a suggestion during a World Leaders Forum at Columbia University in September 2013 that relocation might be an option for his people, he was reflecting a sometimes difficult to understand sentiment held among the vast majority of Marshallese who still bear witness to the legacy of the US nuclear testing program in the Pacific (Rudiak-Gould 2013: 15–39). Sitting next to him and in measured follow on comments, the Marshall Islands' lead climate negotiator Tony deBrum explained further that although "the prospect of planned displacement of communities and peoples is repugnant" and constitutes "an admittance of defeat before we are actually defeated," nevertheless, he concluded, "it is prudent for leaders to keep it [migration] in mind as it is a matter of survival." (deBrum 2013). Such duty-bound pragmatism similarly compels me to explore adaptation limits as themselves opportunities for progressive climate migration planning. Given the specific

M. H. N. Stege (✉)
Executive Council, Maloelap Atoll Local Government,
Taroa Maloelap, MH 96960, Marshall Islands
e-mail: mark@atollconservation.org

W. Leal Filho and J. Nalau (eds.), *Limits to Climate Change Adaptation*,
Climate Change Management, https://doi.org/10.1007/978-3-319-64599-5_21

geological context of my treatment of adaptation limits, I will interchangeably refer to such limits as 'atoll habitability thresholds.'

Our discussion will be taken up in three parts, beginning first with an overview of atoll geology and societies so that the reader may have a better appreciation for the 'home' that this ecosystem represents to approximately 500,000 atoll dwellers worldwide. Building on this foundational knowledge of the atoll setting, I next synthesize decades of published research calling for the development of habitability threshold theory and—as case study of habitability threshold methods and procedures—develop the evidence base for flood risk as one practicable threshold or adaptation limit therein. Given the lack of empirical evidence quantifying magnitudes of climate change that constitute adaptation limits worldwide (Klein et al. 2014), and more specifically the paucity of data needed to conduct flood hazards science analysis in atoll environments (Hwang 2014), I draw on internationally recognized standards of flood risk management so as to introduce the methods and procedures by which to develop an atoll-appropriate flood risk threshold in the Marshall Islands. Emphasis will be placed on opportunities to engage atoll communities and incorporate indigenous knowledge into the flood risk modeling methodology by way of the *Reimaanlok* framework. The third part of this chapter will develop a case for the advancement of habitability threshold theory, methods and procedures using synthesized multiple-evidence strategies across different knowledge systems, pointing out several important benefits to adaptation limit research and best practice as a result. Hopefully this will help enable atoll communities to self determine the point(s) at which 'home' becomes uninhabitable, and thereby provide a timely contribution within the context of an evolving international discourse surrounding climate-induced migration planning today.

Atoll Geology and Societies

Atolls are largely the product of tiny tropical marine organisms called polyps, which themselves have a habitability threshold that is confined to certain depths within the warm tropical Pacific (Fig. 21.1) and Indian oceans and to a lesser extent the Atlantic (Stoddart 1965).

They rise like a crescendo of live coral positioned atop a long dormant underwater volcanic foundation dropping thousands of meters deep.[1] In certain locations along the upper reef platform, narrow stretches of land have been built up by wave and current-driven sedimentation and erosion products of the surrounding coral reef system to form sand cays and vegetated islets. These are all that remain above water during high tide, and represent the only potentially habitable areas of an atoll.

[1]This discussion on atolls is restricted to 'subsiding atolls,' and do not include 'emergent atolls' being lifted above sea level by tectonic forces, nor those single coral islets interspersed among atolls such as the 5 islets amongst 29 atolls within the Marshall Islands.

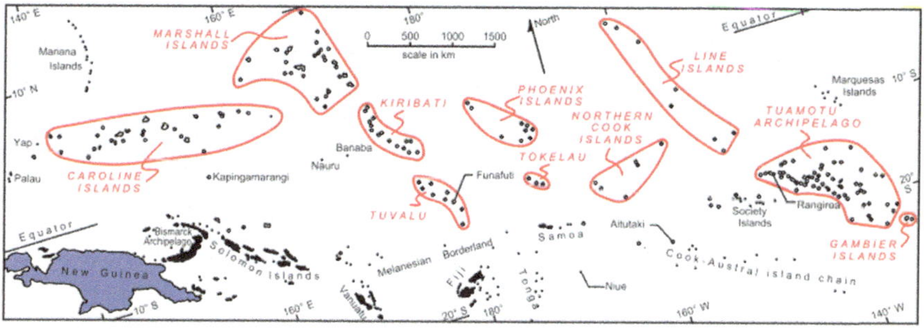

Fig. 21.1 Atoll provinces (in *red*) of the tropical Pacific Ocean. (Reprinted with permission, Dickinson 2009)

It was Charles Darwin who, while observing some of these unique landforms near the western coast of South America, first theorized how polyps could form barrier reefs grown atop subsiding volcanic edifices (Darwin 1842: 109). Since then reef core and radiocarbon evidence from multiple atoll sites have confirmed Darwin's theory, and expanded on the important role that sea-level fluctuations due to tectonic- and glacial-controlled forces have played in episodic coral reef growth and discontinuity relative to sea level along millennial timescales (Dickinson 2009: 5–7; Woodroffe and Webster 2014: 193). Of particular note was the last episode of coral reef growth and discontinuity, which is uniformly evidenced among atolls starting around 8000–10,000 BP with the inundation and submergence of present day atoll platforms due to sea level rise during the Last Interglacial. Coral growth appears to have lagged behind rising sea level during this time, until eustatic sea level stabilized at the 'mid-Holocene highstand' around 5000–6000 BP (Woodroffe and Webber 2014: 241–243). A period of explosive 'catch up' growth followed, during which different reef structures 'caught up' to sea level at different times and where margins of the atoll more exposed to wind-driven seawater nutrients generally grew fastest. Sea levels remained stable at the mid-Holocene highstand between 3000 and 5000 BP, a period marked in atolls today by the development of conglomerate platforms which formed after vigorous 'catch up' reef growth had ceased and lateral reef consolidation flourished (Dickinson 2009: 7).

Some have suggested that the period of coral growth relative to rising sea levels provides evidence that atoll islands might "rise in step with a rising sea" predicted due to anthropogenic climate change (Pala 2014: 496; Kench et al. 2014). Indeed this might be the case, especially if corals can be equally effective in coping with increasing ocean acidity and temperatures also predicted due to the same. But when considering the human dimensions of habitability on atolls in the face of sea level rise, as this chapter does, one must be able to distinguish between the geologic persistence of atolls as evidenced during the mid-Holocene highstand against human persistence in the face of increasing inundation and submergence. This is especially evident when one considers the millennial aggradation of an atoll, as described above, in which the habitable atolls we have today became submerged (i.e. uninhabitable) around 8000–10,000 BP and then 'caught up' around 2000–3000 BP, juxtaposed to the human timescales in which the decision of migration will need to be made. Clearly the impacts of climate change that compel atoll societies to migrate will occur on human rather than geologic timescales.

Near consensus within the scientific community further holds with "very high confidence" that by the end of this century global sea levels will have risen by 1.2 m (IPCC 2013: 1140) and even higher beyond the 21st century (Joughin et al. 2014; Rignot et al. 2014). All this on top of several meters of natural sea level variability due to fair-weather ocean dynamics leaves little doubt that coral atolls will ultimately become, again, uninhabitable due to sea level rise. These "irrefutable, as well as shocking" realities of climate change, as pronounced by US Secretary of State John Kerry in his address to the 2013 Pacific Islands Forum, necessitate a coordinated and scalable response. Planned migration from low-lying atolls will be necessary.

There are altogether four Indo-Pacific atoll nations who will face the realities of mass climate migration, namely Kiribati, Maldives, Marshall Islands, and Tuvalu. Based on archeological records, it appears that the beginning of human experiences on present day atolls coincides with the period when sea levels fell to present position around 2000–3000 BP (Tracey and Ladd 1974; Dickinson 2009; Kayanne et al. 2011; Weisler et al. 2012). Atoll societies, in essence, are the descendants of seafarers with distinct customs, life skills, and languages suited for human habitation in atoll archipelagos. In addition to the abundance of marine resources they discovered, they introduced with them different varieties of breadfruit (Petersen 2006), pandanus (Miller et al. 1956), and a salt-resistant strain of taro (Weisler 2001)—altogether capable of supporting dense populations in otherwise unhospitable margins of land.

These four atoll populations have since grown to a combined 500,000 people, with a projected growth to over 700,000 by 2050. There are various distinguishing characteristics among them in terms of population size, distribution, age, and colonial and religious affiliations; distinctions in their relative life expectancy, education, and income are also evident in their respective human development index (HDI) scores (See Table 21.1).

Despite these historical and contemporary differences, there are notable similarities among Marshallese, i-Kiribati, Tuvaluan, and Maldivian societies due to their similar natural environments. One of the most evident is an atoll society's

Table 21.1 Compiled demographic information for the four atoll nations

Atoll nation	Population in 2014	Projected to 2050	Urban/rural distribution[c]	Median age[d]	Colonial/Religious affiliation	HDI score
Kiribati	100,786	163,266[a]	44/56	23.9	British/Christianity	0.607[e]
Maldives	345,023	474,660[b]	40/60	27.4	British/Islam	0.698[e]
Marshall Is.	52,555	61,217[a]	72/28	22.6	American/Christianity	0.738[f]
Tuvalu	9,860	13,858[a]	50/50	25.2	British/Christianity	0.711[f]

Sources [a]SPC, [b]Maldives (2006), [c]UN (2011), [d]CIA (2016), [e]UNDP (2011), [f]UNDP (2013)

relationship to land. Specifically, the scarcity of habitable land in an atoll greatly heightens its value. One American anthropologist referred to a Marshallese's attitude towards their land as "the very root of their worldly existence" (Mason 1987: 4). This essentially indigenous attitude to our atolls is further substantiated in the linguistic perspective of Marshallese sociologist Kristina Stege et al. (n.d), who explains:

> Every newborn [is] born into the society with primary rights to land parcels held collectively by members of its mother's bwij [or lineage group]. In Marshallese the word bwij comes from the word bwijen, signifying either the navel or the umbilical cord, which nourishes life. The related term for land, bwidej, links ideas on maternal nourishment with the notion of earth itself.

For all that our ancestral domain holds dear, the contemporary push to seek education, health, and economic opportunities has driven approximately three-quarters of all Marshallese to now reside in urban centers or abroad (RMI Census 2011; US Census 2010). This is the social setting in which our conversation on climate-inducted migration takes place, where the biocultural loss of the motherland herself is at stake and atoll societies are already in a migratory flux.

Atoll Habitability Thresholds

Renowned psychologist Abraham Maslow outlined the hierarchy of human needs in the 1950s, but it was perhaps William Alkire (1978) who was the earliest to investigate atoll habitability specifically. His discussion of atoll physiography and ecology included the identification of four major indicators—island size, elevation, rainfall, and typhoon frequency—which he argued influence coral island vegetative growth, and thus human support potential. Lautze's (1997) equally groundbreaking work spelled out the need for more analytically grounded emergency aid interventions in complex emergencies including stress migration. Advancing habitability and migration research further into climate change adaptation, human geographers Barnett and Adger (2003: 321) dialed into atoll habitability analytics by highlighting "a need to explore empirically the effects of climate change at the local level" including changes in mean conditions for coral reef health and changes in extreme weather events (e.g. drought, storm surge, and typhoons) above periodic natural rates of occurrence. Later publications also advocate for broader thematic areas for measuring atoll habitability such as food security (Barnett 2011) and environmental security (O'Brien and Barnett 2013). Similar sentiments were also offered in the lead up to post-2015 UN Sustainable Development Goals, as they relate to Small Island Developing States (SIDS):

> Land is one of the most valuable natural resources of small islands and is closely linked to freshwater, another important resource. The degradation of these resources could pose serious constraints on the sustainable development of the affected SIDS. There may be tipping points where the degradation of vital land and water resources rapidly accelerates and threaten island habitability (UNEP 2014: 27).

In addition to the physical basis of atoll habitability, Barnett and Adger (2003: 330) also highlight the importance of perceptions of adaptation efficacy wherein greater and greater impacts from climate change undermine adaptation efforts, "leading to change in domestic resource use and decreased assistance from abroad." Clearly multiple thresholds are needed to account for the various biophysical and cultural parameters of atoll habitability, and to achieve such a goal would require empirical understandings of what constitutes a healthy and safe atoll environment for human habitation as well as culturally sensitive understandings of what constitutes the day-to-day functioning of a viable atoll society.

Therefore, in response to the widely recognized need for more nuanced investigations into the incremental loss of atoll habitability in the face of climate change, and with the same duty-bound pragmatism that prompted Marshall Islands lead climate negotiator Tony deBrum to declare the repugnance and at the same time acknowledge the prudence of keeping planned displacement in mind "as a matter of survival," (deBrum 2013), this chapter will present the case for one habitability metric in particular: a flood risk indicator by which atoll communities may be able to determine and communicate their own atoll's habitability and 'when it is time to move.'

Flood Risk Threshold

Atoll communities are inherently exposed to considerable flood risk, with potential flooding scenarios being almost omnidirectional as flooding can occur laterally from either lagoon or ocean (and in some instances simultaneously) as well as vertically from underground well water uplift during extreme high tides.[2] Wave-driven flooding is especially common when a storm surge, swell, or tsunami coincides with periods of elevated water levels typical during high tide, an El Niño Southern Oscillation (ENSO) event, and a transient eddy. The frequency and intensity of such episodic flooding events are predicted to increase exponentially with sea level rise because elevated water levels diminish wave-energy dissipation along fringing reefs and reef flats prior to waves reaching atoll shorelines (Merrifield et al. 2014; Storlazzi et al. 2011; Woodroffe 2008). To make matters worse, it is also predicted that increased sea surface temperature and ocean acidity will weaken coral structures that buttress atoll shorelines, giving waves further free reign before reaching inhabited shorelines. (Nicholls et al. 2007). Suffice it to say, as an atoll community's exposure to flood risk increases exponentially in the coming decades, its inhabitants' safety will become ever more threatened, thus

[2]Flooding during heavy rainfall is not considered significant due to an atoll's porous soil. It is possible that tropical cyclones could produce wave and rain flooding simultaneously, but the wave flooding would be so significant during these instances that rainfall flooding would be negligible.

raising the question: *how much flood risk is too much?* The answer that atoll communities arrive at underlies the flood risk threshold presented here.

The present method for measuring flood risk is to conduct a hydraulic analysis to quantify an area's susceptibility to flooding across spatial and temporal scales. In the United States, the identification, publishing, and updating of flood-prone land areas is mandated within the National Flood Insurance Act as part of the National Flood Insurance Program (NFIP). These areas are called Special Flood Hazard Areas (SFHA), and are essentially areas under the Base Flood Elevation (BFE) within which floodwater is calculated to pool or surge—depending on the category of the SFHA—during a 100-year storm.[3] In order to be covered within the NFIP, local communities must adopt and enforce flood plain ordinances and private property owners within a SFHA must pay insurance premiums corresponding to the elevation of their first floor relative to the BFE and the category of the SFHA in which they reside (FEMA 2014). In addition to informing insurance rates for homeowners and businesses, these flood maps also provide developers and land use planners foresight when proposing and approving housing eligible for funding under US federal housing programs (HUD 1994).

Although the need to account for projected sea level rise is not lost on the NFIP (FEMA 1991), its base flood maps presently account for only historic water levels. In contrast, Barnett et al. (2014: 1103) tested a community-based, sea level-responsive flood risk management framework in the township of Lakes Entrance in Australia by offering a 'pathway' or "sequence of linked strategies" corresponding to three progressively disruptive flooding regimes. The 5000 or so inhabitants of this six-river estuary in South-Eastern Australia have historically experienced flooding during high tides coupled with storms and strong rainfall. Should sea level rise exasperate this sufficiently enough to increase instances of 'nuisance flooding' to greater than five days per year, the Lake Entrance local government would take measures, per the early steps of the 'pathway,' to prepare for retreat of critical infrastructure and dwellings. A second flood risk threshold would be the occurrence of two 100-year storms within a single year, at which time actual retreat to higher ground surrounding Lake Entrance would commence. The third and final threshold would occur at the breach of a barrier dune surrounding the community's coastline, leading to permanent inundation of low-lying areas of the town. With this final threshold achieved, according to the proposed 'pathway' approach, adaptive management determined by future generations would be employed.

Similar to how State and local communities in the US and the city of Lake Entrance in Australia are determining *how much flood risk is too much* by way of a

[3]A '100-year storm' is also called a 1%-annual chance-flood. It refers to a storm level or peak that has a one in a hundred, or 1%, chance of being equaled or exceeded in any year. Other terms that express the same idea, such as 1% annual exceedance probability (or 1% AEP), are sometimes preferred because they avoid the common misconception that a '1 in 100 year storm' can only occur once every 100 years; or that it is 'safe' for another 100 years after experiencing such an event.

SFHA or progressively disruptive flooding regimes, so too can an atoll community facing climate-induced migration come to better examine, understand and communicate an adaptation limit to address the question of 'when it is time to move.' Unlike in the US and Australia, however, a lack of empirical data necessary to conduct hazards science analysis is presently limiting research into atoll flood risk (Hwang 2014). Should resources become available to address these gaps, determinations on what datasets to gather and how to gather and analyze them will present opportunities for both in situ actors such as local community members who manage cultural and biological resources, and external, ex situ actors such as national and international policy makers and non-local researchers, to share in the responsibility of designing and implementing the resulting atoll flood risk analysis tool. Both sets of actors must build in enough time and appropriate conditions for iterative reflection to achieve an ideally collective determination between migrant and recipient communities that displacement-inducing thresholds or adaptation limits have been exceeded. In validation of Barnett and Adger's (2003: 321) reference to the "culturally-specific adaptive capacity of…resilient atoll societies," these ideal conditions and processes whereby both in situ and ex situ actors are allowed time and appropriate conditions for iterative reflection presently exist in the Marshall Islands as part of an 8-step process known as *Reimaanlok.*

The *Reimaanlok* Framework

Reimaanlok is an adapted community-based resource management planning framework within which the 23 municipalities comprising the Republic of the Marshall Islands (RMI) may choose to infuse a data-driven and culturally-sensitive planning process into the cadence of daily atoll life. The *Reimaanlok* process was originally developed in the mid-2000s by in situ and ex situ actors comprising an interagency, non-government, and higher education working group known within the Marshall Islands as the Coastal Management Advisory Council (CMAC). It was primarily developed as a mechanism to expand the RMI Protected Area Network, though its name means "looking forward" or "looking to the future," and over its now 10-year application in the RMI it has become increasingly recognized as at the forefront of contemporary coastal zone management and climate adaptation planning, particularly where decision-making about the use of natural resources occurs primarily at the local level (Baker et al. 2010; Govan 2011; Grantham et al. 2011).

In its purest form, the *Reimaanlok* is an eight step process that, when triggered by an atoll community's leadership (Step 1), includes a scoping and budgeting exercise (Step 2), site visits by *Reimaanlok* facilitators to build awareness on the need for resource planning by the target atoll community (Step 3), followed by the gathering and analysis of various socio-ecological data parameters (Step 4) in order to design (Step 5) and ultimately legislate (Step 6) an integrated atoll resource management plan inclusive of programs to ensure ongoing monitoring and adaptive management (Step 7) and local commitment retention (Step 8). Given the specific

Table 21.2 Checklist of inputs needed to determine an atoll flood risk threshold

a. Digital elevation model
b. Vertical datum and High Water Mark
c. Numerical wave model
d. Sea level rise projections
e. Historic vegetation line comparisons

needs and unique circumstances of atoll municipalities, the *Reimaanlok* facilitation consortium known as CMAC may follow these eight steps in a linear or iterative process. This helps foster a sense of trust and shared purpose within the community and of the *Reimaanlok* facilitators, so that the process itself is an empowering experience for atoll communities and a vehicle for national cohesion and shared purpose among members of CMAC, all of whom seek to make sense of a dizzying array of international development goals, regional commitments, and national policies and action plans.

For the purposes of this chapter, we will focus our attention on *Reimaanlok* Steps 3 and 4 which together represent the appropriate conditions for iterative reflection among in situ and ex situ actors to design flood risk models specific to local scales and values. Such an atoll-appropriate flood risk threshold would require five basic data inputs (Table 21.2). As will be examined, each data input provides opportunity for a synthesis of scientific and indigenous knowledge, which together enhance the quality of the tool itself as well as its perceived salience among the atoll communities, and potential recipient hosts, which it serves.

Digital Elevation Model (DEM)

Creating a DEM for flood risk analysis typically involves using light detection and ranging equipment or LiDaR to achieve <10 cm spatial information of shallow water bathymetric and nearshore topographic features, but this method is cost prohibitive for the vast majority of isolated atoll communities. A more feasible method might be to use remotely piloted systems (RPS) and 3D photogrammetry. Advances in the last five years in sophisticated computer vision, robotics, and geomatic engineering allow cm-level resolution and accuracy products that can be generated with off-the-shelf, low-cost equipment (Colomina and Molina 2014: 80). Similar to LiDaR, the "RPS + photogrammetry method" is capable of extremely high accuracy and point density but at a fraction of the time and cost, making this requisite data input more accessible to *Reimaanlok* practitioners and under resourced atoll communities. Another low cost, high-resolution method utilizes conventional ground survey equipment, however the aerial system covers large area with significantly greater efficiency.

Per the *Reimaanlok* process, prior to undergoing scientific assessments, indigenous knowledge holders within an atoll community are provided ample time during *Reimaanlok* Step 3 site visits to share stories, songs, and oral histories.

Oftentimes the cultural knowledge touch on special or sacred places relevant to the topography and bathymetry of the site. In northwestern Marshall Islands community of Wotho Atoll, for instance, interviews with Wotho indigenous knowledge holders revealed notable landscape features undistinguishable in satellite imagery including an elevated ridge leading towards the large intertidal cul-de-sac of Wotho, and sink holes in the middle of the island (Stege et al. n.d). According to legend, the ridge is the deceased body of a slain giant named *Likuriōpjen*, while the sinkholes were dug up by a tiny bird named *Annañ*. These consultation exercises and site visits served to gather important planning information for later RPS + photogrammetry survey mission, as well as reinforce existing rights and tenures of traditional governance regimes at the individual atoll level. In other words, the synthesis of scientific and indigenous knowledge enhanced the quality of the tool itself as well as its perceived salience among the atoll communities.

Vertical Datum and the High Water Mark

Consistent and accurate measurements of flood risk require establishing both a community's vertical datum and its High Water Mark (HWM). Determinations on these two spatial reference points present a rich opportunity for the synthesis of scientific and indigenous knowledge when designing the atoll flood risk threshold analysis tool.

Establishing a vertical datum requires use of survey-grade Global Positioning System (GPS), which in the case of the Marshall Islands is available at the Lands & Survey Division within the Ministry of Internal Affairs. The benchmark may be demarcated by a stake driven into reliably dry soil or better yet a bolt inserted into concrete on a community landmark, and there should be at least 2–3 each established per community in case of disturbances to the benchmarks due to human activities or natural coastal geomorphology (Coastal Management Advisory Council 2012). Community participation during this component of flood risk threshold design is needed to ensure minimal disturbances of the benchmarks due to human activities well into the future. In terms of establishing the HWM, it is important to note that no consensus has been reached on its definition worldwide, with corresponding capture methods varying among different definitions (Liu et al. 2014). In most cases, the HWM measurement is based on tidal observations over the National Tidal Datum Epoch which is 19 years, however in remote atoll locations where there are often no tide gauges available much less over periods of 19 years, the community-based *Reimaanlok* process recommends establishing the HWM based on community consultations indicating the level of the highest spring tide (i.e. "the tide came up to this tree"). Again, 2–3 stakes driven into the soil or sand are needed to demarcate the HWM separating land and water along its coastline.

Together, these two spatial benchmarks will situate all surrounding land and infrastructure relative to highest spring tide, and provide a baseline on which each flooding scenario that the community is vulnerable to due to high waves and sea level rise can be added, as elaborated on in the next two dataset components below.

Numerical Wave Model

Developing a wave model for an atoll entails developing wave-driven flooding scenarios for swell, storm surge, and tsunami emanating from different directions (e.g. Northwest) and in unison with variable tidal (e.g. high tide) and climatic water level contributors (e.g. ENSO). The various water level contributors must be parameterized within wave-driven flooding scenarios, and later calibrated against observed flooding events representative of each of these wave-driven flooding scenarios, so as to supply hindcast flooding information that can be used to validate the model. This work offers an opportunity for community participation in two important ways. First, centuries-old Marshallese wave charts may be used to help identify historic oceanographic trends alongside more modern wave modeling techniques (Genz et al. 2009: 234). The second instance in which local knowledge can be leveraged is in documenting wave-driven flooding scenarios during and after actual flooding events via eyewitness, photographic, and video evidence (Hoeke et al. 2013: 130). Making use of these two opportunities to participate in the development of the numerical wave model both validates the wave model for developers and enhances the readability and salience of the model among atoll inhabitants.

Sea Level Rise Projections

Sea level rise (SLR) will affect the atoll flood risk model in two important ways. First, the return rates of various wave-driven flooding events will shorten, and secondly, the amplitude and duration of such events will increase as overtopping floodwaters extend further inland and high water levels linger longer. As such, SLR will become an increasingly pivotal data input in an atoll community's determination of unacceptable flood risk, and so it is important that atoll communities be able to distinguish SLR from natural sea level variability. The process of distinguishing between the two and applying this knowledge presents an important opportunity for community participation.

Atoll societies hold a rich language, experience, and traditional knowledge system related to gravitational tidal fluctuations that offer a pedagogical opportunity to fashion the scientific concepts of sea level variability. Marshallese communities, for instance, refer to this tidal variability as *bōkā*. More nuanced aspects of the science behind sea level variability can be "scaffolded" onto this indigenous knowledge system, so that communities may be able to discern the differences in temporary increases or decreases in flood risk due to ENSO, for instance, (Chowdhury et al. 2006: 214) as opposed to a sustained increase in risk due to SLR.

Historic Vegetation Line Comparisons

More commonly known as shoreline change analysis, this data input entails comparing historic positions of the vegetation edge of islands over time using historic aerial imagery and satellite imagery to quantify erosion and accretion trends. This method of analysis is highly valuable for determining flood risk locations on an atoll as well as flood resilience, with recent analysis results suggesting "a more optimistic prognosis for the habitability of atoll nations" (Kench et al. 2015: 515). It should be noted, however, that limited local knowledge or worse mischaracterization of shoreline changes can introduce misguided findings. For instance, one study failed to underscore the critical role that a 3-km airport construction project played in causing a trend of shoreline accretion for Majuro Atoll (Ford 2012). Therefore, of the various data inputs that comprise an atoll flood risk threshold determination tool, this is one of the most illustrative of the need to employ a synthesis of scientific and indigenous knowledge.

Efforts are currently underway to introduce such participatory shoreline change analysis capabilities within the CMAC consortium using available historic aerial photographs and satellite imagery housed on the *Reimaanlok* geospatial database known as ConservationGIS. Once set up, atoll communities will be able to digitize its collective memory of shoreline changes over time, thereby engendering an evidence-based conceptualization of flood risk exposure and therefore a rational approach to flood risk management. This particular data input of the flood risk threshold tool is highly intuitive to atoll inhabitants, and therefore an important source for justifying resource management measures and interventions that help prolong the habitability of their atoll homes.

Discussion

As evidenced in the above examination of data inputs within the context of an atoll-appropriate flood risk model, the answer to the original question posited to atoll government officials on whether we will recognize in advance the point at which atoll areas become uninhabitable can be considered using synthesized multiple-evidence strategies across different knowledge systems. There are several important benefits and insights that emerge in adaptation limits research and best practices when in situ and ex situ actors in this way share in the responsibility of designing and implementing determinants of adaptation limits. For one, it engages existing national resource management frameworks—such as the *Reimaanlok*—in contributing much needed empirical data and insight to the process of determining adaptation limits. This directly addresses the lack of empirical evidence needed to quantify magnitudes of climate change that constitute adaptation limits (Klein et al. 2014) in atoll environments. The paucity of data on atoll flood risk is particularly surprising, given the amount of attention given to highlighting their extreme vulnerability to sea level rise.

This can be partly attributed to the costly process in which high-resolution elevation data has been developed in the past, but more cost-effective solutions are now available and are being employed by *Reimaanlok* practitioners within the CMAC consortium (Stege et al. n.d). Moreover, the digital outcome of the various data gathering initiatives comprising Step 4 of the *Reimaanlok* is an invaluable and ever-growing database. Already a key tool for in situ and ex situ actors collaborating within the CMAC consortium to support atoll communities, it can also be used by the atoll communities themselves to compile a robust collective identity that is also portable, while also participating in the application of various skill sets needed to keep the digital database updated. In this way, the *Reimaanlok* database provides a platform that enables Marshallese to proactively address the climate challenges that our atoll ecosystems and society are encountering today and well into the future.

Leveraging existing data-rich national frameworks will bring forth additional insightful spinoff benefits within adaptation limit research and best practices. In the case of atoll countries, this will be particularly timely as wholescale climate displacement is not likely to happen for decades to come, and during which time atoll dwellers must continue the cadence of daily atoll life including the articulation and fulfillment of specific development objectives within in our still-thriving and still-habitable atoll homes. In the case of adaptation limits related to flood risk, adaptation limits may become mainstreamed into national frameworks thus causing atoll communities to discover adaptation 'pathways' (Barnett et al. 2014) that include sometimes difficult to champion resource management measures and interventions such as marine protected areas that, in the long run, help us thrive and prolong the habitability of our atoll homes. In the decades to come and owing to an increasing global discourse on climate-induced migration planning wherein the prognosis of atoll habitability has been referred to as "optimistic" (Kench et al. 2015: 515), this will become increasingly important as atoll communities will need to navigate the pitfalls of risk perception outpacing reality, leading to the intensification of outmigration from our still-habitable atoll homes. The value of adaptation limit management tools such as atoll-appropriate flood risk models inclusive of their perceived salience among atoll communities will be become increasingly significant.

In an effort to develop empirically-based adaptation limit models, there is also a risk that the capacity gap between outside experts and community actors will broaden due to increasing information accessible to climate researchers though inaccessible or difficult to access among researched communities. Within the context of communities facing climate displacement, this may inadvertently lead to the disempowerment of those who may already feel disempowered by the seemingly insurmountable challenge which climate change presents. While it is recognized that the need to increase local people's skills and capacity in relation to climate change impacts and appropriate adaptation activities for long-term sustainability remains urgent, (Pelesikoti et al. 2013) outside experts developing research and policy instruments for climate migration must further apply innovative methods to make research processes behind these instruments participatory, such as done within the *Reimaanlok* framework. Indeed, outside experts are in the best position to mitigate the risk of inadvertent disempowerment of community actors by prioritizing an applied research approach to adaptation limit data gathering and management tools design. The sustainability of their work will benefit as adaptation limits require continuous monitoring often most cost effectively conducted by local actors (Meentemeyer et al. 2015).

Lastly, the empirical and participatory approach demonstrated above to adaptation limit methods and procedures is a timely contribution within an evolving international dialog on climate-inducted migration planning. The most acknowledged effort to date is the Nansen Initiative, which has set forth as its goal to build consensus among interested governments on a protection agenda addressing the needs of people displaced across borders. Leading up to the Paris Accord in late 2015, over 100 governments had affirmed broad support of the Nansen Principals and the Nansen Workplan (Nansen Initiative 2015). While its initial emphasis has been on identifying standards for admission, stay, and status for climate-displaced peoples, there is growing recognition of the need to develop guidance for how to operationalize clause 14 (f) of the Cancun Adaptation Framework, which "Invites all Parties to enhance action" on "Measures to enhance understanding, coordination and cooperation with regard to climate change induced displacement, migration and planned relocation, where appropriate, at the national, regional and international levels" (UNFCCC 2010: 5). Concurrently, another climate-induced migration planning initiative called the Peninsula Principles has echoed the need for actors to emphasize local level participation and international facilitation in the planning and development of master relocation plans, which would determine the habitability and feasibility of any relocation site based on "specific, geographically appropriate, standard criteria including habitability of the land…" (Displacement Solutions 2013: 23). Leading climate-induced policy experts have also outlined success factors for the development of such standards, including having "sufficient lead-time to enable careful, participatory planning processes, appropriate land acquisition and ensuring sustained and sufficient financing to resettle individuals in a way that improves rather than deteriorates living standards" (Warner et al. 2013: 9).

Clearly there is broad recognition that multi-level coordination and participatory planning will be critical success factors in an anticipatory climate-induced migration planning scenario. In contribution, this chapter has made a case for the development of data-driven atoll habitability threshold management tools across different knowledge systems, where in the development of habitability threshold theory and design offers numerous intuitive premises for participatory planning. Furthermore, a demonstration of participatory planning as it can occur within habitability metrics development has been introduced and outlined, in terms of the various data inputs needed for atoll communities and their external partners to accurately understand and communicate the temporal and spatial parameters of flood risk on an atoll including 'when it is time to move.' Additional research rooted in duty-bound pragmatism and using the existant *Reimaanlok* ecosystem monitoring framework and flood risk modeling procedures outlined in this chapter are needed to provide additional contributions (Stege et al. n.d). As a poetic leader from the Marshall Islands put it when addressing the UN Climate Summit in New York City (Jetnil-Kijiner 2014):

We deserve not just to survive.
We deserve to thrive.
Take us along on your ride.
We won't slow you down.
We'll help you win the most important race of all.
The race to save humanity.

Conclusion

Looking towards the future, or to *reimaanlok*, pragmatic poetry in a sense encapsulates the habitability narrative that this chapter is proposing to be applied in a future of "increasing human migration from low to high elevation islands and continental sites" (Leong et al. 2014: 554–555). We must be 'pragmatic' in that such habitability thresholds or adaptation limits must be based in empirical and evidence-based knowledge systems so as to be realistic, and we must use 'poetry' in that these thresholds or limits must also draw from humanist and non-empirical knowledge systems so as to be salient and relevant to displaced peoples. This approach is currently applied via the *Reimaanlok* framework in the atolls of the Marshall Islands which are "unquestionably the most precarious landscapes for [human] settlement" (Weisler 1999: 1) regardless of climate change, and where humans have lived life "on the edge" (Weisler 2001: 109) for so long already. Just as atoll communities recognize those ecosystem services which make life on these already extreme human habitats possible today, they also serve to intuitively communicate "habitability thresholds," per se, to non-atoll dwellers when those ecosystem services and the cadence of atoll life diminish beyond both locally and internationally recognized standards. Our indigenous experiences speak to the dignity and value as atoll people, and in so doing also offer insights and best practices into adaptation limits which may prove useful across various other ecological or 'home' settings where other communities are facing the prospect of planning for climate-induced migration.

Acknowledgements First and foremost, to the people of Maloelap Atoll and Wotho Atoll, thank you for your honesty and humor. Thank you also to my advisors at Columbia University for your support and guidance while I prepared my thesis which became this chapter, especially Benjamin Orlove and Michael Gerrard for your patience and generous intellects. I also thank Ambassador Tony deBrum, Minister Mattlan Zackhras, and Councilman Jisam Kaisha for your enduring leadership. In addition, I'm so appreciative for the inspirational work by the Coastal Management Advisory Council, the Micronesia Challenge Measures Working Group, Glen Joseph, Holly Barker, Leslie Stein, Vicente Diaz, Murray Ford, Dirk Spennemann, Julie Walsh, Peter Rudiak-Gould, Maxine Burkett, Jon Barnett, Scott Leckie, Donna Davis, James Miller, Joseph Muhlhausen, Tina Stege, Eleanor Sterling, Jennifer Newell, Kathy Jetnil-Kijiner and the JoJiKuM team. And to every future climate migrant and host throughout the world, my unending gratitude for your courage and strength as we *reimaanlok* together.

References

Alkire, W. H. (1978). *Coral Islanders. Worlds of man studies in cultural ecology*. Arlington Heights: AHM Publishing. 164 pp.

Baker, N., Beger, M., McClennen, C., Ishoda, A., & Edwards, F. (2010). Reimaanlok: A national framework for conservation area planning in the Marshall Islands. *Journal of Marine Biology, 2011*.

Barnett, J., Graham, S., Mortreux, C., Fincher, R., Waters, E., & Hurlimann, A. (2014). A local coastal adaptation pathway. *Nature Climate Change, 4*(12), 1103–1108.

Barnett, J. (2011). Dangerous climate change in the Pacific Islands: Food production and food security. *Regional Environmental Change, 11*(1), 229–237.

Barnett, J., & Adger, W. N. (2003). Climate dangers and atoll countries. *Climatic Change, 61*, 321–337.

Coastal Management Advisory Council. (2012). A Facilitators Guide to Implementing the Reimaanlok Conservation Planning Process. Available at http://seagrant.soest.hawaii.edu/sites/default/files/publications/reimaanlok_fieldguide.pdf.

Chowdhury, Md R, Chu, P., & Schroeder, T. (2006). ENSO and seasonal sea-level variability—A diagnostic discussion for the US-Affiliated Pacific Islands. *Theoretical and Applied Climatology, 88*(3), 213.

CIA. (2016). The World Factbook. Available at https://www.cia.gov/library/publications/the-world-factbook.

Colomina, I., & Molina, P. (2014). Unmanned aerial systems for photogrammetry and remote sensing: A review. *ISPRS Journal of Photogrammetry and Remote Sensing, 92*, 79–97.

Darwin, Charles. (1842). *Geological observations on parts of South America* (p. 279). London: Smith Elder and Co.

deBrum, T. (2013). Question and Answer Session. Response given at the World Leaders Forum: Christopher Jorebon Loeak, President of the Republic of the Marshall Islands. September 25, 2013. Columbia University, New York, NY. Available at https://www.youtube.com/watch?v=dU_hVpduok4.

Dickinson, W. (2009). Pacific atoll living: How long already and until when? *GSA Today*. doi:10.1130/GSATG35A.1.

FEMA. (1991). Projected Impact of Relative Sea Level Rise on the National Flood Insurance Program. FEMA Federal Insurance Administration. October 1991. Available at http://papers.risingsea.net/federal_reports/flood-insurance-and-sea-level-rise.pdf.

FEMA. (2014). National Flood Insurance Program: Flood Hazard Mapping Website. Available at http://www.fema.gov/national-flood-insurance-program-flood-hazard-mapping.

Ford, M. (2012). Shoreline changes on an Urban atoll in the Central Pacific Ocean: Majuro Atoll, Marshall Islands. *Journal of Coastal Research, 28*(1), 11–22.

Genz, J., Aucan, J., Merrifield, M., Finney, B., Joel, K., & Kelen, A. (2009). Wave navigation in the Marshall Islands: Comparing indigenous and Western scientific knowledge of the ocean. *Oceanography, 22*(2), 234–245.

Govan, H. (2011). Good coastal management practices in the Pacific: Experiences from the field. Apia, Samoa: SPREP, 2011. 42 p. Available at http://www.icriforum.org/sites/default/files/ICRI%20Best%20Practice%20in%20Coastal%20Management%20-%20Govan%202011_0_0.pdf.

Grantham, H. S., Mcleod, E., Brooks, A., Jupiter, S. D., Hardcastle, J., Richarson, A. J., et al. (2011). Ecosystem-based adaptation in marine ecosystems of tropical Oceania in response to climate change. *Pacific Conservation Biology, 17*, 241.

Hoeke, R. K., McInnes, K. L., Kruger, J. C., McNaught, R. J., Hunter, J. R., & Smithers, S. G. (2013). Widespread inundation of Pacific islands triggered by distant-source wind-waves. *Global and Planetary Change, 108*, 128–138.

HUD. (1994). Minimum Property Standards for Housing 4910.1, 24 CFR Appendix K 200.926. US Department of Housing and Urban Development. Available at http://portal.hud.gov/hudportal/documents/huddoc?id=49101xkHSGH.pdf.

Hwang, D. (2014). Gap Analysis for Implementation of Hazard Science into the Community with Test Applications for Kauai, Hawaii and Majuro, Republic of the Marshall Islands. Prepared for National Sea Grant Office, NOAA. University of Hawaii Sea Grant College Program.

IPCC. (2013). Climate change 2013: The physical science basis. In T. F. Stocker, Qin D., Plattner G.-K., Tignor M., Allen S. K., & Boschung J., (Eds.), *Contribution of Working Group I to the Fifth Assessment Report of the Intergovernmental Panel on Climate Change*. Cambridge: Cambridge University Press, 1535 pp.

Jetnil-Kijiner, K. (2014). Dear Matafele Peinam. UN Climate Summit. Available at http://www.un.org/climatechange/summit/2014/09/watch-marshallese-poet-kathy-jetnil-kijiner-speaking-climate-summit/.

Joughin, I., Smith, B. E., & Medley, B. (2014). Marine ice sheet collapse potentially under way for the Thwaites Glacier Basin, West Antarctica. *Science, 344*(6185), 735–738.

Kabir, H. E., Kim, E. M., Lee, H. Y., Nam, B., & Maeng, J. H. (2013). *Human Development Report 2013*. Doctoral dissertation, Ewha Womans University.

Kayanne, H., Yasukochi, T., Yamaguchi, T., Yamano, H., & Yoneda, M. (2011). Rapid settlement of Majuro Atoll, central Pacific, following its emergence at 2000 years CalBP. *Geophysical Research Letters,* 38.

Kench, P. S., Owen, S. D., & Ford, M. R. (2014). Evidence for coral island formation during rising sea level in the central Pacific Ocean. *Geophysical Research Letters, 41,* 820–827.

Kench, P. S., Thompson, D., Ford, M. R., Ogawa, H., & McLean, R. F. (2015). Coral islands defy sea-level rise over the past century: Records from a central Pacific atoll. *Geology, 43*(6), 515–518.

Klein, R. J. T., Midgley, G. F., Preston, B. L., Alam, M., Berkhout, F. G. H., Dow, K., & Shaw, M. R. (2014). Adaptation opportunities, constraints, and limits. In Field, C. B., Barros, V. R., Dokken, D. J., Mach, K. J., Mastrandrea, M. D., & White, L. L. (Eds.), *Climate change 2014: Impacts, adaptation, and vulnerability. Part A: Global and sectoral aspects. Contribution of Working Group II to the fifth assessment report of the intergovernmental panel on climate change* (pp. 899–943). Cambridge: Cambridge University Press.

Lautze, S. (1997). *Saving lives and livelihoods. The fundamentals of a livelihood strategy*. Feinstein International Famine Center. Medford, MA: Tufts University.

Leong, J. A., Marra, J. J., Finucane, M. L., Giambelluca, T., & Merrifield, M. (2014). Chapter 23: Hawai ‘i and US Affiliated Pacific Islands. Climate change impacts in the United States: The third national climate assessment. *US Global Change Research Program*, 537–556.

Liu, X., Xia, J., Wright, G., & Arnold, L. (2014). A state of the art review on high water mark (HWM) determination. *Ocean & Coastal Management*. Vol. 102 Part A, December 2014, pp. 178–90.

Maldives. (2006). Population and Housing Census Report. Ministry of Planning and National Development. Available at http://statisticsmaldives.gov.mv/nbs/wp-content/uploads/2006/12/analysiscd/.

Mason, L. (1987). Tenures from subsistence to star wars. In R. Crocombe (Ed.), *Land tenure in the atolls* (pp. 3–27). Suva: University of the South Pacific.

Meentemeyer, R. K., Dorning, M. A., Vogler, J. B., Schmidt, D., & Garbelotto, M. (2015). Citizen science helps predict risk of emerging infectious disease. *Frontiers in Ecology and the Environment, 13*(4), 189–194.

Merrield, M. A., Becker, J. M., Ford, M., & Yao, Y. (2014). Observations and estimates of wave-driven water level extremes at the Marshall Islands. *Geophysical Research Letters, 41* (20), 7245–7253.

Miller, C. D., Murai, M., & Pen, F. (1956). *The use of Pandanus fruit as food in Micronesia*. University of Hawaii.

Nansen Initiative. (2015, October 14). *More than 100 governments affirm broad support to better protect people displaced across borders by disasters and the effects of climate change* [Press release]. Retrieved from https://www.nanseninitiative.org/more-than-100-governments-affirm-broad-support-to-better-protect-people-displaced-across-borders-by-disasters-and-the-effects-of-climate-change/.

Nicholls, R. J., Wong, P. P., Burkett, V. R., Codignotto, J. O., Hay, J. E., & McLean, R. F. (2007). Coastal systems and low-lying areas. In M. L. Parry, O. F. Canziani, J. P. Palutikof, P. J. van der Linden, & C. E. Hanson (Eds.), *Climate Change 2007: Impacts, adaptation and vulnerability. Contribution of Working Group II to the Fourth Assessment Report of the Intergovernmental Panel on Climate Change* (pp. 315–356). Cambridge: Cambridge University Press.

O'Brien, K., & Barnett, J. (2013). Global environmental change and human security. *Annual Review of Environment and Resources, 38,* 373–391.

Pala, C. (2014). Will Islands rise in step with a rising sea? *Science Magazine, 345,* 496–497.

Pelesikoti, N., Ronneberg, E., Nakalevu, T., & Leavai, P. (2013). *Report on adaptation challenges in Pacific Island countries.* Apia, Samoa: SPREP, APAN.

Petersen, G. (2006). Micronesia's breadfruit revolution and the evolution of a culture area. *Archaeology in Oceania, 41*(2), 82–92.

Rignot, E., Mouginot, J., Morlighem, M., Seroussi, H., & Scheuchl, B. (2014). Widespread, rapid grounding line retreat of Pine Island, Thwaites, Smith, and Kohler glaciers, West Antarctica, from 1992 to 2011. *Geophysical Research Letters, 41*(10), 3502–3509.

RMI. (2011). National Census 2010. *Republic of the Marshall Islands.* Available at http://prism.spc.int/images/census_reports/Marshall_Islands_Census_2011-Full.pdf.

Rudiak-Gould, P. (2013). Climate change and tradition in a small Island State: The Rising Tide. New York: Routledge. (especially Chapter 5).

Solutions, D. (2013). The Peninsula principles: on climate displacement within states. Displacement Solutions.

SPC. (2014). Secretariat of the Pacific Community Statistics for Development Data Portal. Available at https://prism.spc.int/.

Storlazzi, C. D., Elias, E., Field, M. E., & Presto, M. K. (2011). Numerical modeling of the impact of SLR on fringing coral reef hydrodynamics and sediment transport. *Coral Reefs, 30*(1), 83–96.

Stege, M., Muhlhausen, J., Shumann, G., Menke, B., & McCloud, E. Lessons learnt from experimental flood risk modeling in Wotho Atoll, Marshall Islands: Benefits of participatory mapping and parallel use of technology. Manuscript in preparation.

Stege, K. (2008). An Kōrā Aelōn̄ Kein (These Islands belong to the women): A study of women and land in the republic of the Marshall Islands. In E. Huffner (Ed.), *Land and women: The matrilineal factor* (pp. 1–34).

Stoddart, D. R. (1965). Geography and the ecological approach: The ecosystem as a geographic principle and method. *Geography, 50,* 242–251.

Tracey, J. I. J., & Ladd, H. S. (1974). Quaternary history of Eniwetok and Bikini Atolls, Marshall Islands. In *Proceedings of 2nd international Coral Reef Symp*osium (Vol. 2, pp. 537–550).

UN, Dept. of Economic and Social Affairs, Population Division (2011). World Population Prospects: The 2010 Revision.

UNDP. (2011). Human Development Report 2011: Sustainability and Equity: A Better Future for All. Available at http://hdr.undp.org/en/content/human-development-report-2011.

UNEP. (2014). Emerging Issues for Small Island Developing States: Results of the UNEP Foresight Process. United Nations Environment Programme (UNEP), Nairobi, Kenya. Available at http://www.unep.org/pdf/Emerging_issues_for_small_island_developing_states.pdf.

UNFCCC. (2010). The Cancun Agreements: Outcome of the work of the Ad Hoc Working Group on Long-term Cooperative Action under the Convention. FCCC/CP/2010/7/Add.115 March 2011. Available at http://unfccc.int/resource/docs/2010/cop16/eng/07a01.pdf#page=4.

U.S. Census Bureau. (2010). The native Hawaiian and Other Pacific Islander population: 2010 Census Brief C2010BR-12. May 2012.

Warner, K., Afifi, T., Kälin, W., Leckie, S.,Ferris, B., Martin, S. F. & Wrathall, D. (2013). Changing climates, moving people: Framing migration, displacement and planned relocation.

UNU-EHS Publication Series Policy Brief No. 8, June 2013. Available at http://www.nanseninitiative.org/consolidating-knowledge.

Weisler, M. I., Yamano, H., & Hua, Q. (2012). A multidisciplinary approach for dating human colonization of Pacific atolls. *The Journal of Island and Coastal Archaeology, 7,* 102–125.

Weisler, M. I. (1999). *Atolls as settlement landscapes: Ujae, Marshall Islands.* Chicago: Smithsonian Inst.

Weisler, Marshall I. (2001). Life on the edge: Prehistoric settlement and economy on Utrõk Atoll, northern Marshall Islands. *Archaeological Oceania, 36*(2001), 109–133.

World Bank. (2014). World population statistics. Available at http://data.worldbank.org/indicator/SP.POP.TOTL.

Woodroffe, C. D. (2008). Reef-island topography and the vulnerability of atolls to sea-level rise. *Global and Planetary Change, 62*(2008), 77–96.

Woodroffe, C. D., & Webster, J. M. (2014). Coral reefs and sea-level change. *Marine Geology, 352,* 248–267.

Chapter 22
Conclusions: Overcoming the Limits to Adaptation

Walter Leal Filho and Johanna Nalau

Introduction

One old African proverb says that one needs to know how to engage elephants (so their power and strength can be optimized), how to educate baboons (so that they do not do mischief) and how to avoid snakes (so efforts are not damaged). In terms of climate change adaptation, this African proverb serves to illustrate its potentials and limits. The various contributions amassed on this book have shown the wide range of activities in the field of climate change adaptation. They have also shown some of the limiting factors, which are slowing down progress in implementing adaptation and also in some cases making business-as-usual approaches ineffective and impossible.

Throughout this book, there are many examples of successful approaches, examples of the use of successful methods and the successful application of research results in support of climate change adaptation, as outlined in the introduction. Together, they tackled a number of very important issues. Some of them are:

(a) Governance in climate change adaptation
(b) The need for information, communication and capacity-building
(c) The value of engaging the relevant stakeholders in the design and implementation of strategies and discussions surrounding thresholds and adaptation limits
(d) The relevance of due consideration to gender issues

W. Leal Filho (✉)
School of Science and the Environment, Manchester Metropolitan University, Chester Street, Manchester M1 5GD, UK
e-mail: w.leal@mmu.ac.uk

J. Nalau
Griffith Institute for Tourism and Griffith Climate Change Response Program, Griffith Business School, Griffith University, Nathan, QLD, Australia
e-mail: j.nalau@griffith.edu.au

W. Leal Filho and J. Nalau (eds.), *Limits to Climate Change Adaptation*, Climate Change Management, https://doi.org/10.1007/978-3-319-64599-5_22

(c) The need to consider and reflect on the influence of climate change and dislocation of communities.

These elements, as outlined in the Handbook of Climate Change Adaptation (Leal Filho 2015), need to be properly understood since many of them play a key role in setting the current limits to adaptation.

Overall, the papers part of this volume underline the importance of knowledge exchange for coping with climate change, and the examples from the Americas, Europe, African and Asia-Pacific countries display various levels of cooperation, including hands on solutions and integrative approaches by local stakeholders, also including the role played by the provision of academic and scientific support. This is in line with the Sustainable Development Goal 13, which outlines the need for more systematic efforts to cope with climate change. In pursuing them, the engagement of the concerned stakeholders is an essential pre-condition for success. Yet, the extent to which this engagement takes place leaves much room for improvements, and this poses one of the many the major constraints towards implementing adaptation efforts.

The next part of this conclusions section does not intend to repeat the messages and experiences documented and disseminated by each paper. Rather, it offers a bird´s eyes overview of the complexity of climate change adaptation, and list some of the challenges seen today.

The Complexity and Challenges of Climate Change Adaptation

The scientific evidences amassed by the work of the Intergovernmental Panel on Climate Change (IPCC) suggest that a continuous rise in mean temperatures is very likely for the remaining of this century. In addition, there is increasing observational evidence that regional changes in climate patterns have contributed to various alterations in physical and biological systems in many parts of the world (IPCC 2001a, b).

The literature also contains a wide range of works, which illustrate the need for climate change adaptation strategies, which are needed parallel to global climate mitigation efforts. As knowledge of the benefits of early adaptation and the new momentum created by the Paris Agreement in 2015 have grown, climate change adaptation has become an ever more prominent item on the political agendas of both industrialised and developing nations.

However, even though the need for additional policies on adaptation was already acknowledged at the time of the ratification of the UN Framework Convention on Climate Change (UNFCCC) back to 1992, and is now probably one of the most fashionable issues in the international public arena, the implementation of adaptation strategies is still problematic in many contexts. If one looks at the lack of integrated policies for instance, a paradox can be identified: climate change policies are often detached from mainstream ones. For instance, Poland wishes to pursue climate change mitigation and adaptation, yet continues to keep its reliance in coal-fired power stations. The United States has an ambitious climate change mitigation and adaptation plan, but pursues fracking. This paradox needs to be addressed.

One of most prominent issues is the search for appropriate means to increase adaptive capacity, which is known to significantly vary between countries, depending on governance systems (Leal Filho and Knieling 2013), social structure, culture, economic capacity, technological capabilities and—in many cases—the level of environmental degradation they experience. The Intergovernmental Panel on Climate Change IPCC (2001a) emphasized that Africa, for instance, is characterized by a low adaptive capacity. Indeed, many African countries are actively engaged in climate change adaptation—as it is the case elsewhere in the world—and have prepared and/or are implementing climate change adaptation programmes (Leal Filho et al. 2015). However, the existence of adaptation strategies does not necessarily equal their actual implementation. This is especially so if the realities of industrialised and developing countries are considered. In broader terms, the current differences seen between "rich" and "poor" countries are mostly a result of the fact that developing countries are surrounded by a range of limiting factors, as outlined in Table 22.1. Together, they represent what can be called the limits to adaptation, the main focus of this book.

The above list is by no means comprehensive, and its purpose is to offer an overview of a few of the issues at hand. Whereas industrialised countries are able to set-up, agree and fund a variety of climate change adaptation strategies, the same cannot be said to the majority of developing countries, which often do not have access to the financial resources and technologies needed to properly adapt. For example, in their study on "Challenges and Opportunities to Climate Change Adaptation and Sustainable Development Among Tanzanian Rural Communities", Dungumaro and Hyden (2010) have found out that challenges are centred on the pervasive poverty, rapid population increase and high illiteracy rates in the country.

Table 22.1 Some of the limiting factors towards climate change adaptation in developing countries

Limiting factor	Consequences
Weak governance systems Leal Filho (2011)	No assurance agreements will be met
Lack of prioritising	Climate change adaptation is victimised by other competing issues
Financial resources	Lack of funding creates dependency on donors, which have their own agendas
Technology access	Limited availability of technologies leave many areas untackled
Lack of training	Staff at government and non-government organisations unable to engage in some crucial areas (e.g. city planning, coastal zone management, etc.)

Similar findings are reported across many of the chapters in this book, to the extent that changing these basic conditions and their interaction can pose a significant adaptation limit where individuals' and communities' values and priorities cannot be realised.

The findings of Dungumaro and Hyden (2010) are in line with the report of IPCC (2001b), which emphasised that adaptive capacity is low in a developing country due to low GDP per capita, widespread poverty, inequitable land distribution, and low levels of education. The same report also identifies that the absence of social safety nets—for instance, after harvest failures—poses a serious problem in the implementation of climate change adaptation strategies. A report on climate change, poverty and adaptation in Bolivia pointed out that poverty and deforestation have lead the country to question its ability to strategically adapt to climate change (Oxfam International 2009). The large uncertainties in climate projections for parts of Africa, a weak evidence base of complex, often non-deterministic, climate–society interactions and institutional issues, were also some problems found by Conway and Schipper (2011). In many cases, the right government framework is not in place, which makes it very difficult to achieve globally set commitments towards protecting their natural resources and the vulnerable sectors of their populations. As illustrated in Table 22.2, climate change adaptation is taking place across multiple levels.

As can be interpreted from Table 22.2, the fact that adaptation is taking place across multiple levels is at the same a curse and a blessing. It can be a curse because it shows the various challenges that need to met towards its efficient implementation. But it can also be a blessing: even though matters related to climate change are global in nature, action towards adaptation can be undertaken at the local and community level, in order to yield the expected benefits. Here comes the blessing: action take at the local level is visible and more likely to lead to tangible outputs.

Unfortunately, many of the current climate change projects are sector-focused and hence the benefits deriving from them often only benefit a particular group (e.g. farmers). There is also a great need for innovative approaches when handling climate change adaptation initiatives. As demonstrated by Leal Filho in the book

Table 22.2 Levels of implementation of climate change adaptation

Level of implementation	Features
Government	Preparation of policies and frameworks meant to apply at the country level
Regional	Implementation of policies and initiatives aimed at the regional level
Local	Implementation of policies and initiatives at the local level, mostly by means of projects
Community	Implementation of policies and initiatives at the local level by means of projects involving specific local stakeholders
Cross-sectoral	Implementation of policies and initiatives involving stakeholders from various sectors

"Innovation in Climate Change Adaptation", there is a chronic problem in respect of addressing old problems using unsuccessful tools and approaches. For instance a few of these issues include:

1. Misguided donor support: hundreds of millions of dollars are spent by international donors on climate change adaptation projects each year. Even though the majority of the projects are well intended and planned, many projects—particularly large ones—seldom manage to fully reach their objectives. Part of the problem is due to the use of weak monitoring systems. Many donors focus on evaluations at the end of projects, when it is usually too late to take any meaningful action to address issues arising during the project.
2. Lack of long-term perspectives: the usual project cycles entail funding for activities lasting three to four years. Yet, in order to become sustainable and yield long-term benefits, project cycles of seven to eight years are needed. As a result of current practice, most projects are folded without mature outputs, which could be reached if they lasted a little longer.

According to Moser and Ekstrom (2010) there are three key components, which underlie a framework to diagnose barriers to climate change adaptation:

- First, an idealized depiction of a rational approach to decision-making to adaptation is needed;
- Second, a set of interconnected structural elements need include the various actors, and consider the larger context in which they act (e.g., governance), and the context in which they may engage (.e.g. a concern to exposure to climate change in low-lying coastal areas); and
- Third, to overcome identified barriers, a matrix is needed to helps map the influence of specific actors over it.

Local government authorities are at the forefront of managing the impacts of climate change through land-use planning, development consent and asset management and protection (NCCARF 2013). Lack of priority policy and weak effective vulnerability assessment combined with low technical and financial capability are the main problems faced by locals government in their attempts to implement climate change adaptation (NCCARF 2013). The lack of physical evaluation of vulnerabilities (to understand where action is most needed) and social vulnerabilities (to ensure that adaptation decisions are also considered as well as lack of community's empowerment to build resilience to climate change, to reduce dependency especially during extreme events) are also part of the list of problems in the implementation of climate change adaptation (NCCARF 2013).

Decision makers at all levels in Europe currently face several significant challenges when devising appropriate and sustainable adaptation strategies (Juhola 2016). In the summary of the project 'Bottom-up Climate Adaptation Strategies Towards a Sustainable Europe' (BASE) an international team highlighted the uncertainties and insufficient knowledge, the need to improve assessments, and the need to improve mainstreaming of adaptation in policy as main problems in the implementation of climate change adaptation in Europe. Uncertainty and insufficient knowledge is a broad issue, extending beyond the natural sciences to encompass socioeconomic factors, as well as our understanding of the costs and benefits of adaptation measures themselves (BASE 2013).

By considering a wider set of constraints to practical attempts towards adaptation, Measham et al. (2011) based on a case study of three municipalities in Sydney, Australia in 2008, demonstrated that climate adaptation was widely accepted as an important issue for planning conducted by local governments. However, constraints on local adaptation planning have tended to be conceptualised in a relatively straight-forward, mechanical way, emphasising the importance of inadequate information, institutional limitations, lack of resources and a culture of reactive management.

Contributions from This Book

Given these constraints and limits to adaptation noted above, the chapters in this book have provided a diverse lens through which we can expand the definition and notion of the concept of adaptation limits. As noted in the introduction, the IPCC's Fifth Assessment Report considered adaptation limits mostly in terms of defining conceptual linkages between limits and transformational adaptation, including ethical considerations and cross-linkages between concepts. The three main kinds of limits focused on biophysical, sociocultural and economic. Literature to date has been more interested in examining adaptation limits within a particular sector, ecosystem or at specific species level. The chapters in this book have expanded this focus to community level, to psychology, to policy and governance arrangements, to knowledge and its use for adaptation, with case studies across the continents.

Several of the authors have provided interesting insights into economic adaptation limits, which constitute intolerable risks. Greenough described how pastoralists face economic limits in that they cannot keep up the necessary peak livestock number, which enables them to maintain pastoralist lifestyle. Intolerable risk hence emerges when households are not able to provide for their well-being through their herds and have to let go of being pastoralists. However, as Greenough notes, many of them move to cultivate and earn income through wage labour but only until they are able to increase their herd size and have the economic and social security in re-entering pastoralism. Yet, physical and ecological limits also influence whether this move can or cannot be done back to these livelihoods. What this also tells us is that adaptation limits are in some cases mutable: people might reach intolerable risk, as in losing their way of life, but in some cases this can be temporary. Hence, these represent soft adaptation limits, which can be overcome but not necessarily in the present day (Klein et al. 2014).

Ologhe et al. reported also on the economic limits in terms of economic recession in Nigeria, which impedes efforts and funding towards adaptation to climate change in particular amongst the agricultural sector. However, Ologhe et al.'s research reminds us that many of the adaptation limits interact: although economic limits pose a challenge in terms of adequate funding for adaptation actions, socio-cultural limits, such as strong beliefs in particular practices and ways of being, limit the potential of adaptation opportunities among Nigerian farmers. Certainly, as Coultier notes, our "personal cognitive, psychological and creative constraints" and the limit of our imagination, which also influence what exactly limits adaptation to climate change.

Several chapters also provided evidence of the interplay between different levels of adaptation governance and hence different kinds of adaptation limits. The findings from Gbharbaoui's and Blocher's research on relocation in Papua New Guinea and Fiji demonstrate clearly that in order to truly understand limits to adaptation, adaptation limits need to be considered across multiple settings. Hence, there are particular limits which work at individual level but especially in the Pacific islands context, adaptation limits are very much communal and socially constructed, which are influenced by not just the relocating communities but the existing communities in the new place of settlement. Therefore, in the context of community relocation, intolerable risks needs to be understood in a multi-setting, which includes both the place of relocation and the new settlement. Luetz also notes this in his case study on migration in Bangladesh that the new place of settlement for migrants from rural to urban areas can bring forth heightened vulnerability instead of a solution. Luetz sees adaptation limit as "the precise point in time when livelihoods are compromised beyond reasonable hopes of recovery": facing this limit people decide to migrate. Gbharbaoui and Blocher also demonstrate how loss of customary land can present a major limit to adaptation and result in psychological limits to adapt.

Yet, Charan et al. note that relocations of communities are already taking place in the Pacific where particular physical and social limits are being reached. Examples of adaptation limits in the Pacific are increasing although there is frequent discussion in regards to the attribution of anthropogenic climate change versus other factors (Green 2016). But even without full certainty of attribution to climate change, atolls in particular are facing adaptation limits, both soft and hard, which will require decisions to be made in the near term on the thresholds and liveability of places. Stege's chapter on adaptation in the Marshall Islands in particular demonstrates well that there are tools that can be used in particular in atoll settings to start the collection of evidence, based on both indigenous and scientific knowledge) and the discussion as to 'when is it time to move'. Physical and ecological adaptation limits are a reality for atoll islands where there simply is not higher ground to move to. While many atoll countries, such as Kiribati, are looking for other options such as raising the islands, Stege notes that climate adaptation and relocation planning is a near-term reality for these countries. Determining an adaptation limit is therefore a joint exercise involving both in situ and ex situ actors.

The chapters in the book carry further the notion of adaptation limits and constraints' complexity and dynamic nature: while one constraint in itself might not significantly impede adaptation, multiple interacting constraints can lead to a significant adaptation limit, a deadlock where moving the different pieces around might not be possible due to their interconnectedness and dynamic interactions. Gilfillan provided evidence of this in the Cambodian context where many of the constraints around coordination together are seen to constitute an adaptation limit for effective adaptation action. More research is clearly needed to explore this interplay and dynamic interactions between constraints and limits to adaptation.

This book has also provided insights into how adaptation limits and constraints can be understood at the conceptual level. Robinson for instance finds that the use of adaptation constraints and limits overlap and are used interchangeably in the National Communications submitted by Pacific Islands governments to the UNFCCC. Some of the authors in the book focused more on adaptation constraints rather than strictly limits and a few also focused their discussion on adaptation barriers and how to overcome these in particular national policy settings. This also shows that the field is conceptually still in development and is in need of theoretical consolidation, which can enable the vast diversity of adaptation research undertaken to contribute to and expand our understanding of the basic concepts relevant to adaptation.

Conclusions: Moving Forward

The many chapters in this book have shown that adaptation to climate change can be pursued and implemented at various levels, despite the many constraints, which hinder its progress. Yet, the discourse is still strongly dominated by the assessment of future climatic changes and impacts with discussion on adaptation often limited to the identification of generic options (Gagnon-Lebrun and Agrawala 2006). Some industrialised countries do identify existing policies, particularly in natural hazards management that might be synergistic with adaptation to climate change. However, only very few countries currently report on actual implementation of anticipatory measures that consider future climate change. This however is likely to change with the implementation of the Paris Agreement, which calls for all nations to develop national adaptation plans.

To conclude his thoughts on "Political Barriers to Climate Change Adaptation" Dupuis (2011) stated that barriers to adaptation exist in both the developed and developing world, but they appear to be different. If a lack of material resources might still be the main hindrance to the development and implementation of adaptation measures in less developed countries, in developed countries where the need to act is not as obvious, the political feasibility and acceptance of adaptation policies is of greater relevance.

The lesson to be learned here is that, similar to the fact that climate change is a problem created by mankind, climate change adaptation measures may be part of man-led solutions. The some of the current limits posed to adaptation can be overcome by the mutual political will of industrialised and developing countries. Yet, some will continue to exist, such as hard limits, and innovative and radical thinking is needed to identify the kinds of actions that can be taken within the context of limits. The promotion and dissemination of good practice is one of the ways to convincingly show the diversity of the limits to adaptation and likewise the diversity of strategies to address these. There are plenty of lessons from the past showing what works and what does not. Left unattended however, these limits to adaptation will continue to play a negative role in attempts to address the problems seen today, perpetuating the negative trends tomorrow.

References

BASE. (2013). Policy Brief No 1 April 2013.pdf. Retrieved on 30.01.2017 from http://base-adaptation.eu/sites/default/files/BASE%20Policy%20Brief%20No%201%20April%202013.pdf.

Dungumaro, E. W., & Hyden, G. (2010). Challenges and opportunities to climate change adaptation and sustainable development among Tanzanian rural communities. In *Sustainabilidade em Debate*. Brasilia: University of Brasilia.

Dupuis, J. (2011). Political barriers to climate change adaptation. In *IGS-SENCE conference on resilient societies,* Enschede, Netherlands, 19–21 October 2011.

Conway, D., & Schipper, L. F. (2011). Adaptation to climate change in Africa: Challenges and opportunities identified from Ethiopia. *Global Environmental Change, 21*(1), 227–237.

Gagnon-Lebrun, F., & Agrawala, S. (2006). *Progress on adaptation to climate change in developed countries: An analysis of broad trends*. Paris: OECD.

Green, M. (2016). Contested territory. *Nature Climate Change, 6*(9), 817–820.

Intergovernmental Panel on Climate Change (IPCC). (2001a). *Climate change: Impacts, adaptation, and vulnerability*. Cambridge: Cambridge University Press.

Intergovernmental Panel on Climate Change (IPCC). (2001b). Climate change 2001: Synthesis report. In R. T. Watson et al. (Eds.), *A contribution of working groups I, II, and III to the third assessment report of the intergovernmental panel on climate change*. Cambridge: Cambridge University Press.

Intergovernmental Panel on Climate Change (IPCC). (2014). *Intergovernmental panel on climate change fifth assessment report*. Cambridge: Cambridge University Press.

Juhola, S. (2016). Barriers to the implementation of climate change adaptation in land use planning: A multi-level governance problem. *International Journal of Climate Change Strategies and Management, 8*(3), 338–355.

Klein, R. J. T., Midgley, G. F., Preston, B. L., Alam, M., Berkhout, F. G. H., & Dow, K., et al. (2014). Adaptation opportunities, constraints and limits. In C. B. Field, V. R. Barros, & D. J. Dokken et al. (Eds.), *Climate change 2014: Impacts, adaptation, and vulnerability. Part A: Global and sectoral aspects. Contribution of the working group II to the fifth assessment report of the intergovernmental panel on climate change* (pp. 899–943). Cambridge: Cambridge University Press.

Leal Filho, W. (Ed.). (2011). *The economic, social and political aspects of climate change*. Berlin: Springer.

Leal Filho, W., & Knieling, J. (Eds.). (2013). *Climate change governance*. Berlin: Springer.

Leal Filho, W., Esilaba, A., Rao, K. P. C., & Sridha, G. (Eds.). (2015). *Adapting African agriculture to climate change transforming rural livelihoods*. Berlin: Springer.

Leal Filho, W. (Ed.). (2015). *Handbook of climate change adaptation*. Berlin: Springer.

National Climate Change Adaptation Research Facility (NCCARF). (2013). Policy Guidance Brief 5, Challenges of adaptation for local governments retrieved on 30.01.2017 from http://base-adaptation.eu/sites/default/files/BASE%20Policy%20Brief%20No%201%20April%202013.pdf.

Measham, T. G., Preston, B. L., Smith, T. F., Brooke, C., Gorddard, R., Withycombe, G., et al. (2011). Adapting to climate change through local municipal planning: Barriers and challenges. *Mitigation and Adaptation Strategies for Global Change, 16*(8), 889–909.

Moser, S. C., & Ekstrom, J. A. (2010). A framework to diagnose barriers to climate change adaptation. *Proceedings of the National Academy of Sciences, 107*(51), 22026–22031.

Oxfam International. (2009). *Bolivia: Climate change adaptation*. London: Oxfam.